Springer Series in

MATERIALS SCIENCE 145

Springer Series in

MATERIALS SCIENCE

Editors: R. Hull C. Jagadish R.M. Osgood, Jr. J. Parisi Z. Wang

The Springer Series in Materials Science covers the complete spectrum of materials physics, including fundamental principles, physical properties, materials theory and design. Recognizing the increasing importance of materials science in future device technologies, the book titles in this series reflect the state-of-the-art in understanding and controlling the structure and properties of all important classes of materials.

Please view available titles in *Springer Series in Materials Science* on series homepage http://www.springer.com/series/856

Swift Heavy Ions for Materials Engineering and Nanostructuring

D.K. Avasthi
G.K. Mehta
Inter University Accelerator Centre
New Delhi
India

Springer

A C.I.P. Catalogue record for this book is available from the Library of Congress.

ISSN 0933-033X
ISBN 978-94-007-1228-7 (HB)
ISBN 978-94-007-1229-4 (e-book)

Copublished by Springer,
P.O. Box 17, 3300 AA Dordrecht, The Netherlands with
Capital Publishing Company, New Delhi, India.

Sold and distributed in North, Central and South America by Springer, 233
Spring Street, New York 10013, USA.

In all other countries, except SAARC countries—Bangladesh, Bhutan, India, Maldives, Nepal, Pakistan and Sri Lanka—sold and distributed by Springer, Haberstrasse 7, D-69126 Heidelberg, Germany

In SAARC countries—Bangladesh, Bhutan, India, Maldives, Nepal, Pakistan and Sri Lanka—sold and distributed by Capital Publishing Company, 7/28, Mahaveer Street, Ansari Road, Daryaganj, New Delhi, 110 002, India.

www.springer.com

Printed on acid-free paper

Printed in India.

Preface

Ion beams have been of increasing focus in fundamental and applied research in materials science. Development of ion beam facilities along with sophisticated devices for the characterization and analysis on nano scale has given a new direction in nanotechnology and towards tailoring desired properties in materials for their applications in various fields. A large amount of interesting research work in the direction of engineering materials properties with high energy heavy ion beams has been performed. Need was felt to have an overview of the work done in this direction by using high energy heavy ions [Swift Heavy Ions (SHI)].

The prime purpose of the book is to bring out salient features of research in materials with swift heavy ions. It will be of interest to young scholars engaged in research with swift heavy ions to focus towards specific problems in this emerging area. It has evolved from the experience gained by interacting with the large community of young researchers from various universities and research institutions in India and abroad. It was felt that because it is a relatively new area of research which is providing avenues of exploration in various directions, there is possibility of missing the directions required to focus on the emerging challenges in various areas, and that request from them for their contribution for this book may help in focussing their attention towards focussed goals. A consolidated compilation of contributions from the users of the high energy heavy ions for materials studies is being attempted here.

Swift heavy ions have a unique feature of depositing a large energy density in materials which can drive the material far from equilibrium, resulting in modifications in materials, difficult to achieve by other means. Studies on understanding of the modifications produced by swift heavy ions give an insight to the interaction of energetic ions with materials. The irradiation of materials by swift heavy ions can result in different effects in materials such as defect annealing, defect creation, ion track formation, plastic deformation etc. and these can be tuned by ion beam parameters. High energy density provided by SHI in laboratory allows to simulate extreme conditions.

Ion beams have been used for a long time to characterize the materials with techniques developed specifically for the situations which could not be handled with the conventional techniques. The worry that the ion beams being used to characterize the samples will damage the material soon turned into a challenge that the modifications in the properties during irradiation can be monitored in dynamic mode which can provide possibility of controlled change in the properties of the materials. This started the discipline of "Materials Engineering with Ion Beams". It has now emerged as an area with immense scope of development particularly because of the way it is becoming possible to create ion beams with specific and diverse characteristics giving rise to new avenues to exploit. Considering the fact that the low energy ions have been in the field for a long time it was decided to concentrate on the high energy heavy ions.

This book is an effort to present a scenario of research in materials science with SHI in a way which could be useful for the students planning research in ion accelerator based materials science. We would like to emphasize that the material is prepared with the help of researchers in this area and is based on their publications and existing materials. The challenge was to try and convert the large number of research publications in the field into a kind of a text to motivate the students to plan their experiments with a focus on engineering some well defined property of some materials to make its application possible.

First three chapters provide an overview of the field, challenges in materials and state-of-the-art developments in ion beams, very brief description of ion matter interaction and survey of the wide field of applications of ion beams for the characterization and analysis of materials. The next two chapters are on the SHI induced ion beam mixing and SHI for synthesis and modification of nanostructured materials. The last chapter focusses on engineering the properties of materials with SHI, where various types of materials are considered. It covers different types of materials. It has not been possible to cover all the materials which have been worked upon using SHI irradiation.

This exposition of experimental work owes much to the persons who provided the information about the importance of ion beam interaction in changing the properties of materials in unique ways. We had significant inputs from the present and past research scholars who had been at IUAC for their PhD. We gratefully acknowledge Amit Kumar (now at LPMS, Toulouse, France), Y.K. Mishra (now at Kiel University, Germany), S.K. Srivastava (now faculty in IIT Kharagpur), S. Ghosh and Rajendra Singh (now faculty in IIT Delhi), D.C. Agarwal, R. Singhal and Yogita Batra. There are many to acknowledge but special role was played by Drs Naresh C. Mishra, Ajay Gupta, K.M. Varier, Anand Pathak, R.G. Sharma, Ashok Kumar, A.C. Chaudury, Ratnamala Chatterjee, S. Dhamodaran, Avinash Pandey, Utpal Joshi, T. Som, D. Behera, D. Mohanta and Maulik Patel, to name a few. The manuscript could only be completed because of the support from the colleagues in the Nuclear Science Centre [now Inter University Accelerator Centre—IUAC] who have been active collaborators with the users from various institutions in experiments conducted with the Pelletron accelerator in New Delhi. The discussions at various occasions with scientists at IUAC e.g. D. Kanjilal, A. Tripathi, S. Chopra, Ravi Kumar, D. Kabiraj, Fouran Singh, K. Asokan, S. Mookherjee, S.A. Khan, P. Kulriya and I. Sulania had been fruitful.

The authors acknowledge the help rendered by Ms. Srashti Gupta and Mr. Jai Prakash in this project.

Appreciation, constant interest and encouragement by Dr. Amit Roy, Director IUAC, needs special mention. The financial assistance provided by the Department of Science & Technology and the help from the staff of the Inter University Accelerator Centre is gratefully acknowledged. In the end we would like to mention that it was late Dr. C.P. Srivastava of DST who gave the required push to take up this project.

February 2011

D.K. Avasthi
G.K. Mehta

Contents

1

Ion Beams for Materials Engineering—An Overview

1.1 INTRODUCTION

Nuclear accelerators have provided revolutionary advances in several disciplines. Starting with the reputation of atom smashers which demonstrated that atoms are not indivisible they provided facilities to explore nuclear properties. The exploration went from macroworld into femto and beyond with fantastic developments in nuclear accelerator technology. Then came a period where the word nuclear became a serious suspect, and the nomenclature changed from nuclear to particle accelerator. The accelerators continue to accelerate the technology in multi-dimensions. Until sixties, these accelerators were dedicated to research in nuclear and particle physics. Research using beams from particle accelerators has told us almost everything we know about the basic building blocks of matter, and about nature's fundamental forces. Efforts are now on to recreate conditions that have not occurred since shortly after the Big Bang. Accelerators unravel nature's deepest mysteries. They are central to the effort to unravel the mysteries of dark matter and dark energy.

Physics developed essentially along the energy axis i.e. particles were given increasingly higher energies by developing nuclear accelerators—a trend which seems to continue, as evident from the efforts in CERN to produce light particles of TeV energies. However, a considerable progress is also made on the mass axis, by accelerating heavy particles, as nuclear matter is being probed in depth with heavy ions to explore nuclei far from the line of stability. High energy heavy ions are also sought after for research in various other fields.

Nuclear accelerators—besides their role in basic nuclear, atomic and elementary particle physics—provide revolutionary advances in several disciplines and invariably open up completely new perspectives in various fields. Energetic ions were usually thought to introduce disorder in solids, normally an undesirable phenomenon. Irradiation studies showed that the energetic ions create defects and thus modify the properties of the materials. Radiation damage to materials used in nuclear reactors has been of great concern [1-4]. Ion beams from accelerators provide unique ways to study the radiation damage to materials. It is necessary to understand the fundamental mechanism of radiation damage to determine what changes in the constituents and structures of the material will reduce the damage. Detailed investigations of this type have helped in predicting how changes produced in materials due to ion irradiation could improve its properties of interest.

In 1947 Brattain and Bardeen in Bell labs showed that a crystal of Ge doped with some atoms which could accept electrons could amplify signals. This along with the development of techniques to make pure crystals led to the transistor era. Doping the materials in a controlled way was the challenge. Particle accelerators provided the way. The multi-billion-dollar semiconductor industry relies on ion beams from accelerators to dope desired atoms in semiconductors. Ion implanters were developed for the purpose of shallow-impurity doping into semiconductors. The ion implantation technique then dominated conventional thermal diffusion methods in the semiconductor-IC processes. The initial and running costs of ion accelerators for implantation were considerably high but they were found to be indispensable because of the excellent controllability of doping levels, which is required for the industrial processes involved in the fabrication of transistors and integrated circuits. Support from the semiconductor industry produced concerted research efforts in doping technologies. Although the position and distribution of the implanted ions could be accurately predicted, the defect distribution, their stability and effect on the properties were of concern. Studies on these lines provided appropriate understanding of the changes in materials properties due to ion matter interaction. Semiconductors, once exotic, are now commonplace because of the unbelievable things achieved with ion beams in understanding and controlling defects. Developments in ion beam technologies continue to provide the solutions for the bottlenecks in the path of miniaturization of electronic systems and have now started the miniaturization of systems used for chemical and bio-systems by providing nano fluidic channels fabricated by nano focussed ion beams.

It was in the sixties that the importance of ion beams was recognized for materials studies. Rutherford Backscattering Spectroscopy (RBS) and Channeling provided a quantitative technique for applications in surface studies, thin films, element profiling and interface analysis [5-7]. Ions can kick out (recoil) the atoms being struck through elastic nuclear interaction. Detection of these recoils with appropriate detectors provide a good method for elemental concentration depth profiles, known as Elastic Recoil Detection Analysis (ERDA). It provides excellent technique for depth profiling light elements [8-12]. On-line and dynamic measurements with ERDA can provide efficient ways to engineer the properties of materials, particularly ones which depend on the distribution of light elements [13, 14].

Proton Induced X-Ray Emission (PIXE) provided a technique to determine the elemental make-up of materials [15]. This method of trace elemental analysis proved to be of immense importance in diverse areas such as forensic science, medical diagnosis, environmental studies, geochemical prospecting etc. [16-21].

These techniques were providing unique ways to analyze the materials. However, there was always a concern that the nuclear techniques cannot be non-invasive and that the ion irradiation will affect the materials and will modify the properties of the materials. The changes in the properties of the sample being characterized were analyzed. Capabilities of ion beams for the modification of the properties of materials started surfacing and it was realized that there are certain desirable modifications of the properties of materials, which can only be done with ion beams. This led to the field of "Ion Beam Induced Modification of the Materials Properties".

Diamond has been the most sought-after material as it is considered to be the best material for semiconductors but very difficult to make. It is known that high temperature and pressure can convert graphite into diamond as done by nature but it is difficult to achieve. Considerable efforts were made in laboratories to make Diamond-like Carbon [DLC] films and make them more and more diamond-like by appropriate controls. DLC films were studied using Elastic Recoil Detection Analysis [ERDA] in on-line fashion [22] and the possibility of engineering DLC films by on-line monitoring its hydrogen content in dynamic fashion was demonstrated [23].

The bottleneck in the development of the techniques connected with ion beams has always been the cost of ion beam facilities. So the question, which naturally arises, is whether these techniques can be commercially viable. The answer is provided by the fact that no body had imagined in sixties that electronics industry would not survive without an ion implanter. Many thousands of accelerators, most of them only room-sized or smaller, serve as essential tools for biomedical and materials research, for diagnosing and treating illnesses, and for a growing host of tasks in manufacturing devices, sensors etc. Now the remarkable advances in ion source technology and associated developments in the accelerator field are opening possibilities of the table-top ion beam facilities.

The use of ion beams at the bio-medical interface is generating novel and exciting research. In parallel, interesting possibilities are emerging with the use of ion beams in several other disciplines. Bio-effects induced by ion implantation has opened a new branch in the field of ion beam applications in the life sciences and applications in biotechnology, biomedical engineering etc.

There have been phenomenal developments in ion beam technology providing possibilities of tailoring the properties of materials as per needs and developing new materials. The field of materials science is going through revolutionary developments, particularly after the realization that the properties of the materials critically depend on the size of its constituents, leading into nanotechnology. The thrust of nanotechnology is in the direction of the creation of functional materials, devices and systems through control of matter on the nanometre length scale and exploitation of novel phenomena and properties (physical, chemical, biological) at that length scale. Ability to synthesize nanoscale building blocks with control on size, composition etc. and assembling into larger structures with designed properties will revolutionize materials manufacturing. The field is interdisciplinary but everything starts with materials science. Developments involve challenges which include:

- Novel synthesis techniques of nano-materials
- Characterization of nanoscale properties
- Developments of nanostructures
- Basic science studies at nano-level
- Fabrication of nanosystems
- Large scale production of materials
- Application development

Broadly the ion beam, depending on the energy regime, can play different roles in materials science such as synthesis, modification and characterizations

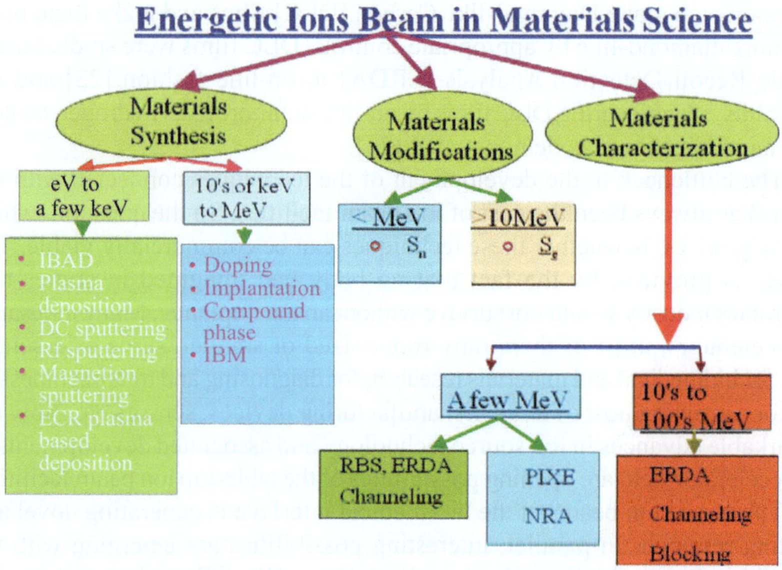

IBAD: Ion Beam Assisted Deposition
IBM: Ion Beam Mixing
RBS: Rutherford Backscattering Spectrometry
ERDA: Elastic Recoil Detection Analysis
PIXE: Proton Induced X-ray Emission
NRA: Nuclear Reaction Analysis

Figure 1.1: A schematic of the role of ion beam in materials science. In plasma based sputtering set up, the +ve ion in plasma are attracted by cathode. These positive ions strike the cathode to sputter the material, which is deposited to form a film. Thus the plasma based deposition utilizes the ion induced sputtering process for sputter deposition. [© Springer - *Hyperfine Interactions* (2005) **160:** 195-106, Nanostructuring by Energetic Ion Beams, D.K. Avashti, Fig. 2, with kind permission from Springer Science+Business Media BV.]

as shown in Fig. 1.1. Low energy from a few keV to a few MeV is useful in synthesis of materials. A few MeV to hundreds of MeV is employed for the materials characterization. Ions of all possible energies are useful in the modification and tailoring the properties of materials; understanding of this aspect can help in engineering the properties of materials. Ion beams are providing unique ways to analyze, synthesize and control the formation of nanomaterials, produce nanostructures and showing ways for doing actual manufacturing in nanoscale, making it possible to fabricate nanoelectro-magnetic systems [NEMS].

Increasing utilization of the opportunities offered by ion beams has also led to tremendous developments in ion beam technologies. This is now opening new horizon needing a proper appreciation of the scope offered by the ion beam facilities in the area of materials engineering with a focus on the nano region. There is a need of experimentation to learn emerging new mesoscopic physics exhibiting quantum confinement effects, which will provide directions for the futuristic technologies.

1.2 CHALLENGES IN MATERIALS SCIENCE AND ENGINEERING

There are two routes toward the advances in materials science through nano approach. On one route, nanoscale not only reduces the scale of handling but also yields new characteristics and functions. On the other route, nanostructures are created from non-equilibrium processes and new characteristics are discovered. Ion beams are providing unique ways on all fronts and playing a critical role in the development of different types of materials. In order to survey possibilities of engineering materials properties with ion beams, challenges in various types of materials are surveyed here.

Nanomaterials

Nanomaterials have structured components with at least one dimension less than 100 nm, such as layers, thin films or surface coatings, nanowires and nanotubes, colloids and quantum dots (nano particles of semiconductor materials). Materials made up of nanometre-sized grains are called nano-crystalline materials. Development of nanostructured materials are of immense interest due to their exotic properties and emerging exciting physics.

Synthesis of nanomaterials requires development of techniques to create nanoparticles having a controlled size and also having controlled morphology usually related to the surface structure. Methods are being developed to control or modify the surface structure and to arrange and integrate nanoparticles on a given substrate. Nanotubes, one of the most attractive groups of nanomaterials, need to be synthesized with well-controlled methods to obtain semi-conducting or metallic nanotubes selectively.

Energetic ions play a crucial role in the development of nano materials. Ions of different energy regimes have different roles in growth of nano particles. Low energy ions (typically up to a keV) have been in use for the growth of nano particle thin films. A few hundred keV ions from ion implanters are used for growth of nano particles in a matrix. Highly charged slow moving ions and focussed ion beams are now providing exotic ways for creating nanostructures. High energy heavy ions (swift heavy ions) provide unique ways for the growth of nanostructures [24, 25] and for their controlled modification [26-29].

Superconducting Materials

High Tc superconductors [HTSC] have been in the field for decades. A limit on technological applications of HTSC comes from the intrinsically low critical current density and from the ease with which magnetic vortices can move. Swift heavy ions produced controlled columnar defects provide efficient pinning centres which improves critical current density by order of magnitude [30-32]. Efforts to engineer materials with higher critical temperatures continue. It is found that the pinning centres can also be achieved in nanostructured materials [33]. Iron based superconducting materials have opened a new horizon.

Magnetic Materials

Magnetic materials are key industrial materials that are widely used in the telecommunication, electric power and automobile industries. Development of thin film high-performance magnetic materials are required for future applications in Micro/Nano electro-mechanical systems (MEMS/NEMS) and other small portable components. Due to the rapid increase of the size of data communication, the frequency used in portable electro-communication devices has reached the GHz regime. Write heads for high-density magnetic recording require materials of higher magnetic flux density. Magnetic recording technology needs new materials to increase the aerial density of magnetic recording to the range of 1 Tbit/in^2.

Nanoparticles of magnetic substances—metals, alloys and oxides—acquire noteworthy new properties and their magnetic characteristics are significantly altered relative to those of the corresponding bulk materials. Magnetic nanoparticles are active components of ferro fluids, recording tape, and flexible disk recording media. There are exciting future applications of magnetic nanoparticles in spintronics—a new paradigm of electronics that utilizes both the intrinsic charge and intrinsic spin of electrons for ultra-high-density data storage and quantum computing [34].

Magnetic nanoparticles are well suited for biomedical applications, as they can be fabricated in sizes that range from several to hundreds of nanometres making them smaller than or comparable to important biological entities such as cells (10-100 μm), viruses (20-250 nm), proteins (3-50 nm), and genes (10 nm wide and 10-100 nm long). Further, they can be surface functionalized with biological molecules that enable selective interaction with various target bio-entities. Moreover their motion can be manipulated, controlled by an external magnetic field and can absorb energy from a time-varying magnetic field and dissipate it at the cellular level in the form of heat.

Magnetic thin film materials can be tailored by ion irradiation, since the magnetic properties of multilayers depend sensitively on the mutual interfaces. Modification of these interfaces by ion irradiation leads to a local modification of the magnetic anisotropy, the exchange bias or the interlayer exchange coupling. Also structural phase transitions can be induced by ion irradiation. The ordering in intermetallic alloys can be induced by the ion irradiation, to tailor the magnetic anisotropy in the FePd films, which can be exploited for magnetic recording [35]. In addition to pure radiation effects, doping effects can be exploited to achieve a pure magnetic patterning. Pure magnetic microstructures can be easily fabricated by local ion irradiation without changing the substrate topography [36].

Semiconducting Materials

We have silicon, undoubtedly the most important material of electronics. Silicon is cheap, robust, easy to process and an insulating layer on it can be created by just growing an oxide layer on it. But it has problem, it has indirect band gap. What are the requirements we have for electronic materials and how do these requirements translate to basic material properties, the manipulation of these properties, and the technologies required to produce the intricate structures that perform that electronic function? Search for new semiconducting materials and the improvement of the existing materials by band gap tuning etc. is an active field of research. One can engineer the material for simultaneously tuning the band gap and the lattice constant to avoid lattice mismatch with the substrate. Ion beams provide ways to engineer the required properties. It is desirable to have thin layer of crystalline semiconductor on an insulator like SiO_2. Such a structure is also referred to as Si on insulator (SOI). Industries look forward for such SOI structures. These SOI structures can be achieved by a process called ion cut/smart cut by ion beam and is discussed in Section 1.3.

Biomaterials

A biomaterial can be natural or man-made, that comprises whole or part of a living structure or biomedical device which performs, augments, or replaces a natural function. It is essentially a material that can be adapted for a medical application. Biomaterials may have a benign function, such as being used for a heart valve, or may be bioactive with a more interactive functionality and used for a more interactive purpose such as coated hip implants. It was first demonstrated by an experiment with high energy accelerator of Yale University that the life time of hip joints can be increased significantly by swift heavy ion irradiation. Biomaterials are also used in dental applications, surgery, and drug delivery (a construct with impregnated pharmaceutical products can be placed into the body, which permits the prolonged release of a drug over an extended period of time). Biomaterials must be compatible with the body. It should not be toxic, unless specifically engineered to be so, for example drug delivery systems that target cancer cells and destroy them. These materials provide a challenge that they should be made in such a way that they can be placed in intimate contact with living structures without any harmful effects [37, 38].

Ecomaterials

Ecomaterials are produced from raw materials taken from the environment by interacting with the geo-system and bio-system. Challenge is to find techniques for developing materials in which physical, chemical, thermal and/or functional properties are improved by minimizing damage done to the natural environment. From the viewpoint of sustainable development, consumption of materials and energy, emission of toxic gases and waste associated with materials processing should be reduced to ease our impact on the resource circulation system.

All materials and their properties are to be considered from the viewpoint of environmental concerns. For example, how can a required property be obtained with less environmental load? How can materials be given improved recyclability? How can maximum performance be acquired with least consumption of materials?

Ceramic Materials

These are inorganic non-metallic materials which are formed by the action of heat. In other words, desirable properties of these materials are normally achieved through high temperature treatment. Technically ceramics can be classified into three distinct categories: Oxides: Alumina, zirconia; Non-oxides: Carbides, borides, nitrides, silicides; and Composites: Particulate reinforced combinations of oxides and non-oxides. Each one of these classes can have unique properties. The friction and wear behaviour of ceramic materials is of utmost importance which can be improved by ion irradiation. Advanced ceramics are now used in the medicine, electrical and electronics industries [39].

Composite Materials

Composite materials are engineered materials made from two or more constituent materials with significantly different physical or chemical properties which do not dissolve or merge completely into one another although they act in concert. There are two categories of constituent materials: matrix and reinforcement. The matrix material surrounds and supports the reinforcement materials by maintaining their relative positions. The reinforcements impart their special mechanical and physical properties to enhance the matrix properties. A synergism produces material properties unavailable from the individual constituent materials. Most commercially produced composites use a polymer matrix material.

A nanocomposite is a multiphase solid material where one of the phases has one, two or three dimensions of less than 100 nanometres (nm), or structures having nano-scale distances between the different phases that make up the material. The mechanical, electrical, thermal, optical, electrochemical and catalytic properties of the nanocomposites differ markedly from that of the component materials.

Metal nanoparticles provide opportunity of applications in many fields by dispersing and ordering them within the dielectric matrix. The chemical method of fabricating the nanocomposites provides a way of changing the bulk of the host matrix as per the requirement. Such types of material have potential application in biosensors, lubricants, paints, medical science, etc. The use of ion beam techniques provides a way to not only modify the layer near the surface but up to some thickness also as required by selecting the ion beam parameters. It provides fabrication of monodispersed nanostructure with high purity near the surface of the material. By selecting the dose and type of ion one can control the refractive index of the layers of the host polymer matrix. Focussed ion beam synthesis of the nanostructure material within the polymer matrix is an emerging field for the fabrication of functionalized material and devices.

Organic Materials

Polymers are one of the simplest organic solids. A polymer is a large molecule (macromolecule) composed of repeating structural units typically connected by covalent chemical bonds. Well-known examples of polymers include plastics and proteins. Biopolymers such as proteins and nucleic acids play crucial roles in biological processes. Synthetic polymer materials such as nylon, polyethylene and silicone have formed the basis for a burgeoning polymer industry. They have been employed in a variety of biomedical applications ranging from implantable devices to controlled drug delivery. Polymers are recognized as important elements of nanotechnology, indispensable materials in medicine and biotechnology, and important for solving environmental, safety, energy and other related problems [40].

Tracks in polymers produced by high energy heavy ions are cylinders of density smaller than the bulk. There are chemical modifications and specific defects induced by radiation in the tracks.

Metamaterials

Metamaterial is a material that gains its properties from its structure rather than directly from its composition. These are artificial materials engineered to provide properties which may not be readily available in nature. Unusual properties are also produced in conventional materials by processing them at nanoscales. However, a distinguishing feature of metamaterials is that they are specifically designed to fulfil a certain objective and to fit the desired application.

Metamaterials can be made with unusual properties such as negative refractive index, infinite inertia etc. Focussed ion beam [FIB] nanostructuring is a convenient tool for rapid prototyping of metamaterials. The FIB is used as a micro-machining tool, to modify or machine materials at the micro- and nanoscale. FIB micro machining has become a broad field of its own, but nano machining with FIB is a field that still needs development.

Biomimetic Materials

These are materials which can be developed using inspiration from nature. Critical analysis of the way things develop around us provides ways to design composite materials, or material structures. Natural structures have evolved many inspiring examples that have been used by man. Common examples are the honeycomb structure of the beehive, the fibre structure of wood, spider silks, bone etc.

Efforts are to be directed to mimic advantageous surface properties or processes of living organisms for materials science and technology by utilizing and combining state-of-the-art nano and micro fabrication, soft lithography, self assembly, and ion beams to achieve the objectives. Challenge to materials scientist is to learn about the material structures within organisms and to exploit that natural technology into scientific and technology context [41].

Materials for Sustainable Energy

Ever increasing need of energy requires development of materials in areas of (i) H_2 storage materials, (ii) materials for producing H_2 by photo splitting of water, (iii) solar cell technology, (iv) thermoelectricity, (v) supercapacitor etc. Ion beam tools can play a role in specific cases, for example (a) to create new patterned surface for back electrode and (b) for synthesizing a controlled plasmonic layer in photovoltaic cells to improve efficiency. The possibilities of enhancing solar cell efficiency has been suggested recently using plasmonics [42].

New Materials

Science and technology has to provide techniques to develop new materials. "Smart" or "intelligent" materials will play an important role in this development. Smart materials are materials that have one or more properties that can be significantly changed in a controlled fashion by external stimuli, such as stress, temperature, moisture, electric or magnetic fields. Smart materials become intelligent when they have the ability to respond intelligently and autonomously to dynamically-changing environmental conditions. Window glass for example passes both light and heat. One will like to develop smart glass which passes only light but not heat. The challenge is to make intelligent glass which passes both light and heat as long as the temperature is below some threshold but blocks heat beyond that temperature, which can then provide a way to air-conditioning the houses.

The intelligent material is one which can fulfill combined functions of sensor, processor, an actuator and a feedback function. The development of intelligent materials is important for real-time monitoring of accidents and development of drug delivery systems. The ion engineering technology can support the development of intelligent materials through the synthesis of new materials and tailoring the properties in the desired direction.

The Challenge

Materials science is now getting close to the situation where one hopes to produce materials with properties tailored to the requirement. The challenges are: How mechanically strong can material be while remaining lightweight? How resistant it can be to radiation? How electrically and thermally conductive? How transparent? How magnetic? How sticky? How biocompatible? ... Can materials be made to modify themselves in response to a changing environment? Can we learn from nature how to make substances that repair themselves?

Powerful permanent magnet produces at its surface a field of ~1.6 T, small compared with the field of a solenoid. Why can't we make permanent magnet with order of magnitude larger field? One needs high density of atoms with strong magnetic dipoles, such as iron or nickel, in a host structure of the type which can force the dipoles to point in a common direction. Quantum confined atoms are pointing in the direction of providing nanomagnets produced this way. The production of high magnetic field in electromagnet is limited by the strength of materials, they must withstand the magnetic pressure required to create a high energy density in magnetic fields. Field up to ~20 T can be produced in superconducting magnets. Present limit due to the strength of the materials is ~50 T [43]. Nanocomposites are bound to provide materials with large strength.

Nature produces extreme conditions such as extraordinary pressures and temperatures in the interiors which gives rise to materials such as diamond. The extreme phenomena are central to many of the fascinating grand challenges of science—behaviour far from equilibrium. Can ion beams create this type of non-equilibrium situation which can synthesize diamond from graphite?

Biological machinery seems to work in ways dramatically different from those of synthetic devices. It is necessary to learn from biosystems to engineer the devices. Biological mechanisms hinge on physical behaviours that exist only at the nanoscale. Biological systems should be considered as the proof of concept for what can be achieved with nanotechnology. All natural phenomena occur away from equilibrium. Can we devise systems to work in nonequilibrium state? Ion beams provide ways to take the system in nonequilibrium states of different forms. The challenge is to let the system equilibrate after nonequilibrium has reached the desired state. A real understanding of nonequilibrium phenomena is required. Presently most of the current understanding of physical and biological systems is based on equilibrium concepts.

Our body is made of materials which have self healing properties. The challenge is to develop such materials. For example, it should be possible to develop materials for constructing nuclear reactors with materials that can self-heal following radiation damage.

Trends in materials science have been to control the nanoscale structure of materials which leads to better properties and more functionality. Attention now is on beyond simple materials, to fully functional nanoscale devices.

1.3 ION BEAM BASED/ASSISTED PROCESSES

Developments in ion beam technology provide ions of remarkably controlled characteristics permitting their use in various possible ways. Some of these are described here.

Ion Implantation

Ion implantation is a process by which ions of a chosen material can be injected into another material at a desired depth. Virtually any element can be injected into the near-surface region of any solid material. Ion-implantation enables the precise placement of ions into the solid based on the energy and angle of implantation. The ions penetrate the surface up to a depth determined by the energy and the mass of the ion thereby providing a way to change the physical properties of the solid in a controlled way.

Ion implantation process provided ways to tailor the conducting properties of materials and thereby initiated the transistor revolution in electronics. It is a non-equilibrium method for introducing atoms into solids, and is applicable to almost any solid object. Limitations due to diffusion, precipitation, segregation, solid solubility etc. can be avoided in this method. Normally the interest in ion implantation is in modifying some specific property of the solid by the addition of a chosen ion. For example, chosen impurity can engineer the electrical properties of the crystal (doping), a process which revolutionized the semiconductor device technology.

The ions introduce both a chemical change in the target and structural modification. It is especially useful in cases where the chemical or structural change is desired to be near the surface of the material. In higher concentration the implanted impurities can form alloys as in the case of oxygen ions on silicon to form buried SiO_2 layer for the silicon-on-insulator technology.

High dose implantation can create a solid state super saturation of the implanted ions in an inside layer close to surface. Subsequent thermal processing or further irradiation can induce the implanted material to precipitate as discrete nanoparticles. The versatility of the process arises from the fact that essentially any element can be implanted into virtually any selected host material. It provides a powerful technique for synthesizing nanometre-scale clusters and crystals embedded in near-surface region of a variety of hosts. The condition for making buried nanoparticles of implanted species is that the implanted species should not be miscible with the host matrix/substrate. The resulting nanocomposite materials can have unique optical, magnetic and electronic properties because of two major effects: (1) Increase in surface energy contributions associated with the large surface to volume ratio of small particles and (2) Unique characteristics of electrons in confined systems. The first effect largely determines the thermodynamic properties of particles or nano-composites, e.g., melting point, bulk modulus, solid phase transition etc. Both the surface properties and electron confinement combine to produce novel electronic properties that can be manifested in a wide range of effects such as: nonlinear optical susceptibility, intense photoluminescence, altered band structures, super magnetism etc.

Ion Beam Induced Epitaxial Crystallization (IBIEC)

Ion implantation provides an excellent method for putting the desired dopant material in the semiconductor. Unfortunately, each impinging ion before settling in the solid displaces many atoms of the solid (target atoms) damaging the crystalline lattice, referred to as the radiation damage, which needs annealing treatment. The annealing temperatures greater than 500°C are generally needed. With the reduction in the size of the devices to submicron scales it has become a challenge to regain the lattice structure at lower temperatures to avoid undesired diffusion of the dopants. Ion beams have also provided a way to tackle this problem. It can be achieved in silicon at ambient temperatures by simultaneous irradiation with energetic ions [44]. This recrystallization process achieved by ion-atom interaction is termed as ion beam induced epitaxial crystallization (IBIEC). High energy ions have emerged as a powerful tool to provide epitaxial crystallization of the semiconductors, referred as Swift Heavy Ion Beam Induced Epitaxial Crystallization (SHIBIEC) [45]. This process is discussed in detail in Appendix 1.

Sputtering

An energetic ion incident on a solid interacts with the atoms causing electronic excitations and nuclear collisions. Surface atoms are emitted if the energy transfer in such collisions is high enough to overcome surface binding energy. This removal process of surface atoms by energetic ions is called sputtering. Depending on projectile energy loss, different scenario of sputtering occurs. Emission of atoms caused by atomic collision cascade due to nuclear energy loss is called "Nuclear Sputtering" [46, 47]. The ejection of atoms due to the electronic energy loss is referred as the "Electronic Sputtering". Nuclear sputtering and electronic sputtering are completely different processes. Simple and complex molecules can be ejected intact into the vapour phase when a material is electronically excited by incident particles.

Some ion-induced phenomena depend on the internal (potential) projectile energy, particularly if this potential energy greatly exceeds the kinetic projectile energy. The stored potential energy in a highly charged ion (HCI) can be quite high and produce sputtering in insulators, which is known as the "Potential Sputtering" [48]. Swift heavy ion beam based sputtering process is discussed in Appendix 2.

The sputtering plays crucial role in creating nanostructures at surface, which is discussed in more details under Ion Beam Assisted Self Organization.

Ion Beam Milling

Sputtering with ion beam in a controlled way provides an excellent way for milling the surface as the energetic ions remove material atom by atom. With nanometre focussed ion beams [FIB] one can get extremely precise milling. Computerized control of the ion beam provides a way to calculate the time for the ion beam to dwell on each area to remove the material and bring the surface to desired specifications.

Ion Beams for Material Deposition

(a) In "Ion Beam Deposition (IBD)" the atoms or clusters to be deposited are ionized and accelerated before impinging on the substrate surface. One can do mass selected ion beam deposition (MSIBD) [49] and also energy selected ion beam deposition (ESIBD) [50].

Metal clusters with about a thousand atoms can be ionized, typically with a single charge and accelerated to keV energy. The average energy per atom in the cluster is about a few eV. Owing to the low energy carried by each atom the damage produced on the substrate is very small. It produces very dense and smooth thin films.

In MSIBD, ion beam of given element with very small energy dispersion is obtained. The chosen element gets incorporated into the substrate forming a layer. At low energy, sputtering is not significant and sticking probability is high providing films with good adhesion.

A mass and energy selected ion beam source for deposition (M & E-SIBD) can produce ions with a mass up to 2048 amu and with energies of 5–500 eV by selecting from a Penning discharge chamber with a quadrupole mass filter and an electrostatic lens system.

(b) Ion Beam Assisted Deposition (IBAD) is a thin film deposition process that combines evaporation with concurrent ion beam bombardment in a high vacuum environment. The evaporant (or coating) material is produced using a thermal or high power electron beam, laser ablation etc. The individual coating atoms or molecules condense and stick on the surface of the substrate to form the coating. Simultaneously energetic ions (100–2000 eV) are produced and directed at the surface [51]. The concurrent ion bombardment improves adhesion, and permits control over film properties such as morphology, density, stress level and crystallinity.

(c) In Dual Ion Beam Deposition (DIBD), the sputtered material is deposited on the surface of the substrate, while simultaneously the growing film is bombarded by ions from another source operating at low energies (a few hundreds of eV). Strict control possibilities of ion beam parameters of the two ion sources makes the quality of the films to be excellent in terms of roughness (planarity) and good adherence to the surface [52].

(d) Ion Beam Induced Chemical Vapour Deposition (IBICVD): In the three methods discussed above the thin film prepared by vapour deposition is modified by the effect of the ion beam. In IBICVD, the film is prepared with the help of volatile precursor of the constituent element of the film [53]. Typically CVD consists of thermal decomposition of a volatile metallic precursor. For example, by controlling the atmosphere and temperature of the substrate during the deposition, a metal oxide or nitride thin film can be deposited by using an appropriate precursor, while the volatile molecules of the other atomic constituents of the precursor (e.g., H_2O, CO_2, HCl etc.) are removed. The ions impinging on the substrate produce the decomposition of the molecules of the volatile precursor adsorbed on the surface of the film while simultaneously assisting its growth process. Ion beams provide excellent method because they can be focussed to nm size. The technique for introducing a precursor gas such as organic metal into the neighbourhood of a specimen and irradiating a focussed ion beam onto it enables making three dimensional nanostructures with precision. Versatility of this technique is demonstrated in nanofabrication using the focussed ion beam (FIB) based chemical vapour deposition method by utilizing scattered Ga ions to decompose organometallic molecules of the precursor gas for depositing the metallic element on a surface with the advantage of producing uniform metallic coats on those surfaces of nanostructures which are not directly accessible to the primary beam. The method can be used to provide electrical contacts on inaccessible regions of prototype nanodevices [54].

(e) Atom/Neutral Beam Co-sputtering: Using neutralized ions (atom beams) instead of ion source facilitates sputtering of insulators as the latter has the problem of charging of the insulating sputter target and the space charge obstructs the ion beam from striking the sputter target. A neutral beam sputtering set-up specially designed for the synthesis of nanocomposite thin films using neutral Ar atoms of 1.5 keV energy incident at 45° at the sputter target produces uniform films. The metal fraction in the nanocomposite films can be varied by controlling the relative area covered by metal foils exposed to the Ar beam [55]. Au-silica nanocomposite thin films prepared by atom beam co-sputtering for metal fraction with 1.2 at%, 11 at%, and 20 at% provided a way to make controlled size Au nanoparticles [56]. Si-Ge nanocrystals were made through synthesizing Ge-silica nanocomposites by co-sputtering of Ge and silica with 1.5 keV Ar atoms [57]. The cosputtering with atom beam is an excellent method for synthesis of nanocomposite thin films and nanopatterning of the surface discussed in Section 1.8.

Ion Induced Dewetting

A planar metal film prepared on a dielectric substrate is typically metastable and will dewet from the substrate at elevated temperatures. Ion beam provide novel means of producing localized melting of thin film deposited on substrates. It allows one to pattern metallic surface nanostructures with precisely controlled size, spacing and location. Predefined patterns by focussed ion beam direct-writing can be used as the templates for ordered nanostructures. This is a maskless process with no chemical etching etc., which can be applied to many metallic systems in constructing complex nanostructures for numerous applications.

Patterning of metal nanowires by directed ion-induced dewetting was demonstrated by ion beam irradiation of thin Au and Pt lines on SiO_2 substrates, 60–400 nm wide and 8–25 nm thick which led to the formation of nanowires by directed dewetting [58]. Continued irradiation of these wires further resulted in the formation of a linear array of metal dots. In experiments with the accelerator in Darmstadt, Swift heavy ion induced surface modifications of a 50-nm-thick Fe_2O_3 film on Si substrate exhibited submicrometre size holes due to irradiation induced dewetting. In situ high resolution electron microscope (HRSEM) in Darmstadt allows investigations of the development of individual dewetting holes as a function of the ion fluence, and even the interaction between them can be investigated. To illustrate pattern formation during grazing incidence, 3.6 MeV/amu ^{131}Xe ion irradiation was carried out on a 25-nm-thick NiO film on SiO_2/Si at a tilt angle of 75°. The SEM image sequence recorded during the experiment reveals the development of a lamellae-like structure [59].

Ion Beam Nanosculpting

Ion beam sculpting enables nanometre control of structural dimensions. It allows manipulation of matter on the scale of 10^{-8} to 10^{-9} metres from which many future electronic, chemical and biological advances are anticipated. In one of the applications it is applied to the problem of fabricating a molecular scale hole or nanopore. It is a two-step process to make solid-state nanopores. The first step in ion sculpting is to mill a through hole. Nanometre sized holes can be made using a focussed ion beam (FIB). Next, 'sculpt' the hole with broad area ion exposure. Hole size can be reduced to the desired level by this process of sculpting. This leads to transport of atoms from elsewhere to the edge of the holes. An appropriate feed-back control can be made to provide a way to monitor nanopore fabrication in real time. Detector can be set up to register the number of ions passing through the hole as a function of time [60]. Ion beam nanosculpting can provide a new tool for studying ion induced materials phenomena on nanoscale. It has been used to make an electronic detector consisting of a single nanopore in a Si_3N_4 membrane, capable of registering single DNA molecules in aqueous solution [61].

Films on substrates can be delaminated at room temperature by ion irradiation. Som et al. [62] reported MeV He^+ ion induced delamination of diamond films on a Si substrate. In situ monitoring of hydrogen, during irradiation, using ERDA can be used to estimate the threshold ion fluence for exfoliation to occur.

Ion Cut/Smart Cut

Ion cut is a unique technique provided by ion beams for the integration of dissimilar materials which enables the transfer of very fine layers of crystalline material onto a mechanical support. The process involves essentially cutting a thin layer from the bulk substrates on to the host substrate. Normally protons are used having energy in the range of 50–200 keV, to induce a mechanically fragile zone under the surface of the donor wafer. The implanted wafer is then bonded to a handle wafer and the obtained pair is subjected to thermal annealing at a temperature in the 200–500°C range. During annealing, the interaction of the implanted species with the radiated surface acts as an atomic scalpel producing extended internal surfaces parallel to the bonding interface. This leads to the splitting and transfer of a thin layer with a thickness roughly equivalent to the implantation depth. The range of the proton determines the thickness of the layer. This process is commonly known as ion-cut or smart-cut process.

Silicon-on insulator (SOI) wafers are used as the starting material for Ultra Large Scale Integration (ULSI) device structures. A typical SOI system consists of a thin layer of single–crystal silicon supported by an underlying insulator (for example SiO_2 and sapphire). Ion-cut process is now primarily used for fabricating SOI wafers which have become the foundation for ULSI devise structures, which was created to obtain delamination of thin layer in a thick substrate [63]. It turned out to be very effective in the production of SOI wafers because it enables greater uniformity of the top Si layer while preserving its good crystal quality. It involves four steps: (1) Wafer A capped with a dielectric layer (for example Si O_2) is imparted a thin layer of sufficiently dense hydrogen ions (proton implantation), (2) Wafer A is then bonded with wafer B, (3) The bonded wafer is then split and annealed. The wafer A is then split in two parts, yielding an SOI structure and the remainder of the wafer A which can be used again, and (4) The SOI layer is then polished to produce a high quality surface.

The smart-cut process enables to have practically any type of mono-crystalline layer on any type of support. Proton implantation enables delamination of a thin film from a thick substrate. The wafer bonding enables different multilayer structure to be achieved by transferring this delaminated layer on to the second substrate. This crude looking but smart–cutting process helps ensure that the wafers are accurate in thickness to less than 4 nm. Feng and Huang [64] have studied the fundamental physics and mechanics of smart cutting. It is possible to produce three dimensional patterns by, instead of implanting protons into the whole wafer, directing them to localized regions by specialized metal masks.

Ion-cut process is now being extended to make it into the process of "Ion Cut Synthesis". This can be achieved by simultaneous nanostructure synthesis and layer transfer. Formation and blistering of a layer of luminescent GaN-rich nanostructures, using N ion implantation into GaAs, followed by high temperature rapid thermal annealing can provide a way. Ion-cut synthesis is expected to enable the development of low-cost manufacturing methods.

Ion Beam Mixing

Ion beam mixing (IBM) is a very powerful process which can mix materials at the interface between two layers under the influence of incident ions [65, 66]. Consider a substrate with a film of some material deposited on the surface. Ion implantation can be used to achieve mixing atoms of the two materials. In the process of ion implantation, the ions incident on a solid are incorporated over a range of depths determined by their initial energy. It gives rise to mixing but is not ideal for the desired mixing of the atoms of the two materials because in the mixed region unwanted ions are implanted. Ions with higher energy pass through the film into the substrate and are not implanted in the region of interest. Ions passing through the material do interact with the atoms of both the substrate and the film which causes mixing of the atoms of the film and substrate atoms. Materials under ion irradiation undergo atomic rearrangements giving rise to atomic mixing. In low energy regime, IBM is primarily a result of ballistic effects in which the kinetic energies of impinging ions are transferred to target atoms on collision.

Ballistic ion beam mixing can be classified into two basic types: recoil mixing and cascade mixing, which happen simultaneously as a result of ion bombardment. In recoil mixing, atoms are relocated by single collision events and the ion itself gets implanted in the material. This mode of mixing, however, is not the dominant process in ion beam mixing. The relocated atoms of the material undergo further collisions producing a collision cascade. A collision cascade is a set of energetic collisions of atoms induced by an energetic particle in the material. If the energies in a collision cascade are higher than the threshold displacement energy of the material the collisions can permanently displace atoms from their lattice sites. The fact that lots of atoms are displaced by a cascade means that ions can be used to deliberately mix materials, even for materials that are normally thermodynamically immiscible which is referred to as cascade mixing. Collision cascades produce defects in the film and the substrate (or both the layers for bilayer systems). These defects assist in diffusion across the interface. This radiation enhanced diffusion (RED) also contributes to the Ion Beam Mixing. The nature of collision cascades can vary strongly depending on the energy and mass of the recoil/incoming ion and density of the material. The degree of mixing of a film depends on the ion mass, fluence and the total dose.

The non-equilibrium nature of irradiation can also be used to drive materials out of thermodynamic equilibrium, and thus form new kinds of alloys. Higher energies are preferred for ion beam mixing process. Energy of the ion is chosen depending on the thickness of the film and such that the nuclear energy loss is maximum at the interface [67].

Ion beam mixing (IBM) finds applications in the generation of new phases, novel materials and enhancing the adhesion of the film to substrate. Low energy ion beams have been very widely used for decades for such applications. The ion energy and species are chosen in such a way that they impart a large amount of energy to the atom at the interface of thin film and substrate by the process of elastic collisions. The electronic energy deposition beyond a certain threshold can cause the movement of atoms which can lead to mixing. Swift heavy ions have vast potential in the field of modification of materials by extremely large energy transferred through the electronic excitation but negligible damage due to elastic collisions. Ion beam mixing with swift heavy ions requires about two orders of magnitude less fluence than that with low energy ions. Progress in the area of materials engineering with swift heavy ions induced mixing is discussed in Chapter 4.

1.4 MATERIALS MODIFICATIONS WITH ION BEAMS

Interaction of the ions with the material needs to be understood in detail in order to analyze the ion induced changes in the materials. The interaction is with both electrons and the nuclei of the target material. At low energies the incident ion looses its energy predominantly by elastic collisions with the nuclei of the solid. It is referred to as the nuclear energy loss. Materials modification due to nuclear energy loss is caused by the diffusion properties of the implanted species and on the defect formed, their probability to annihilate, to agglomerate and on their ability to induce stable or metastable phase transitions. At higher energies ions start having inelastic collisions leading to excitation/ionization of electrons, which is referred as the electronic energy loss. The transition from the initial electronic excitations to structural modification and macroscopic changes in irradiated material involves a series of physical processes occurring on largely different time scales. These include excitation of electrons by a projectile, thermalization of the electronic subsystem, electron-to-ion energy transfer, relaxation of the excited material and the accompanying structural transformations. In this case the ability to induce material modifications is mainly determined by the efficiency and the rapidity of the energy transfer from the target electrons to the target lattice. Very spectacular effects are observed when extremely high densities of energy deposition are reached. Non-linear and threshold effects as function of deposited energy density are observed. As radiation induces non-equilibrium states of matter, new materials can be created with novel properties.

With the availability of high energy ions for materials studies interest emerged in studying the electronic loss associated changes in the materials [68]. A few examples of attempts made in the direction of modifying the properties of materials with high energy heavy ions are mentioned here.

Performance of tribological coatings depends on adhesion, which requires bond between the two materials having little chemical affinity. Significant improvement was observed in micro-hardness of Cr coatings on stainless steel irradiated with 75 MeV ions at 4.5×10^{13} ions cm^{-2} [69].

Study of magnetic and magneto transport properties of Fe/Tb and Fe/Cr multilayers with 80 MeV Si ions indicated that interface roughness increased without intermixing [70].

The minority carrier life time can be reduced significantly by introducing deep level defects in Si and switching characteristics of diodes can be improved [71, 72].

The microstructure and properties of polycrystalline diamond thin films were studied with 100 Mev I^{7+} ions using elastic recoil detection analysis (ERDA). Films containing lower amounts of hydrogen showed better resistance to radiation damage [73].

The specific and unique feature of ion beams is that it can deposit a large energy density which can drive the material to far from equilibrium, which can result in unusual unexpected effects on materials. The large energy density in nature exists, for example, in earth core in form of molten core. The ion beam can produce transiently molten phase along the ion track in case of high energy (~1 MeV/nucleon) heavy ions. Another example of large energy density in nature is: the impact of meteorite at the earth and such large energy density has created new phases of materials at the impact region. Similar situation exists in materials under impingement of energetic ions where phase transitions due to temperature and/or pressure are created. A schematic given in Fig. 1.2 shows such possible effects. The pressure can arise due to expanding molten liquid (transiently) causing pressure to the surrounding.

There have been surprises in this field of ion beams in materials science. Some examples are given here:

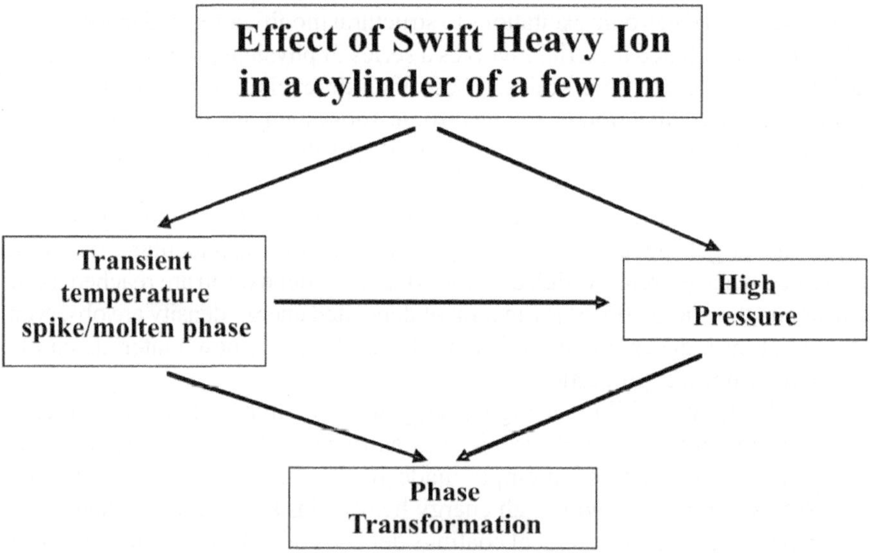

Figure 1.2: A schematic to show how SHI can create phase transformation. SHI can cause a temperature spike or a pressure spike (discussed in Chapter 2), which can result in phase transformation. The volume expansion during transient melt phase can cause pressure to the surrounding annular region.

(i) The amorphous and glassy materials undergo plastic deformation under SHI. The effect of SHI was just like hammering. The phenomenon is called hammering effect [74].
(ii) The metals, considered to be unaffected by ion beams, showed ion track with SHI [75].
(iii) Defect creation by ion beam was well known but the defect annealing and ion beam induced crystallization attracted attention of researchers and industries [76].
(iv) Polymers, known to get huge defects in the form of latent tracks by ion beams, exhibited crystallization under SHI irradiation [77].
(v) SHI at low fluences causes defect annealing of CNTs [78].
(vi) Phase transformations are observed from one crystalline to other crystalline phase form under SHI irradiation in several materials (HfO_2, ZrO_2) [79, 80] and TiO_2 [81].
(vii) There have been unusual phase transformation of material under simultaneous effect of pressure and SHI irradiation [82].
(viii) Phase transformation in BISCO (superconductor) was shown under SHI irradiation [83].
(ix) Ejection of large complex molecules occur as electronic sputtering under SHI irradiation [84].
(x) Ejection of large Au clusters favouring the possibility of thermodynamic liquid gas type phase transition under energetic ion bombardment [85].
(xi) Unusually large electronic sputtering was observed in thin Au films and sputtering yield was found to be higher for small thickness [86].
(xii) Electronic sputtering from LiF occurs in the form of jet [87].
(xiii) Change of shape of particles from spherical to elongated along the beam direction under ion impingement [88]. No other method allows the control of shape of buried or core shell particle.
(xiv) Collective rotation of nanocrystal under SHI irradiation [89].
(xv) The crystalline plane of Ni encapsulated in CNT gets tilted under SHI irradiation [90].
(xvi) Nanoscale line-like features observed in Cu_3N/glass film irradiated with 200 MeV Au^{15+} ions [91].
(xvii) Nanotracks were shown to act as transport routes for enhanced and reversible hydrogen diffusion in swift heavy ion irradiated Pd–Pr layers [92].
(xviii) Ferromagnetic behaviour is observed in irradiated C_{60} [93] and enhancement in ferromagnetic signal in SHI irradiated Pd-C nano-composite [94].
(xix) Self affine structures in low energy ion irradiation are known. Recently similar self aligned nanostructures are demonstrated by SHI irradiation of ZnO [95].
(xx) SHI irradiation of $Gd_2Zr_{2-x}Ti_xO_7$ gives amorphous track surrounded by crystalline but disordered shell [96].

(xxi) Nanosize defects are observed in LiF crystals under SHI irradiation [97].

(xxii) An Asymmetric Polymer Nanopore synthesized by etching of single ion exposure of polymer was demonstrated for Single Molecule Detection [98].

(xxiii) CNTs can be welded by low energy ions and fullerene can be welded as dimmer or polymer [99, 100].

(xxiv) Flux Pinning in HTSC by SHI was demonstrated [101].

1.5 POSSIBILITIES OF TAILORING THE MATERIALS PROPERTIES WITH ION BEAMS

Ion implantation has been exploited for decades in widely diverse fields both as a powerful research tool for investigating solid state material processes and properties, and as a means of controllably modifying the electrical, physical, chemical, mechanical and optical properties of solid surfaces. The ion-matter interaction processes lead to the modification of the composition, structure and surface topography of materials.

Flexibility provided by ion beams to manoeuvre the changes in the properties of the materials through the ion matter interaction is due to the fact that we can have the ions of energy of almost zero (fraction of eV) to very high energies (GeV), and can also control the ion fluence over a wide range covering 10^7 ions cm^{-2} to 10^{14} ions cm^{-2} in case of swift heavy ions and orders of magnitude larger in case of low energy ions.

Immense possibilities have emerged with ion beams for engineering the properties of materials. The opportunities here are in the development of new materials and nanostructures. Ion beams can be employed in a reasonably controlled way to make changes in the properties of the material. However, although the position and distribution of the implanted ions can be predicted, the defect distributions, stabilities and changes in the properties are not adequately understood and needs serious efforts. Developments in the direction of the use of ion beams towards nano-scale materials are giving rise to exciting possibilities but there is much to be researched, found, understood and exploited.

Amorphous materials under MeV ion irradiation can undergo anisotropic deformation. For example, silica contracts in the ion direction, while it expands perpendicularly. This ion beam induced deformation provides a unique method to tailor the shape of a wide variety of nanostructure materials. One can control the pore shape and pore structure of mesoporous materials etc., discussed in Chapter 5.

Availability of ions of wide range of energy is being exploited to explore the possibilities in this vibrant and exciting field where developments are taking place on a whole range of multidisciplinary frontiers particularly in the fields of nano-materials and related areas. These are summarized in Table 1.1.

Table 1.1: Developments in the field of nano-materials with ions of different energy ranges

❖ Soft Landing through Ultra Low Energy Ions	Surface structuring without any collision with these ions is opening new horizons
❖ < 10 keV Low Energy Ion Beam	Ion assisted nano patterns/nano-ripples Example: Self Organized Quantum Dots [SOND] ~75 eV-1.8 keV Ar ions with normal incidence produce "Hexagonally ordered Quantum Dots", with characteristic length of the array depending on the squareroot of ion energy. Ripple Structure: Ion beam incident at an angle produces a ripple structure on the surface, which depends on the angle of incidence.
❖ 10 keV-10 MeV near Medium Energy	Ions provide most versatile means to introduce an element "A" in surface region of "B" and obtain supersaturated solution of the two elements. This leads to: Synthesis of nanoparticles through ion implantation and Ion Beam Mixing in bilayers. Ions alter tribological properties on a micro scale, alter optical and other properties and form nanocomposites nucleating nano dots. Patterning nanostructures by ion induced dewetting
❖ > 10 MeV (SHI)	Swift heavy ions generate track shaped damage, a few nm in radius which can be a few mm in length. Tracks are modified materials contained in well defined columns. By varying the parameters of ion beams one can engineer the property of the material inside the track. The track can be etched to make holes and filled with the desired materials.

The table gives a brief account of the developments in the fields of nanostructures and nanomaterials synthesis using ion beams accessible from existing particle accelerators in different energy ranges from keV to hundreds of MeV, and special facilities for creating ultra low energy heavy ions. It is found that unique manipulations in the properties of the materials is possible as we go to higher energies if we can understand how the ion matter interaction changes as a function of energy.

There are plans to have a beam of heavy ions in the order of 10 GeV/amu from the FAIR accelerator complex at GSI equipped with a raster scanning beam delivery system which can ensure beam quality and shape homogeneous large-field irradiations over a wide range of particle fluences. At such high energies there are no systematic measurements and thus it is not possible to predict the effect of these ions only from an extrapolation from lower energies. At energies higher than 1 GeV/amu the radiative energy loss becomes relevant in addition to the electronic energy loss.

1.6 WHY ION BEAMS FOR MATERIALS ENGINEERING?

Starting with the creation of the revolution in semiconductor technology, ion beams were making impact in characterization of the samples and were being appreciated as non–destructive techniques for analyzing samples of various kinds. Changes caused in the material being investigated by the ion matter interaction were analyzed to focus on the possibilities of modification of the properties of the materials of different kinds. With the advances in this direction and availability of ion beams of all possible energies it became possible to tailor the properties of the materials. This along with the development of the understanding of ion matter interaction led to the idea that one can in fact engineer the properties of materials in a planned way. The question naturally arises as to why ion beams for materials engineering.

An attractive point of ion beams for materials science is the non-equilibrium conditions produced by the ion matter interactions. Energy of 1 eV corresponds to a temperature of 10^4 K. Of course, the comparison is not directly applicable to materials effects, since the energy stored is rapidly relaxed. However, the highly non-equilibrium states create unusual material processes and novel material properties. These are never obtained under thermal equilibrium conditions. Non-equilibrium materials science is still at a nascent state, but with a promise of unlimited possibilities for the future.

Appropriate modelling of the non-equilibrium phase diagram produced by ion irradiation can provide predictive power to ion-based techniques. It has been shown how one can, in this way, go so far as to design systems with interesting physical properties that may lead to applications in microelectronics or magnetism [102].

There has been evolution in this field with high energy heavy ions during the last two decades. A sizeable scientific community of materials science has been researching with ion beam facilities originally devoted to nuclear science studies. The interest of this community has been on the highly excited states of matter induced by swift heavy ions [SHI]. Importance of other types of beams such as cluster beams, highly charged ions etc. for materials engineering is now gradually being realized.

Ion beam facilities provide tremendous flexibility due to the possible variations in ion parameters and the sensitivity of the ion matter interaction on these parameters. Energy of the ions can be produced from almost zero velocity to GeV energy with the following variations:

- Ions: of all elements
- Charge states: 1^+ to 92^+ (in principle) Singly charged negative ions
- Current density: few ions cm^{-2} to 10^{18} ions cm^{-2}
- Beam: Continuous or Pulsed (up to ~50 Picosec)
- Focussed Beams: Nano, Micro, ..., or, Broad beams with uniform intensity
- Beam in high vacuum to atmospheric pressure
- Programmed scanning and dwell times

There are opportunities in the area of development of new materials and nanostructures. Ion beams of isotopically pure species can be employed in an energetically tailored environment combined with the remarkable in-situ and favourable conditions of dynamic controls. A large array of parameters such as the dose, the dose rate, the target temperature, the velocity of ions, the initial state of the system, synergetic effects with the previous state etc. determine the final outcome of the interaction of ion beam with matter. The subsequent post-processing provides even more flexibility in the creation and manipulation of new material. Currently most of the work seems to be empirical requiring much more research and characterization before it can mature as technique to be exploited industrially. The subject is very vibrant and exciting.

There has been significant activity in the direction of simulating the expected results of the ion matter interactions under different conditions. These efforts are helped by a great deal of enormous experimental data which is becoming available, thanks to the availability of various types of ion beam facilities being developed and increasing interest of scientists in the emerging challenges in this field of engineering new materials, and proceeding in the direction of making smart materials and thinking about ways to make intelligent materials.

1.7 SPECIALLY CONFIGURED ION BEAMS

There have been phenomenal developments in accelerator technology providing ion beams of any desired characteristics. It is making significant impact in the futuristic technology development through the engineering of materials with charged particles. A brief description of the contribution of the specially configured beams to the development of materials science and new materials is given here.

Low Energy Highly Charged Ions

Low energy highly charged ions (HCI) provide a technique to produce nanostructured surfaces. With the shrinking of semiconductor devices, nano dimensional features and structuring become increasingly important. For fast ions, the kinetic energy is not only dissipated close to the surface but also in deeper layers of the material. Slow (< 1 keV) highly charged ions as opposed to fast ions are a new tool for gentler structuring of surfaces at the nanometre scale. The substrate is modified only at and slightly below the surface, opening the possibility of controlling electronic properties at the nanometre scale.

The potential energy in a Highly Charged Ion (HCI) can be quite high. The stored potential energy is equal to the energy spent in removing a large part, say q, of their Z electrons (Z being the projectiles nuclear charge). This potential energy becomes very large for high values of q (the ion charge state). Low energy very highly charged ions (HCI) can extract a large number of electrons from surfaces. As the ion approaches the surface, target electrons are captured in high energy levels of the projectile and re-emitted by Auger effect. Consequently, the number of electrons extracted can be higher than the charge state of the projectile [103]. Upon surface impact this potential energy is available for inducing various inelastic processes while the ion regains its q missing electrons to again become fully neutralized. The ion deposits its potential energy in a short time (typically about 100 fs) within a small area (typically less than 1 nm^2). In the course of HCI neutralization at the surface a multiple excited neutral particle with empty inner shells is formed, which is known as "Hollow Atom". Hollow atoms are short-lived multiple-excited neutral atoms which carry a large part of their Z electrons in high energy levels while inner shells remain (transiently) empty. For impact on insulator surfaces the potential energy contained by hollow atom may also cause the release of target atoms and ions via potential sputtering and the formation of nanostructures on a surface.

Negative Ions

Negative ions play a crucial role in materials engineering. To fabricate a two-dimensional structure embedded in an insulator, medium-low energy ions (several tens keV) are effective. In the case of positive ions, Coulomb repulsion due to surface charging impedes precise implantation, especially in case of insulators. For instance, if surface current leakage is neglected, positive-ion irradiation for several seconds builds up a surface voltage of several tens kV, which is comparable with the acceleration voltage, and repels further incident ions by Coulomb repulsion. It can cause decrease in the incident ion energy and thus can effect the implanted species depth profile. Consequently, negative metal ions have a decisive advantage, where slight positive charging by negative ion bombardment is neutralized by surrounding electrons of low energy. The charging-free conditions enable us to make precise and efficient atomic injection into a solid. This merit further provides us with a good tool to kinetically control nanoparticle precipitation, including morphology control, because we can change the ion flux over a wide range without surface charging effects. The insulating matrix are most common substrates of interest.

High Current Ions

Increasing the ion number density provides a variety of material phases resulting in new controllability. "High current density" implies a high number density of ions/atoms per unit time. For materials engineering, not only the high energy but also the high current density is important factor. It provides high energy deposition and high atomic flux per unit time, which give a variety of rate dependent processes such as: radiation-induced diffusion, enhanced sputtering, ion-beam mixing, annealing and so on. In more extreme cases, radiation-induced melting or evaporation takes place. The "high current" here is of a level of 1 mA or 1 mA/cm^2 as a density. Current (~ mA) itself is small but energy is high, tens of kV to several MV. The current density of 1 mA/cm^2 is comparable with the atomic number density in gaseous growth methods, such as vacuum evaporation for thin film making, and each particle carries on huge energy which produces spectacular effects.

Pulsed Ion Beams

High-intensity pulsed ion beams (HIPIB) provide capability for surface engineering based on rapid thermal processing of the top few microns of metal, ceramic and glass surfaces. The processing is based on the beam-material interaction: remelting and/or ablation of a top layer on the irradiated surfaces (extreme surface heating effect); subsequently, the molten state may be frozen at an ultra-fast re-solidification rate after termination of the ion beam pulse. Short pulse lengths can create partial melting with subsequent quenching. The depth of treatment is controllable by varying the ion energy and species. This Ion Beam Surface Treatment (IBEST) process can produce enhancement in the hardness, corrosion, wear and fatigue properties of surfaces by rapid melt and resolidification. Efficient deposition of the energy in a thin surface layer allows melting of the layer with relatively small energies (typically 1-10 J/cm^2) and allows rapid cooling and resolidification of the melted layer by thermal diffusion into the underlying substrate. Typical cooling rates ($\sim 1 \times 10^{14}$ K/sec) are sufficient to cause amorphization and fine grain layer formation and the production of new micro structures including nano-crystalline and meta-stable phases, as seen in the case of magnesium alloy [104].

Surface smoothing and reconstruction of titanium alloys and ZrO_2-Y_2O_3 coatings have been observed as one of the typical outcome under high-intensity pulsed ion beam irradiation. It is demonstrated that the changes in surface morphology may significantly contribute to the improvements of overall performance of the materials [105].

Cluster Beams

The unique capabilities of specially configured ion beams for nano-fabrication are now opening new horizon. The uniqueness comes not only from the atomic dimension of the ion but also from the details of the energy loss of the ion in matter. The availability of energetic cluster beams holds the promise of substantially broadening these capabilities for nano-fabrication. However, the energy loss of cluster ions in matter is neither studied in detail experimentally nor understood theoretically. For example, it is still not known whether the differential energy loss of a cluster ion is the sum of the differential energy losses of the constituent ions considering that the inter-atomic distances in a cluster are much larger than atomic dimensions. The question to be answered experimentally is whether the differential energy loss of a carbon dimer is twice that of a carbon ion at the same ion velocity or is the same as that of a magnesium ion at the same ion velocity.

An energetic ion interacting with a solid loses energy through electronic excitation of target atoms and scattering by target atomic nuclei. If an energetic ion containing two atoms (dimer) enter into a solid they can travel a short distance without being completely separated. Within this distance in the solid, the closeness of the two components of the dimer ion can cause modifications of the response of the electrons in the solid, in comparison to the passage of monomer ions. This coherent dynamic response was indirectly observed in C^{+}_{2} implantation into a semi-insulating GaAs crystal [106]. In a semi-insulating material electronic excitation can cause damage. An enhanced damage production was observed in the GaAs crystal surface region for C^{+}_{2} implantation compared to C^{+} implantation, where in both cases same energy/mass of the ions and the same atomic fluence was used. The near-surface modified layer was studied by X-ray reflectometry, RBS, Channeling and TEM.

Ionized cluster beams are considered as a powerful and versatile tool for modification and processing of surfaces and near-surface layers as an alternative to ion implantation and ion assisted deposition.

Ion beam technologies have attained an advanced stage of development. However, nowadays fabrication of nanometre scale structures needs much more controllable and versatile tools on an atomistic scale. Requirements for ion implantation are becoming especially challenging, for example, in areas of shallow doping (shallow junction formation), low damage/high-rate etching and smoothing, ultra thin film deposition and special surface relief growth. Cluster beams provide special conditions mentioned below:

(i) Elastic scattering dominates from first atomic layer of the substrate

An ionized cluster is characterized by low charge to mass ratio. With clusters consisting of up to a few hundreds or thousands of atoms it is possible to transport a large amount of material with just a single charge. Due to the large size and weak bonding of atoms the cluster-surface impact is fundamentally different from that of monomers. The cluster size usually exceeds a target unit cell size. Hence, nuclear stopping (direct collisions with target atoms) dominates for clusters from the first atomic layer of the substrate.

(ii) High density energy deposition

Multiplication of this low energy by the number of atoms and dividing by very small surface collision area leads to high energy density deposited by the cluster. Depending on implantation energy and cluster size this can cause nonadiabatic phenomena and nonstationary quantum phase transitions. High-density energy deposition allows the formation of smooth films and fabrication of nanostructured patterns [107]. At low energies it is possible to grow ultrathin films and special surface by cluster assembling. One would like to have energy in the realm where the energy per atom of the accelerated cluster is below the binding energy of constituents in the substrater. At energies above this one gets high sputtering yield by cluster bombardment which produces high-etching rates and surface smoothing to a level unachievable using monomer ions. The high sputtering effect gives an opportunity to use it as an effective technology for surface smoothing and cleaning or etching.

When the energy of a cluster constituent atom is more than the penetration threshold energy of a target material, the cluster breaks down under the energetic impact and constituents penetrate into the substrate or form a film. They can also rebound or scatter from a surface, especially at high angles of incidence.

(iii) Shallow implantation to form nm thick layers

Clusters generate multiple-collision effects with lower energies. An acceleration of, for example, 100 atoms cluster to 10 keV provides 100 eV energy to each constituent atom. The possibility to decrease implantation energy to eV-range per atom makes cluster beams very attractive for shallow implantation providing effective doping of nm-scale-thick layers.

New generation of transistors needs to have junction of ~30 nm which requires ions of a few keV energy. The beam current drops for such ions due to the space charge effects which limits low energy implantations. Low energy implantation of $B_{10}H_{14}$ clusters is shown to provide the answer [108].

Cluster-solid interaction is non-linear which gradually increases with cluster size. The nonlinearity arises from the fact that the cluster atoms influence each other during the penetration into the target. Since the cluster breaks down into single atoms quite rapidly after impact (tens of fs), one can expect, for example, an overlapping of the collision cascades originating from individual cluster atoms. Thus, the possible difference between the stopping of a dislodged atom of a cluster and a single atom can make some difference in the projected range of cluster constituents compared to single ion implantation of the same species. MD and Monte Carlo simulations show a decrease in the stopping power per atom in clusters compared to the single atom. In other words, the penetration of clusters is larger than for the corresponding ions at the same incident velocity. It was suggested that so-called clearing-the-way effect, where the "front" atoms of the cluster push target atoms out of the way, could take place. The mean range and straggling are found to be an increasing function of cluster size for the same implantation energy per atom as shown by MD simulations for gold implantation into copper [109, 110].

Cluster impact can cause surface erosion or crater formation [111]. Along with craters, hillocks (nm-size protrusions) are found in cluster implantations of various substrates. It is found that size selected Ar clusters implanted in Si with ~18 keV energy show hillocks formed inside the craters.

The study of cluster–surface impact, soft landing of clusters on the surfaces, and cluster implantation is important for the futuristic technology.

Microfocussed Heavy Ion Beams

Focussed heavy ion beams are developed to achieve targeted irradiation to engineer the properties of materials in configured ways. Focussed microbeam, unlike a collimated one, is not restricted to a single location on the accelerator exit window. This provided a way to develop high resolution lithography technique using focussed ion beams. Proton beam [p-beam] writing is a direct-writing process that uses a focussed beam of MeV protons to pattern-resist material using the beam scanner system. The direct-write technology using focussed 50 keV electron beams (e-writing) has been providing efficient way to fabricate structures up to ~100 nm level. The proton beams provide a better technique because they travel in a more or less straight path apart from a small amount of end-of-range broadening where nuclear energy loss becomes prominent. The probability that a proton interacts with atomic electrons is a few orders of magnitude larger than for nuclear scattering and because of the high mismatch in mass between the proton and the electron, proton-electron interactions do not result in any significant deviation in the trajectory of a proton from a straight line path. On the other hand, focussed electron beam spreads rapidly as it enters the resist material. Proton micro-beams can produce 3D structures with high aspect ratios. The sub-micron lateral resolution and well defined range of proton microbeam provides a way to make multilevel structures in polymers. Sub-micron structures with a depth of tens of microns and aspect ratios approaching 20 have been achieved [112]. The primary physical characteristics of focussed mega-electron-volt (MeV) proton beam, that of high penetration with little lateral beam spreading, makes p-beam writing the only true direct write, three dimensional fabrication technique at the nanoscale. The ability of p-beam writing to create smooth three-dimensional channels in polymers allows the direct writing of microfluidic devices which can sort cells, DNA, and large proteins. P-beam writing in Si results in a pattern damage profile that can be used in a variety of ways [113]. Technologies are being developed to produce and integrate microstructures and microelectromechanical systems (MEMS).

Microbeams opened new horizons in bioscience and related technologies. Study of "Bystander effect" with a single alpha particle hitting a cell is providing information which will help in radiation therapy. It is possible to study whether the DNA is affected in the cell when the alpha particle hits the cytoplasm of the cell but not the DNA [114]. A magnetic-coil-based fast deflector, placed between the two quadrupole triplets allows deflecting the beam to any position in the field of view to observe the cells during irradiation. Microbeams are ideally suited to the study of so-called 'non-targeted' phenomena that are now known to occur when living cells and tissues are irradiated. Non-targeted effects are those where cells are seen to respond to ionising radiation through pathways other than direct damage to the DNA [115].

Micro-PIXE based trace element analysis of cell is being done by exposing individual cell to protons and analyzing the X-rays to look for correlations, if any, of trace elements with diseases such as Alzeimer etc. [116].

Apart from the depth resolution of ion beam analysis techniques, one can have spatial resolution governed by the beam diameter of the microbeam. Such a spatial selectivity is very interesting in fault finding in electronic circuits, IC chips etc. In principle one can perform 3D microscopy. One such example is that of 3D hydrogen microscopy in diamond performed using a microbeam and ERDA technique in conjugation with a magnetic spectrometer [117].

Nano-Focussed Ion Beams

Nanofocussed heavy ions are providing novel routes in the field of materials engineering. Presently accessible focussed ion beam [FIB] system generate gallium (Ga) ions with the beam energy of typically 30 keV, focussed to ~7 nm with ~20 nA beam current. FIB system was first used by Hitachi in 1985 for integrated circuit (IC) cross-sectioning for failure analysis [118]. These beams in nanometre range have been of immense importance in the semiconductor industry where they are used to carve structures with dimensions measured in billionths of a metre, repair defects in masks used for photolithography, isolate and analyze elements of integrated circuits, etc. Currently FIB systems are widely used in nanofabrication, ion lithography, functional structure prototyping etc.

Importance of these nano focussed ions is now realized in diverse disciplines. FIB has emerged as one of the most important tools for the scientific research and industrial applications involving nanotechnology. FIB can shape materials with microscopic precision. They can shape nanoscale materials either by adding atoms to a structure or by shaving them off. Besides the nanoengineering possibilities with these nanofocussed ion beams, there is another point which provides scope in a new direction. In ion beam based engineering of the materials there are applications where one feels limited by the flux of the available ion beams. Due to limited ion flux, the localized excitation induced by single ion impacts are de-linked in time. In nanofocussed ion beams, multiple ions can hit a nanometric zone in a time span, which can be made less than the lifetime of excitation induced by a single ion. The current provided by FIB is not large but because it is focussed in nanodimension the current density is very large. Thus, before the excitation induced by one ion has decayed down, another ion can hit the same zone and induce further excitation. The consequence of such multiple ions hitting the same zone in a short time span can be manyfold. High current density can induce dramatic changes in the properties: magnetic, ferroelectric, optical, etc.

Present FIB systems use Liquid Metal Ion Source [LMIS] with metal gallium because of its low melting point and low vapour pressure. Gallium metal is placed in contact with a tungsten needle and heated. Gallium wets the tungsten, and an intense electric field (> 10^8 V/cm) causes ionization and field emission of the gallium atoms. Ga ions are then accelerated and focussed onto the sample by electrostatic lenses. LMIS produces high current density Ga ion beams with very small energy spread. Other materials, mostly alloys, tried with LMIS have not been too successful.

Focussed Ion Beams [FIB] make it possible to implant single ions a few nm apart. This level of control and the development of technology associated with FIB facilities has opened a new horizon. When ions strike surface of the sample they sputter atoms from the surface. Because the ion beams can be focussed to small sizes their sputtering capability can be used as a nano-machining tool.

When the beam strikes the sample, secondary electrons and secondary ions are emitted from its surface. The electron or ion intensity can be monitored with appropriate detectors to generate an image of the surface. As with any analytical technique, FIB analysis also has its drawbacks. Major problems include damage to the surfaces from Ga ion implantation, and the fact that some milling can occur during the imaging process. This milling slowly degrades the quality of the images. Secondary electrons are used for most imaging applications. Dual-beam FIB systems combine FIB and electron beam. Ion beam can be used for milling and the electron beam for imaging.

Ion beam milling can be performed at an enhanced rate with the assistance of small quantities of gas, termed as Gas assisted etching (GAE). These gases are introduced via gas needles that are positioned near the area being milled. Gas-assisted ion milling offers several advantages, including selective milling of certain types of materials such as metals, insulators or carbon-based compounds. When gas is introduced near the surface of the sample, the sputtering yield can selectively increase depending on the chemistry between the gas and the material. During the gas-assisted process, gas is adsorbed onto the surface, where it reacts with the ion beam and the surface material, producing volatile compounds that are pumped away.

In a similar process to gas-assisted milling, FIB is also used to deposit material via ion beam induced deposition. FIB-assisted chemical vapour deposition occurs when a precursor gas is allowed to chemisorb onto the sample and the beam scans an area. The gas decomposes into volatile and non-volatile components and the non-volatile component remains on the surface as a deposition. In these applications, the beam parameters and gas flow are optimized for the most efficient equilibrium between the cracking of the precursor gases and the milling action of the beam. Complex structures can be built, layer by layer, by spraying appropriate gas on the sample surface and dissociating the gas molecules adhered to the surface by the energetic ions. The volatile components which are mainly hydrocarbons leave the sample, leaving behind the metallic element which is then firmly attached to the surface.

One of the basic structures, which are implemented in many MEMS devices, is a resonator in the form of a cantilever or a pillar. Nano size cantilevers and pillars on a suitable substrate can be easily fabricated by FIB or FEB (Focussed Electron Beam) which offer the ability to detect very small motions thereby improving the ability of mass sensing down to attograms (10^{-18} gm). A cantilever fabricated with FIB has been prototyped and patented to act as a miniature bomb-detector by having a sensor attached to the cantilever top. The sensor part can be tagged to pick up particular types of proteins and can be used for diagnosis.

The focussed ion beam facility provides basic research possibilities in nanoscale physics. In one experiment classical laws governing electron emission and transport in the presence of fields and space charges were studied for nanoscale vacuum electronic systems. The classical Child Langmuir's law, which is known to hold for space charged limited (SCL) flow of current for electrodes placed in vacuum (I proportional to $V^{3/2}$), where I is the current and V is the applied voltage between the electrodes, is investigated in the quantum regime. The voltage is found to scale as $V^{1/2}$ rather than $V^{3/2}$ for the classical case [119, 120].

Present FIB systems deliver tens of nanoamperes of current of Ga ions. These beams are very useful for milling and deposition, or can image the sample with a spot size on the order of a few nanometres. But Ga can contaminate the surface. Erosion and contamination makes Ga poor ion for imaging technique. Materials engineering also needs other types of ions.

There is a need to develop technology so that different type of ions can be focussed to nanometre spots for imaging, milling and deposition. Attempts are being made to develop ion beams from a variety of different atomic sources. The ability to use lighter elements such as helium or lithium could be particularly useful for imaging applications, such as ion microscopy. Ion microscopes have the potential to provide sharper images than electron microscopes, as long as the ions are not so heavy that they damage the surface. As circuits get smaller it becomes increasingly important to control the precise placement of single metal atoms within semiconductor materials to give them the desired electrical properties for developing the next-generation nanodevices.

Nanostructures by Focussed Ion Beams (FIB)

FIB milling has proven to be a powerful tool to fabricate nanoscale structures. It is used in the development of nanoelectronics [121], magnetoelectronics [122], for high resolution modification of MFM tips etc. Nanofocussed ion beams provide simple method to pattern metallic surface nanostructures with precisely controlled size, spacing, and location. Predefined patterns by focussed ion beam direct-writing can be used as the templates for the self-organization of ordered nanostructures [123]. This approach represents a maskless process that combines the top-down and bottom-up patterning methods, and no chemical etching or pattern transfer steps are involved. Nano patterns can be generated using focussed ion beam (FIB) induced adhesion, by selective irradiation of thin metallic films grown on substrates by focussed ion beam followed by peel off. After peel off of the irradiated thin film, the ion beam scanned portions are retained on the substrate, creating nano patterns. The method is suitable for materials of which the adhesion to the substrate can be improved by ion bombardment [124].

Huang et al. [125] produced aligned nanoneedles in Fe/GaAs (001) epitaxial thin film when irradiated with a Ga^+ ion beam normal to the sample surface at an ion dose of 3×10^{17} cm^{-2}. These nanoneedles, with their axes perpendicular to the original sample surface, are composed of GaAs nanorods with Fe nanoislands situated at the top. The crystalline nature of the Fe thin film is disrupted by the ion sputtering, becoming nanocrystalline Fe islands. It is the thin Fe layer that is responsible for the formation of the nanoneedles. The ion sputtering of the Fe layer roughens the sample surface due to the uneven sputtering rate of the Fe layer. The removal of the Fe layer exposes the substrate to the incident ions, which then sputter the substrate leading to the formation of the nanoneedles.

The focussed ion beam is a powerful tool for engineering at nanoscale. It allows high flexibility of the nanostructure designs that can be realized. The slow process is the main drawback of the FIB. This technology is presently best suited for prototyping nanostructures.

1.8 ION BEAM ASSISTED SELF ORGANIZATION

It is possible to guide the self-organization process in a material by imposition of appropriate conditions through ion-matter interaction processes. In self-organization a system develops some kind of structure by itself, that is, without selective intervention from outside. Ion beam sputtering i.e. the removal of material from a surface due to the impact of energetic ions is a surface processing technique which can induce self-organization process resulting in nanostructures with different shape and high order arrangement can be evolved. The formation and self-organization of "ripples" and "dots" spontaneously appear during uniform irradiation of surfaces. Creation of nanoripples by ion irradiation at certain range of angles has an analogy with self assembled sand dunes in the desert due to wind. Glancing incidence ion beam irradiation on a surface induces self ordering of atoms. This is a field with immense potential in creating materials of desired properties for futuristic technologies. We first consider the process of self organization before bringing in the ion beams.

The basic concept of self organization is that the dynamics of a system tends to increase the inherent order of a system. There are processes that can be described as self-organization, e.g. phase transitions, spontaneous magnetization, crystallization, liquid crystal, Bose-Einstein condensation, molecular self-assembly, etc. It is now believed that there are universal laws (arising from fundamental physics and chemistry) that govern the growth through self organization in biological systems.

The monolayer formation of alkanethiols on gold surfaces is an example of self-organization. Alkanethiols are similar to lipids. They possess a hydrocarbon tail group (1–4 nm length) but instead of head group designed to be water soluble, they have a head group designed to specifically bond to a gold surface, such as a thiol group. Because of the strong interaction between the thiols and gold, when a gold surface is placed in a dilute solution containing alkanathiols, a monolayer of these attaches itself to the gold surface via the thiol group. This technique allows us to control the surface properties of materials. It is possible to incorporate a wide range of functional groups into the individual molecules which can be used to further modify the properties of the surface. For instance by making the functional group a methyl group, the surface becomes hydrophobic, so that water runs off the surface. Replacing the methyl groups with negatively charged phosphate groups makes the surface hydrophilic and wettable with water. The surface is sensitive to the acidity of the solution and thus can be used to measure pH changes. It is possible to influence the electronic nature of the substrate material. These are some of the wonderful possibilities with chemistry.

Ion Beam Assisted Self-organized Nanostructure Formation

Synthesis of nanoparticles on the surface can be achieved with low energy ions through ion induced surface modification. This technique can be utilized in two ways. One can either deposit the film on selected areas and irradiate the whole surface or make the ion beam raster over selected areas, thus leading to the formation of nanoparticles in the desired regions.

keV Ion Induced Formation of Nanostructures

Low energy ions can create periodic arrays of nano-ripples and nano-dots on the surface. The ion energies from a few hundred eV to a few keV have been very effective for ripple creation. The basic mechanism of the evolution of surface morphology leading to nanostructures is a complex phenomenon, which includes roughening by sputtering and smoothing by diffusion process, and the final morphology is determined by the balance between them. The periodic ripple can be formed either perpendicular or parallel to the projection of the ion beam direction in the surface plane depending on the incidence angle. An off-axis incident ion beam on solid can produce a ripple structure. The first theoretical approach describing the process of ripple formation due to ion bombardment was developed by Bradley and Harper [126]. Stochastic model was developed by Cuerno et al. to explain the surface morphology [127]. Nonlinear continuum theory was later developed [128] to understand the experimental observations not predicted by the linear theory. In this formalism surface diffusion process was included. Diffusion constant has two terms: thermally activated surface diffusion and the ion induced surface diffusion. Ion induced process predicts an increase of the ripple wavelength with ion energy, but the thermal diffusion process, assumed to be dominating at high temperatures, predicts a reduction of ripple wavelength with increasing energy. Increase of ripple wavelength with energy was observed [129]. Chini et al. [130] showed that bombardment of Si with 50–140 keV Ar ions results in the increase in the ripple wavelength with ion energy if the beam is scanned, and decreases with energy without ion beam scanning.

The effect of pre-existing random roughness on the evolution of ripple structures was investigated by Mishra et al. [131]. There is considerable reduction in initial roughness of the film surface at the early stages of sputtering.

For large scale surface structures, angle-dependent first order sputtering is responsible for ion beam smoothening, while for smaller microscopic features, different relaxation mechanisms dominate for smoothing of the surface. At the later stages of sputtering, the curvature dependent erosion instability sets in leading to the development of either coherent ripples or faceted structures depending on the degree of virgin film roughness and bombarding angle. The gas cluster ion beam (GCIB) provides excellent technique for surface smoothening of materials [132-134].

Ion sputtering produces self-organized structures resulting from the balance between the erosion induced by the ion collisions and diffusion of surface atoms displaced after the impact. General mechanism involved is surface curvature dependent sputtering yield and surface diffusion [135].

Metallic surface can be patterned into metallic nanostructures by utilizing keV ion beam induced dewetting and Rayleigh instability [136]. Dewetting studied by high energy ion irradiation indicate that it can be used for well controllable sub-μm patterning of thin solid films [137].

Glancing incidence ion beam assists in self-organization of a periodic arrays of different configurations depending on the material. A periodic wire arrays can be created on a CaF_2 (111) substrate with low energy ions. Preferential erosion of fluorine by the ion beam creates a surface enriched in calcium. The calcium self-organizes in elongated island structures of preferential width and separation. Batzill et al. [138] demonstrated that the CaF_2 (111) substrate irradiated with 4.5 keV Ar^+ ion beam along a fixed azimuth produces nanowires with ~10 nm periodicity and wire lengths of at least several micrometres oriented along the azimuthal direction of ion beam incidence.

keV Atoms Induced Formation of Ripples and Nanodots

Atom beam has been used to create ripples at surface on Si crystal surface and the dependence of the ripple wavelength on atom fluence and on the orientation of Si surface was studied [132].

Ion/atom bombardment is known to enhance surface diffusion and affect the surface morphology to produce nanostructures. The fluence and angle dependent studies done with 1.5 keV Ar atom beam with a flux of 14.8 mA/cm^2 are shown in Figs 1.3 and 1.4.

The formation of ripples and dot structures on sputtered InP (100) were characterized with SPM. The ripple wavelengths and roughness increases with the fluence [138-140]. A transition from nanodots to ripples was observed. The formation of nanodots at lower angles of incidence is shown in Fig. 1.3. The development of ripple structures at higher angles of incidence is shown in Fig. 1.4.

MeV Ion Beam for Synthesis of Embedded Quantum Dots

Ion beam irradiation can induce a formation of long-range ordered quantum dot arrays in amorphous silica matrix. Twenty alternating ($Ge+SiO_2$) and SiO_2

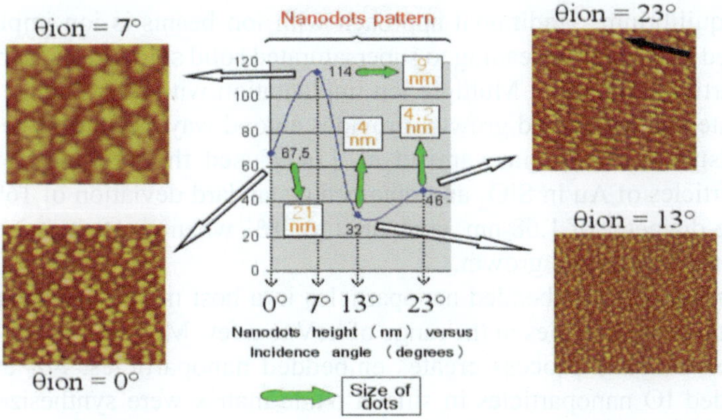

Figure 1.3: The formation of nanodots at different angles of incidence of the atom beam (0°, 7°, 13° and 23°). [From I. Sulania, A. Tripathi, D. Kabiraj, M. Lequeux and D.K. Avashi, *Advanc. Mat. Lett.*, **1** (2010) 118, copyright VBRI Press, with permission from VBRI.]

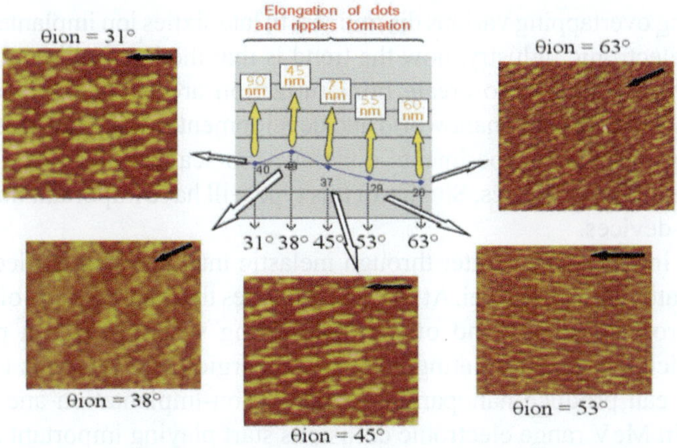

Figure 1.4: The development of ripple structure as the angle of incidence of the atom beam is changed from 31° to 38°, 45°, 53° and 63°. [From I. Sulania, A. Tripathi, D. Kabiraj, M. Lequeux and D.K. Avasthi, *Advanc. Mat. Lett.*, **1** (2010) 118, copyright VBRI Press, with permission from VBRI.]

layers irradiated by 3 MeV O ions with the dose of 1×10^{15} at/cm^2 under the angle of 60° induces the nucleation of amorphous Ge quantum dots in well-ordered chains and the inclination angle of the chains corresponds with the irradiation direction. After the annealing quantum dots crystallize resulting in a 3D array of crystalline Ge quantum dots embedded in SiO_2 matrix [139].

Buried Nanoparticles by Ion Implantation

Ion beams offer possibilities for synthesizing nanoparticles under extreme physical conditions and in highly localized regions, involving processes far from equilibrium. Traditional approach with ion beams is ion implantation followed by annealing leading to supersaturated solid solution and subsequent nanoparticle formation. Multiple ion implantation with intermediate anneals to isolate nucleation and growth, provides a good way to control the particle size distribution. Ramaswami et al. [141] used this technique to form nanoparticles of Au in SiO_2 and obtained a standard deviation of 16% on an average diameter of 1.68 nm, compared to 28% when no attempt is made to isolate nucleation and growth.

Formation of embedded nanoparticles in a host matrix can be achieved using ion beam energies in the range of keV to a few MeV. Ion-beam-induced collision cascade process creates embedded nanoparticles. For example, embedded IO nanoparticles in silicon oxide matrix were synthesized using 320 keV In ions and the size of which was varied by annealing [123].

1.9 SUMMARY AND PERSPECTIVE

Ion beams are now emerging as almost indispensible tools for mesoscopic structuring overlapping various disciplines. In late sixties ion implanters became must of electronic industry, now the trend is that the ion facilities being used by electronic industry to create miniaturization are being re-discovered by scientists for research in nanoworld and development leading towards futuristic technologies. A few experiments done at nano-scale level have given clear indications of new physics. Studies of this type will have important implications for nano-devices.

Ions interact with matter through inelastic interaction with electrons and elastic scattering with nuclei. At very low energies there are cascade of collisions which produce some kind of nanopatterning which helps in producing nanoparticles in thin insulating matrix. In energies of a few tens to hundred keV one can produce nanoparticles through ion-implantation and ion-beam mixing. In MeV range electronic collisions start playing important role in the process of engineering the properties of materials. Then at higher energies track technology starts playing role. Control of the way ion is interacting with the material allows unlimited possibilities of creating materials with unique properties. Ion beams are very versatile tools because of very large number of controllable parameters: energy (eV-GeV), charge, current density, pulsing up to ~pico-sec., focussing up to nanometre range etc. Considering the importance of these parameters facilities are being developed for controllable cluster beams, Ultra Low Energy Ion Beam (ULEIB) for soft landing of ions on the surfaces etc. providing new openings/handle to engineer the properties of the materials. Now the excitement is due to the availability of nanofocussed ion beams. Besides localization in nanorange it can give rise to synergetic effects of individual impacts. Multiple ions can hit a nm zone in a time span which can be made less than the life time of excitation induced by single ion. High current density can create local defect injection, chemical composition modification, and phase transformation. It gives rise to dramatic changes of magnetic, ferroelectric, optical, electric and magnetic properties.

Ion beams are opening various possibilities in the direction of tailoring the properties of materials towards certain functional goals. Basically FIB patterning techniques rely on a sequential process which can be used in manufacturing. FIB instruments combine high resolution patterning by local irradiation and imaging by Scanning Ion Microscopy. This can be turned into a process of unattended–automatic batch operation having self diagnostic and repair capability.

There seem to be unlimited possibilities for nanostructuring materials. To what extent we can control the parameters is important. In order to examine this point we have to concentrate on the fact that the ion-matter interaction at the nano level is not yet fully analyzed. The crucial aspect of nanophysics is that scaling does not hold. At the nanoscale, the properties of materials can be easily ordered by small structural, compositional and size variations. Because of the small energy differences involved, these properties are very sensitive to ion beam interactions.

Ion beams can be used to tailor the structure and properties of nanosystems with high precision. Use of ions to develop futuristic technologies has started opening up not only challenges which can be handled through multidisciplinary approach but has emphasized the need to look at the ion atom interaction physics afresh through the subject of nanoscience. There is no reason to believe that irradiation should have the same effects on nano-systems as on bulk solids.

Present bottle-neck in further developments in this field appears to be the nonavailability of nanofocussed ion beams other than Ga produced with LIMIS. Development of Multi-Element Focussed Ion Beam [ME-FIB] will open up this undeveloped field of non-equilibrium materials science for experimentation and provide new avenues for futuristic technologies. There are efforts world wide to develop ME-FIBs based on different types of ion sources. Scientists in the National Institute of Standards and Technology (NIST) of US have developed a magneto-optical trap based ion source (MOTIS), using laser cooled neutral atoms, which gives very small (~0.02 eV) energy spread [142, 143].

Different types of plasma sources like filamant based, Radio Frequency and Microwave plasma sources are being investigated for FIB applications. Based on the source developed at Lawrence Berkley National Laboratory (LBNL) [144], Scipioni et al. studied the filament-based multicusp ion source for FIB applications [145]. In filament driven plasma source, a hot filament electrode is used for plasma generation and the plasma is confined using a multicusp device.

Jiang et al. at LBNL, Berkeley have developed a mini RF driven multicusp ion source, operated in inductively coupled discharge mode, with an external antenna, for focussed ion beams [146, 147]. Smith et al., from ANU, Australia have developed an inductive coupled plasma (ICP) source for high current FIB applications [148]. Another type of plasma based ME-FIB system is being developed at IIT Kanpur using a microwave driven multicusp plasma source. The system can generate numerous species of focussed ion beams including inert gas ions like Ar, metallic ions like Mn etc. The large beam currents from the plasma source enable high speed milling capabilities for large volume milling applications [149, 150].

FIB fabrication technology can provide solutions for industrial use by providing multiple focussed ion beamlets, since a single nanofocussed ion beam being scanned on surface is inefficient for industrial applications. Some efforts in this direction are in progress in IIT-Kanpur using microwave generated multicusp plasma which can provide high density uniform plasma in a compact cross section [151].

A major project in this direction is underway by the name CHARPAN [Charged Particle Nanotechnology] in Europe [152]. The production process and device is developed in this project of Mask–Less Patterning (MLP) for nanomanufacturing technology, based on hundreds of thousands of finely focussed ion beams with <10 nm resolution.

The research in this fast developing area of use of ion beams of energetic particles for modifying/engineering the properties of materials is constantly opening new directions. It exploits advances in nanoscale materials to develop unique microstructures and material properties. Swift heavy ions have been providing new directions for both the modification of the properties of the materials and their characterization. This area also includes efforts to develop the underlying physics for the behaviour of materials whose properties have been engineered at the nanoscale level. Based on a wealth of experimental results using ion beams of diverse characteristics it is clear that the developments will provide new directions towards exploiting ion-matter interaction to tailor desired properties in materials. The ongoing research is likely to open new avenues for harnessing irradiation on the nanoscale.

Structure and chemistry at nanoscale have long been known to affect the properties of materials. In terms of function, nanoscale features impact mechanical, optical, magnetic, and many other bulk properties. Extremely small grain sizes, nanometre-thickness films, nanodimension precipitates, nanostructured surfaces etc. can have profound impact on the properties of all types of materials. With the constant advances in our ability to characterize, control, and model these nano-dimensioned structures and associated materials properties, an entirely new aspect of materials research has begun. The ability to enhance properties by controlling the synthesis of materials with nanodimensional components provides opportunities which need controllable tools, whereas in many situations ion beams seem to emerge as the only paths to achieve the objectives.

Ion beam-based science and technology has emerged over the years as an area of research having far reaching consequences in both fundamental understanding of materials properties as well as providing a platform for numerous exciting futuristic and some immediate applications. Many physical phenomena that occur in nano-metre regime and femtosecond time domain exist under extreme physical conditions. Ion beam-based techniques offer many exciting possibilities because they produce conditions which are far from equilibrium.

REFERENCES

1. I. Cook, *Nature Materials*, **5** (2006) 77.
2. M.K. Patel, D.K. Avasthi, P.K. Kulriya, S. Kailas, J.C. Pivin, A.K. Tyagi and V. Vijayakumar, *Nucl. Instr. and Meth.*, **B 268** (2010) 42.
3. M.K. Patel, V. Vijayakumar, D.K. Avasthi, S. Kailas, J.C. Pivin, V. Grover, B.P. Mandal and A.K. Tyagi, *Nucl. Instr. and Meth.*, **B 266** (2008) 2898.
4. A. Sarkar, P. Mukherjee and P. Barat, *Journal of Nuclear Materials*, **372** (2008) 285.
5. W.K. Chu, J.W. Mayer and M.-A. Nicolet, "Backscattering Spectrometry". Academic Press, New York (1978).
6. L.C. Feldman, J.W. Meyer and S.T. Picraux, "Materials Analysis by Ion Channeling". Academic Press, New York (1982).
7. A.K. Jain, V.N. Kulkarni, K.B. Nambiar, D.K. Sood, S.C.L. Sharma and P. Mazzoldi, *Radiation Effects and Defects in Solids*, **63** (1982) 175.
8. J. L'Ecuyer, C. Brassard, C. Cardinal and B. Terraeault, *Nucl. Instr. and Meth.*, **149** (1978) 271.
9. T. Som, S. Dhar, N. Banerji, K. Ramakrishnan and V.N. Kulkarni, *Bull. Mater. Sci.*, **19** (1996) 73.
10. D.K. Avasthi and W. Assmann, *Current Science*, **80** (2001) 1532.
11. D.K. Avasthi, D. Kabiraj, A. Bhagwat, G.K. Mehta, V.D. Vankar and S.B. Ogale, *Nucl. Instr. and Meth.*, **B 93** (1994) 480.
12. D.K. Avasthi, S. Hui, E.T. Subramaniyam and B.R. Mehta, *Vacuum*, **47** (1996) 1061.
13. V.N. Kulkarni, V.K. Jain and A.K. Shukla, *Machining Science and Technology*, **9** (2005) 289.
14. D.K. Avasthi, W. Assmann, H. Huber, H.D. Mieskes and H. Nolte, *Nucl. Instr. and Meth.*, **B 142** (1998) 117.
15. T.B. Johansson, R. Akselsson and S.A.E. Johansson, *Nucl. Instr. and Meth.*, **84** (1970) 141.
16. K.M. Varier, G.K. Mehta and S. Sen, *Phys. News*, **11** (1980) 1.
17. K.M. Varier, G.K. Mehta and S. Sen, *Nucl. Instr. and Meth.*, **181** (1981) 217.
18. S. Sen, K.M. Varier, G.K. Mehta, M.S. Rao, P. Sen and N. Panigrahi, *Nucl. Instr. and Meth.*, **181** (1981) 517.
19. P. Sen, N. Panigrahi, M.S. Rao, K.M. Varier, S. Sen and G.K. Mehta, *Journal of Forensic Sciences, JFSCA*, **27** (1982) 330.
20. K.M. Varier, A.K. Nayak and G.K. Mehta, *IEEE Trans. on Nucl. Science*, **NS30** (1983) 1316.
21. A.K. Nayak, K.M. Varier and G.K. Mehta, *Indian Mining and Engineering Journal* (1983) 5.
22. H.C. Barshilia, Somna Sah, B.R. Mehta, V.D. Vankar, D.K. Avasthi, Jaipal and G.K. Mehta, *Thin Solid Film*, **258** (1995) 123.

23. N. Dilawar, S. Sah, B.R. Mehta, V.D. Vankar, D.K. Avasthi and G.K. Mehta, *Vacuum*, **47** (1996) 1269.
24. Amit Kumar, D.K. Avasthi, A. Tripathi, D. Kabiraj, F. Singh and J.C. Pivin, *J. Appl. Phys.*, **101** (2007) 014308.
25. Amit Kumar, D.K. Avasthi, A. Tripathi, L.D. Filip, J.D. Carey and J.C. Pivin, *J. Appl. Phys.*, **102** (2007) 044305.
26. D.K. Avasthi, *Hyperfine Interactions*, **160** (2005) 95.
27. Y.K. Mishra, D.K. Avasthi, P.K. Kulriya, F. Singh, D. Kabiraj, A. Tripathi, J.C. Pivin, I.S. Bayer and A. Biswa, *Appl. Phys. Lett.*, **90** (2007) 073110.
28. Y.K. Mishra, F. Singh, D.K. Avasthi, J.C. Pivin, D. Malinovska and E. Pippel, *Appl. Phys. Lett.*, **91** (2007) 063103.
29. D.K. Avasthi, Y.K. Mishra, F. Singh and J.P. Stoquert, *Nucl. Instr. and Meth.*, **B268** (2010) 3027.
30. S.G. Kaplan, M. Chen, H.D. Drew, M. Rajeswari, R. Liu, T. Venkatesan, D. Kanjilal, L. Senapati and G.K. Mehta, *Physica*, **C32** (1994) 174.
31. S.E. Lofland, S.M. Bhagat, M. Rajeshwari, T. Venkatesan, D. Kanjilal, L. Senapati and G.K. Mehta, *Phys. Rev.*, **B 51** (1995) 8489.
32. A.K. Pradhan, S.B. Roy, P. Chaddah, D. Kanjilal, C. Chen and B.M. Wanklyn, *Phys. Rev.*, **B 53** (1996) 2269.
33. E. Akcoltekin, T. Peters, R. Meyer, A. Duvenbeck, M. Klusmann, I. Monnet, H. Lebius and M. Schleberger, *Nature Nanotechnology*, **2** (2007) 290.
34. Basavaraj Angadi, Y.S. Jung, Won-Kook Choi, Ravi Kumar, K. Jeong, S.W. Shin, J.H. Lee, J.H. Song, M. Wasi Khan and J.P. Srivastava, *Appl. Phys. Lett.*, **88** (2006) 142502.
35. H. Bernas, J-Ph. Attane, K.-H. Heinig, D. Halley, D. Ravelosona, A. Marty, P. Auric, C. Chappert and Y. Samson, *Phys. Rev. Lett.*, **91** (2003) 77203.
36. J. Fassbender, A. Mücklich, K. Potzger and W. Möller, *Nucl. Instr. and Meth.*, **B 248** (2006) 343.
37. F.Z. Cui and Z.S. Luo, *Surface and Coating Technology*, **112** (1999) 278.
38. B. Pignataro, E. Conte, A. Scandurra and G. Marletta, *Biomaterials*, **18** (1997) 1461.
39. Engineering Foundation Conference, **253** (2003) 1124.
40. H. Koizumi, T. Ichikawa, Mitsumasa Taguchi, Yasuhiko Kobayashi and Hideki Namba, *Nucl. Instr. and Meth.*, **B 206** (2003) 1124.
41. M.T. Pham, W. Matz, H. Reuther, E. Richter, G. Steiner and S. Oswald, *Surface and Coating Technology*, **128-129** (2000) 313.
42. H.A. Atwater and A. Pollman, *Nature Materials*, **9** (2010) 205.
43. R.J. Hemley, G.W. Crablree and M.V. Buchanan, *Physics Today* (Nov 2009) 32.
44. J.S. Williams, R.G. Elliman, W.L. Brown and T.E. Seidel, *Phys. Rev. Lett.*, **55** (1985) 1482.
45. P.K. Sahoo, T. Som, D. Kanjilal and V.N. Kulkarni, *Nucl. Instr. and Meth.*, **B 240** (2005) 239.
46. P. Sigmund, *Phys. Rev.*, **184** (1969) 383.
47. R. Behrisch and Wolfgang Eckstein (Eds), Sputtering by Particle Bombardment, Springer 1991.
48. T. Neidhart, F. Pichler, F. Aumayr, H.P. Winter, M. Schmid and P. Varga, *Phys. Rev. Lett.*, **74** (1995) 5280.

49. J. Kulik, G.D. Lempert, E. Grossman, D. Marton, J.W. Rabalais and Y. Lifsitz, *Phys. Rev.*, **B 52** (1995) 15812.
50. O.M. Kuttel, P. Groening, R.G. Agostino and L. Schlapbach, *J. Vacuum Sci. Techn. A: Vacuum, Surface and Films*, **13** (1995) 2848.
51. W. Ensinger, M. Barth, H. Martin, A. Schröer, B. Enders, R. Emmerich and G.K. Wolf, *Rev. Sci. Instr.*, **63** (1992) 2393.
52. F. Alvarez, N.M. Victoria, P. Hammer, F.L. Freire, Jr and M.C. dos Santos, *Appl. Phys. Lett.*, **73** (1998) 1065.
53. V.M. Jiménez, J.P. Espinós, A. Caballero, L. Contreras, A. Fernández, A. Justo and A.R. González-Elipe, *Thin Solid Film*, **353** (1999) 113.
54. S.K. Tripathi, N. Shukla and V.N. Kulkarni, *Nanotechnology*, **20** (2009) 075304.
55. D.K. Avasthi, Y.K. Mishra, D. Kabiraj, N.P. Lalla and J.C. Pivin, *Nanotechnology*, **18** (2007) 125604.
56. Y.K. Mishra, S. Mohapatra, D.K. Avasthi, D. Kabiraj, N.P. Lalla, J.C. Pivin, H. Sharma, R. Kar and N. Singh, *Nanotechnology*, **18** (2007) 345606.
57. Kapil U. Joshi, D. Kabiraj, A.M. Narsale, D.K. Avasthi, T.N. Warang and D.C. Kothari, *Surface and Coating Technology*, **203** (2009) 2497.
58. K. Zhao, R.S. Averback and David G. Cahill, *Appl. Phys. Lett.*, **89** (2009) 053103.
59. S. Amirthapandian, F. Schuchart and W. Bolse, *Rev. Sci. Instr.*, **81** (2010) 033702.
60. Q. Cai, Brad Ledden, Eric Krueger, Jene A. Golovchenko and Jiali Li, *J. Appl. Phys.*, **100** (2006) 24914.
61. Jiali Li, Derek Stein, Ciaran McMullan, Daniel Branton, Michael J. Aziz and Jene A. Golovchenko, *Nature*, **412** (2001) 166.
62. T. Som, S. Bhargava, M. Malhotra, H.D. Bisht, V.N. Kulkarni and S. Kumar, *Appl. Phys. Lett.*, **72** (1998) 3014.
63. M. Bruel, *Nucl. Instr. and Meth.*, **B 108** (1996) 313.
64. X.Q. Feng and Y. Huang, *Int. J. Solid Struct.*, **41: 16-17** (2004) 4299.
65. M. Nastasi and J.W. Mayer, *Materials Science and Engineering*, **R12** (1994) 1.
66. S.K. Srivastava, D.K. Avasthi, W. Assmann, Z.G. Wang, H. Kucal, E. Jacquet, H.D. Carstanien and M. Toulemonde, *Phys. Rev.*, **B 71** (2005) 193405.
67. Sarvesh Kumar, P.K. Sahoo, R.S. Chauhan, D. Kabiraj, Umesh Tiwari, D. Varma and D.K. Avasthi, *Nucl. Instr. and Meth.*, **B 212** (2003) 238.
68. G.K. Mehta, *Nucl. Instr. and Meth.*, **A 382** (1996) 335.
69. Amitabh Jain, S. Loganathan, Jaipal, D. Kanjilal and G.K Mehta, *Vacuum*, **46** (1995) 369.
70. A. Gupta, A. Paul and D.K. Avasthi, *J. Phys Cond. Matt.*, **10** (1998) 9669.
71. S.T. Chavan, P.S. Bhave, Bhoraskar and D. Kanjilal, *J. Appl. Phys.*, **78** (1995) 2328.
72. P.S. Bhave and V.N. Bhoraskar, *Nucl. Instr. and Meth.*, **B 127** (1997) 383.
73. N. Dilawar, R. Kapil, V.D. Vankar, D.K. Avasthi, D. Kabiraj and G.K. Mehta, *Thin Solid Films*, **305** (1997) 88.
74. S. Klaumuenzer, Ming-dong Hou and G. Schumacher, *Phys. Rev. Lett.*, **57** (1986) 850.
75. H. Dammak, A. Dunlop, D. Lesueur, A. Brunnelle, S. Della-Negra and Y. Le Beyec, *Phys. Rev. Lett.*, **74** (1995) 1135.
76. J. Nakata, *Phys. Rev.*, **B 60** (1999) 2747.
77. A. Biswas, R. Gupta, N. Kumar, D.K. Avasthi, J.P. Singh, S. Lotha, D. Fink, S.N. Paul and S.K. Bose, *Appl. Phys. Lett.*, **78** (2001) 4136.

78. A. Kumar, D.K. Avasthi, J.C. Pivin and P.M. Koinrkaor, *Appl. Phys. Lett.*, **92** (2008) 221904.
79. A. Benyagoub, *Eur. Phys. J.*, **B** 34 (2003) 395.
80. A. Benyagoub, *Nucl. Instr. and Meth.*, **B 245** (2006) 225.
81. H. Rath, P. Dash, T. Som, P.V. Satyam, U.P. Singh, P.K. Kulriya, D. Kanjilal, D.K. Avasthi and N.C. Mishra, *J. Appl. Phys.*, **105** (2009) 074311.
82. U.A. Glasmacher, Maik Lang, Hans Keppler, F. Langenhorst, R. Neumann, D. Schardt, C. Trautmann and G.A. Wagner, *Phys. Rev. Lett.*, **96** (2006) 195701.
83. U. Tiwari, N. Sen, A.K. Bandopandhyay, D. Kanjilal and P. Sen, *Europhys. Lett.*, **25** (1994) 705.
84. R.E. Johnson and Bo U.R. Sundqvist, *Physics Today* (1992) 28.
85. P.K. Kuiri, B. Joseph, H.P. Lenka, G. Sahu, J. Ghatak, D. Kanjilal and D.P. Mahapatra, *Phys. Rev. Lett.*, **100** (2008) 245501.
86. A. Gupta and D.K. Avasthi, *Phys. Rev.*, **B 64** (2001) 155407.
87. M. Toulemonde, W. Assmann, C. Trautmann and F. Gruner, *Phys. Rev. Lett.*, **88** (2002) 057602.
88. S. Roorda, T. van Dillen, A. Polman, C. Graf, A. van Blaaderen and B.J. Kooi, *Adv. Mater.*, **16** (2004) 235.
89. I. Zizak, N. Darowski, S. Klaumunzer, G. Schumacher, J.W. Gerlach and W. Assmann, *Phys. Rev. Lett.*, **101** (2008) 065503.
90. Abha Misra, P.K. Tyagi, Padmnabh Rai, Dipti Ranjan Mhahopatra, Jay Ghatak, P.V. Satyam, D.K. Avathi and D.S. Misra, *Phys. Rev.*, **B 76** (2007) 14108.
91. S. Ghosh, A. Tripathi, V. Ganeshan and D.K. Avasthi, *J. Nanosci. & Nanotech.*, **8** (2008) 2505.
92. S. Kala, B.R. Mehta, S.A. Khan and D.K. Avasthi, *Appl. Phys. Lett.*, **90** (2007) 153121.
93. Amit Kumar, D.K. Avasthi, J.C. Pivin, A. Tripathi and F. Singh, *Phys. Rev.*, **B 74** (2006) 153409.
94. P.K. Kulriya, B.R. Mehta, D.K. Avasthi, D.C. Agarwal, P. Thakur, N.B. Brookes, A.K. Chawla and R. Chandra, *Appl. Phys. Lett.*, **96** (2010) 53103.
95. D.C. Agarwal, R.S. Chauhan, D.K. Avasthi, S.A. Khan, D. Kabiraj and I. Sulania, *J. Appl. Phys.*, **104** (2008) 024304.
96. M. Lang, J. Lian, J. Zhang, F. Zhang, W.J. Weber, C. Trautmann and R.C. Ewing, *Phys. Rev.*, **B 79** (2009) 224105.
97. A. Lushchik, Ch. Lushchik, K. Schwartz, E.V. Chenko, R. Papaleo, M. Sorokin, A.E. Volkov, R. Newmann and C. Trautmann, *Phys. Rev.*, **B 76** (2007) 54114.
98. A. Mara, Z. Siwy, C. Trautmann, J. Wan and F. Kamme, *Nano Lett.*, **4** (2004) 497.
99. Kiran Jeet, V.K. Jindal, L.M. Bharadwaj, D.K. Avasthi and K. Dharamvir, *J. Appl. Phys.*, **108** (2010) 34302.
100. N. Bajwa, A. Ingale, D.K. Avasthi, R. Kumar, A. Tripathi, K. Dharamvir and V.K. Jindal, *J. Appl. Phys.*, **94** (2003) 326.
101. R.C. Budhani, M. Suenaga and S.H. Liou, *Phys. Rev. Lett.*, **69** (1992) 3816.
102. C. Chappert, H. Bernas, J. Ferré, V. Kottler, J.-P. Jamet, Y. Chen, E. Cambril, T. Devolder, F. Rousseaux, V. Mathet and H. Launois, *Science*, **280** (1998) 1919.
103. T.M. Tona, H. Watanabe, S. Takahashi, Y. Fujita, T. Abe, S. Jian, N. Nakamura, N. Yoshiyasu, C. Yamada, M. Sakurai and S. Ohtani, *Journal of Physics, Conference Series*, **58** (2007) 331.

104. P. Li, X.G. Han, J.P. Xin, X.P. Zhu and M.K. Lei, *Nucl. Inst. and Meth.*, **B 266** (2008) 3945.
105. T.J. Renk, P.P. Provencio, S.V. Prasad, A.S. Shlapakovski, A.V. Petrov, K. Yatsui, Jiang Weihua and H. Suematsu, *Proc. IEEE*, **92 (7)** (2004) 1057.
106. S.K. Ghose, G. Kuri, Amal K. Das, B. Rout, D.P. Mahapatra and B.N. Dev, *Nucl. Instr. and Meth.*, **B 156** (1999) 125.
107. T. Tada, T. Kanayama, K. Koga, P. Weibel, S.J. Carroll, K. Seeger and R.E. Palmer, *J. Phys. D: Appl. Phys.*, **31** (1998) L21.
108. I. Yamada and J. Matsuo, *J. Mat. Soc. Symp. Proc.*, **427** (1996) 265.
109. K. Meiwes and O. Broer, Metal Clusters at Surfaces. Springer, Berlin (2000) 145.
110. J. Peltola, K. Nordlund and J. Keinonen, *Nucl. Instr. and Meth.*, **B 217** (2004) 25.
111. Y. Yamaguchi and J. Gspann, *Phys. Rev.*, **B 66** (2002) 155408.
112. T. Osipowicz, J.A. van Kan, T.C. Sum, J.L. Sanchez and F. Watt, *Nucl. Instr. and Meth.*, **B 161-163** (2000) 83.
113. F. Watt, M.B.H. Breese, A.A. Bettiol and J.A. van Kan, *Materials Today*, **10** (2007) 20 and the references therein.
114. A. Bertucci, R.D.J. Pocock, G.R. Pehrson and D.J. Brenner, *J. Radiat. Res.*, **50** (2009) 49.
115. M. Folkard, K.M. Prise, A.G. Michette and B. Vojnovic, *Nucl. Instr. and Meth.*, **A 580** (2007) 446.
116. G. Randers-Pehrson, C.R. Geard, G. Johnson, C.D. Elliston and D.J. Brenner, *Rad. Research*, **156** (2001) 210.
117. P. Reichart, G. Datzmann, A. Hauptner, R. Hertenberger, C. Wild and G. Dollinger, *Science*, **306** (2004) 1537.
118. M. Ultaut, Handbook of Charged Particle Optics, CRC Press, Boca Raton, Fl. (1997) 429.
119. Sudeep Bhattacharjee, A. Vartak and Victor Mukherjee, *Appli. Phys. Lett.*, **92** (2008) 191503.
120. Sudeep Bhattacharjee and Tathagata Chowdhury, *App. Phys. Lett.*, **95** (2009) 061501.
121. U. Dotsch and A.D. Wieck, *Nucl. Instr. and Meth.*, **B 139** (1998) 12.
122. Gang Xiong, D.A. Allwood, M.D. Cook and R.P. Cowburn, *Appl. Phys. Lett.*, **79** (2001) 3461.
123. J. Orloff, M. Utlaut and L. Swanson, High Resolution Focused Ion Beams: FIB, and its Applications. Springer Press. ISBN 0-306-47350 (2003).
124. Neeraj Shukla, Sarvesh K. Tripathi, Mihir Sarkar, Nitul S. Rajput and V.N. Kulkarni, *Nucl. Instr. and Meth.*, **B 267** (2009) 1376.
125. Y.Z. Huang, S.G. Wang, C. Wang, Z.B. Xie, D.J.H. Cockayne and R.C.C. Ward, *Appl. Phys. Lett.*, **88** (2006) 103104.
126. R.M. Bradley and J.M.E. Harper, *J. Vac. Sci. Technol.*, **A6** (1988) 2390.
127. R. Cuerno, H.A. Makse, S. Tomassone, S.T. Harrington and H.E. Stanley, *Phys. Rev. Lett.*, **75** (1995) 4464.
128. B. Kahng, H. Jeong and A.-L. Barabási, *Appl. Phys. Lett.*, **78** (2001) 805.
129. A. Datta, Y-R Wu and Y.L. Wang, *Phys. Rev.*, **B 63** (2001) 125407.
130. T.K. Chini, M.K. Sanyal and S.R. Bhattacharya, *Phys. Rev.*, **B 66** (2002) 153404.

131. Y.S. Katharria, S. Kumar, P.S. Lakshmy, D. Kanjilal and A.T. Sharma, *J. Appl. Phys.*, **102** (2007) 44301.
132. P. Mishra and D. Ghose, *J. Appl. Phys.*, **105** (2009) 014304.
133. R.M. Bradley and E. Cirlin, *Appl. Phys. Lett.*, **68** (1996) 3722.
134. J. Lian, L. Wang, X. Sun, Q. Yu and Rodney C. Ewing, *Nano Lett.*, **6** (2006) 1047.
135. W. Bolse, *Nucl. Instr. and Meth.*, **B 244** (2006) 8.
136. M. Batzil, F. Bardou and K.J. Snowdon, *Phys. Rev.*, **B 63** (2001) 233408.
137. P. Kulriya, A. Tripathi, D. Kabiraj, S.A. Khan and D.K. Avasthi, *Nucl. Instr. and Meth.*, **B 244** (2006) 95.
138. I. Sulania, A. Tripathi, D. Kabiraj, M. Lequeux and D.K. Avasthi, *Advanc. Mat. Lett.*, **1** (2010) 118.
139. I. Sulania, A. Tripathi, D. Kabiraj, S. Varma and D.K. Avasthi, *J. Nanosci. and Nanotech.*, **8** (2007) 1.
140. M. Buljan, I.B. Radovic, M. Karlusic, U.V. Desnica, G. Drazic, N. Radic, P. Dubcek, K. Salamon, S. Bernstorff and V. Holy, *Appl. Phys. Lett.*, **95** (2009) 063104.
141. Vidya Ramaswamy, Tony E. Haynes, C. Woody White, Warren J. Moberly Chan, S. Roorda and Michael J. Aziz, *Nano Lett.*, **5 (2)** (2005) 373.
142. J.L. Hanssen, J.J. McClelland, E.A. Dakin and M. Jacka, *Phys. Rev.*, **A 74** (2006) 063416.
143. B.J. Claessens, M.P. Reijnders, G. Taban, O.J. Luiten and E.J.D. Vredenbregt, *Phys. Plasmas*, **14** (2007) 093101.
144. Y. Lee, Q. Ji, K.N. Leung and N. Zahir, *Rev. Sci. Inst.*, **71** (2000) 722.
145. L. Scipioni, D. Stewart, D. Ferranti and A. Saxonis, *J. Vac. Sci. Technol.*, **B 18** (2000) 3194.
146. X. Jiang, Q. Ji, A. Chang and K.N. Leung, *Rev. Sci. Instr.*, **74** (2003) 2288.
147. L. Ji, Q. Ji, K.-N. Leung and R.A. Gough, *Appl. Phys. Lett.*, **89** (2006) 164103.
148. N.S. Smith, W.P. Skoczylas, S.M. Kellogg, D.E. Kinion, P.P. Tesch, O. Sutherland, A. Aanesland and R.W. Boswell, *J. Vac. Sci. Technol.*, **B 24 (6)** (2006) 2902.
149. Jose V. Mathew, A. Chowdhury and S. Bhattacharjee, *Rev. Sci. Inst.*, **79** (2008) 063504.
150. Jose V. Mathew, Indranuj Dey and S. Bhattacharjee, *Appl. Phys. Lett.*, **91** (2007) 041503.
151. Abhishek Chowdhury and Sudeep Bhattacharjee, *J. Appl. Phys.*, **107** (2010) 093307.
152. Elmar Platzgummer, Hans Loeschner and Gerhard Gross, *J. Vac. Sci. Technol.*, **B 26 (6)** (2008) 2059.

2

Ion Matter Interaction

2.1 INTRODUCTION

The ion beam applications, which involve either the modification of the properties of the material or its characterization, stem from the basic interactions of the ions with the target material. The basic physics that governs the interaction of energetic ions with solids has its roots in the atomic and nuclear physics of the last century. The interaction is with both electrons and the nuclei of the target material. In contrast to electrons or photons, energetic ions deposit extremely high localized density of energy to the target. In this way, solids receive in a very short time (~10^{-17} to 10^{-15} s) within a very small volume (~10^{-17} to 10^{-16} cm^3) very high energy density by the impact of just one energetic ion. Consider a high energy heavy ion having an energy loss of ~5 keV/A. It can deposit energy of ~30 k Jule/cm^3 in a cylinder of 5 nm radius in 10^{-16} s. The energy is not high but it is concentrated in such a small area that it yields an extremely large energy density. Such large energy densities are possible only in the vicinity of exploding devices. With the availability of ions from protons to uranium in the energy range of eV to GeV it is possible to have penetration depth of a few nanometre for low energy heavy ions to several tens of micrometres for medium energy (few MeV) light ions. The width of the interaction zone ranges from a few nanometres to hundreds of nanometres. Considering the time scale, the ion takes about a pico second time before it comes to rest in the material.

Techniques used to characterize the material require that the material should not change during the process. On the other hand engineering the properties of the material requires significant changes in the materials properties such as electrical, optical, structural etc. The challenge here is to evolve soft engineering approach, where material is modified in an extremely controlled manner and is characterized during the process of modification. Such a scenario is provided by the way high energy heavy ions interact with the material. At high energy, the ions loose their energy predominantly by interacting with the electrons of the atoms constituting the material, while the nuclei constituting the lattice are not disturbed. The energy locked into electrons can cause modification of the lattice through electron phonon interaction. This provides the soft engineering approach. Engineering the properties will require movement of the nuclei whether it is soft approach or the normal one, which is the displacement of atoms through elastic collisions (nuclear energy loss). For heavy ions (M>4) of energy a few keV/amu, the energy loss occurs predominantly by elastic scattering, which can be used for modifying the properties of the material. For higher energies, above 1-2 MeV/amu, the energy loss occurs predominantly by inelastic scattering process. Materials however are modified when the deposited energy in electrons goes above a certain threshold value, which depends on the materials properties.

Ions travelling through the solid undergo collisions with the stationary target atoms, which deflect the ion from its initial direction. During its transit in the target material, an energetic ion, in addition to loosing energy through various processes, picks up electrons from various shells. Ultimately, a very slow moving ion propagates almost as a neutral atom going through cascade of collisions and stops in the solid.

2.2 NUCLEAR AND ELECTRONIC ENERGY LOSS IN MATERIALS

The incident ion, during its transit in material, continuously lose its energy by elastic and inelastic collisions with atoms in the target materials. The energy loss dE/dX basically has three components as follows [1].

$$dE/dX = (dE/dX)_{\text{nuclear}} + (dE/dX)_{\text{electronic}} + (dE/dX)_{\text{radiation}} \tag{2.1}$$

The energy loss for heavy ions in radiative processes, such as bremsstrahlung and Cerenkov radiation, is very small and will be neglected in our discussions. The energy loss mechanism, at high energies, is the interaction of the effective charge with the target electrons. The nuclear energy loss, $(dE/dX)_{\text{nuclear}}$, dominates at low energies. The established designation "nuclear" is a misnomer since nuclear energy loss is not due to nuclear forces. It is caused by the scattering of the projectile by the screened nuclear potential of the target atoms (Rutherford Scattering). When the recoil energy transferred to target atom is higher than the binding energy, these atoms are released from their chemical environment and can trigger off a chain of recoils, if kinetic energy is big enough. The energy deposition is described by the "stopping power" or dE/dx, which gives the energy loss per unit length of a particle along its trajectory. There is slight difference in stopping power and dE/dx, although both of them represent slowing down of ion in material. The stopping power refers to the property of material to stop the incident ion and is given by $-1/\rho\,(dE/dx)$, where ρ is the density. There are two basic energy transfer mechanisms in the atomic collisions with the target atoms:

(i) Elastic collisions with the nuclei of the atoms, in which energy is transmitted to the target atom as whole. It is referred to as nuclear energy loss S_n, which dominates at low energies (typically below 10 keV/amu), produces cascade of collisions with nuclei before coming to rest and getting implanted. It can be varied from a fraction of eV/nm to a few keV/nm, depending on the mass and energy of ion.

(ii) Through inelastic interaction with electrons, in which the ion causes excitation or ionization of atoms in the target. It is referred to as electronic energy loss (S_e). In this process, the energy is locked into electrons and gets transferred to the lattice through electron-phonon interaction. It dominates at higher energies (typically >100 keV/amu). It can be varied over a wide range of a few eV/nm to a few tens of keV/nm, by varying the mass and the energies of the heavy ions.

Nuclear collisions produce lattice disorder by the displacement of atoms from their positions. Electronic collisions involve small energy loss per interaction and negligible deflection of the ion trajectory. Computer simulation methods have been developed (over the years) to calculate the motion of ions in a medium, which are used to calculate the stopping power theoretically. The basic idea in them is to follow the movement of the ion in the medium by simulating the collisions with nuclei in the medium. The best known simulation programme is TRIM/SRIM (TRansport of Ions in Matter, in more recent versions called Stopping and Range of Ions in Matter), which is developed by Ziegler and Biersack [2] and accessible on the Web [www.srim.org] [3]. Typical values of S_e and S_n estimated from TRIM simulation code are shown in Fig. 2.1(a) for Ag ions of different energies in Si. Both stopping powers increase with increasing energy until they reach a maximum value, and thereafter decrease. The peak in electronic energy loss curve is referred as Bragg peak. The stopping power at energies below Bragg peak region is proportional to velocity of ion ($E^{1/2}$) and is given by LSS theory [4]. The stopping power at energies above the Bragg peak is proportional to $1/E$ and is expressed by Bethe Bloch relation [5, 6]. The electronic energy transfer reaches its peak value at energies, which are orders of magnitude higher than the nuclear stopping maximum. The variation of energy loss (nuclear and electronic) with depth for 100 MeV Ag in Si, is shown in Fig. 2.1(b). It may be noted that the value of S_e is more or less constant within submicron thickness and the corresponding value of S_n is negligible. Therefore the experiments of SHI with thin film provide information on modification of materials, solely due to electronic energy loss of SHI in materials. Although there are possibilities of contribution of S_n in materials modification as discussed in Section 2.4. The nuclear and electronic loss values in materials can be varied by choosing different ions and energies as shown in Figs 2.2(a) and 2.2(b). The possibility of large variations in value of S_e can be seen in Fig. 2.3 where the values of S_e for different ions (C, Si, Ni, Ag and Au) of 1 MeV/amu are shown.

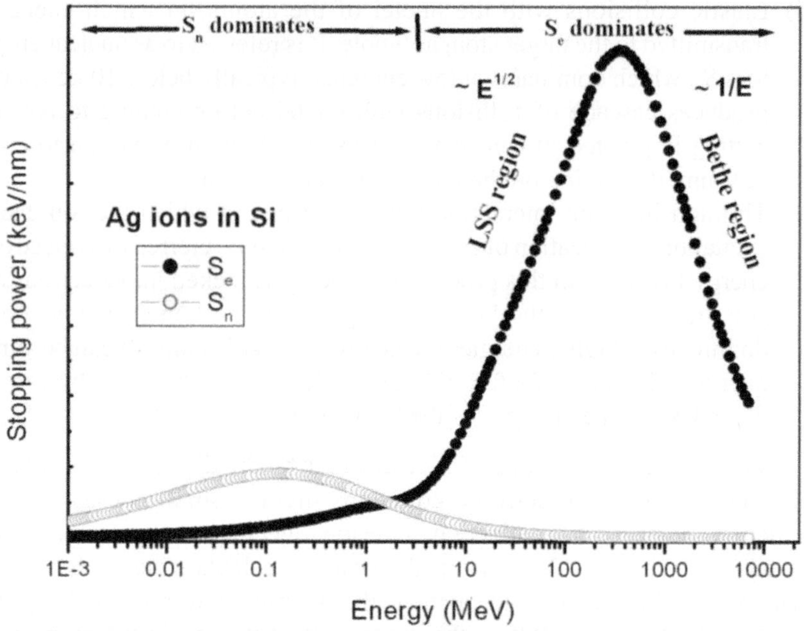

Figure 2.1(a): The values of S_e and S_n for Ag ions in Si, at different energies.

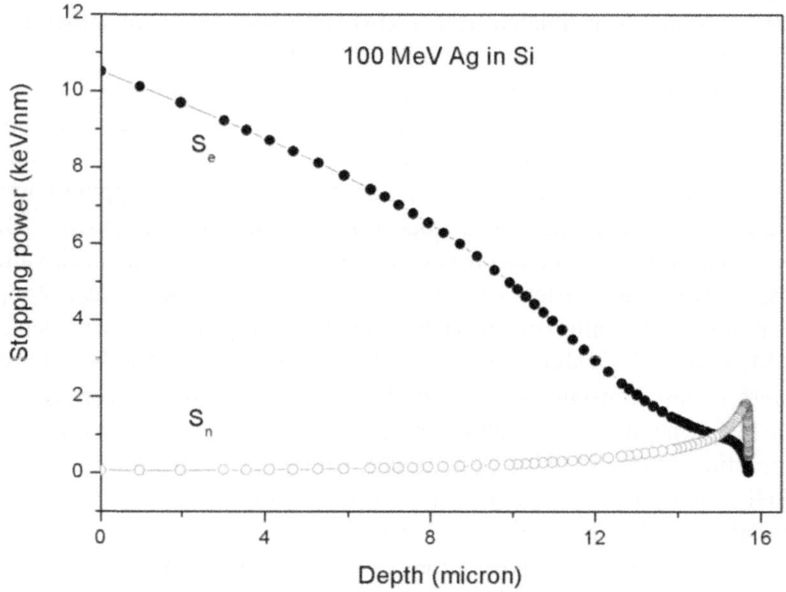

Figure 2.1(b): The values of S_e and S_n are shown from the surface to depth till the range of ion for 100 MeV Ag ions in Si.

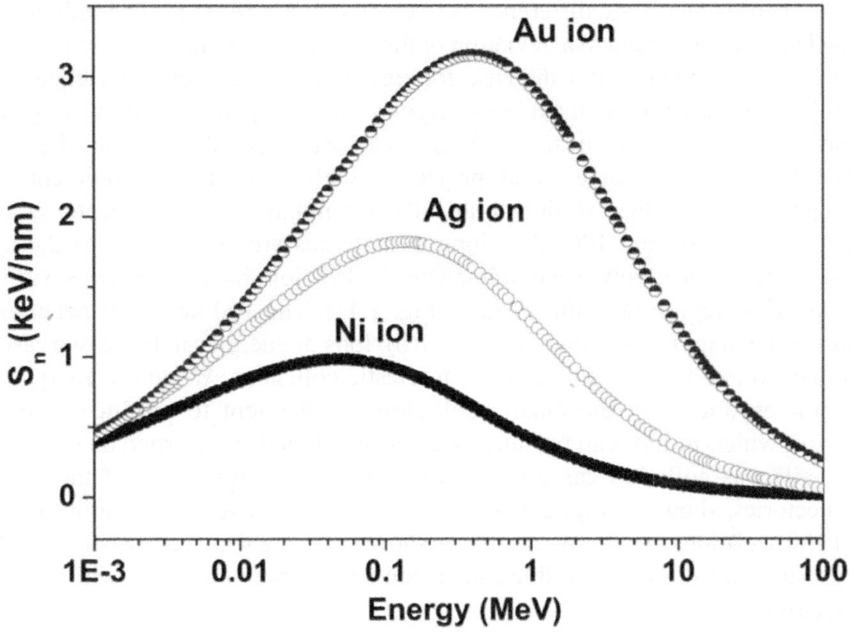

Figure 2.2(a): The values of S_n at different energies are shown for different ions.

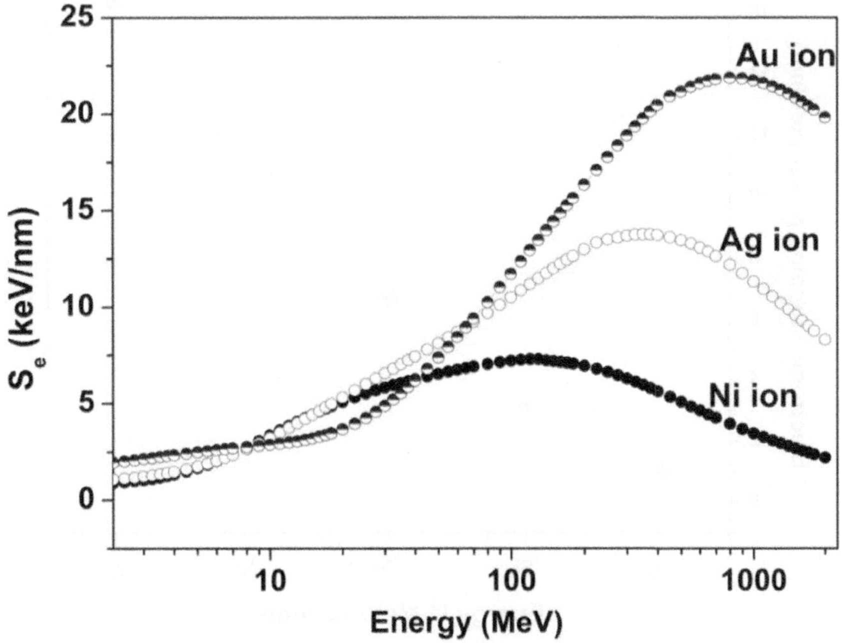

Figure 2.2(b): The values of S_e at different energies are shown for different ions.

There is one basic difference between the two processes. The individual nuclear collisions lead to a deviation of the projectile's flight direction from its original one which is not the case for electronic energy transfer. Hence, for projectiles with dominant nuclear energy transfer, the particles follow a zigzag movement until they come to rest. This gives rise to a spatially extended damage distribution. On the other hand, projectiles with dominant electronic energy transfer, traverse in a straight line. TRIM simulation of the trajectories of 1 keV, 100 keV and 100 MeV ions in materials are shown in Figs 2.4(a), (b) and (c) respectively. Several impacts of 1 keV ions have trajectories within a spherical region near the surface (Fig. 2.4a). The 100 keV ions penetrate inside the material as shown in Fig. 2.4(b). Ions of energies in these two cases do not follow the straight path due to elastic collisions. When the energy of recoil produced by the elastic collision is sufficient to produce further recoil, which in turn can produce another recoil, and so on; such a condition is called as collision cascade. Small branches emerging out of the main trajectories, shown in Fig. 2.4(b) represent collision cascade. The trajectories of 100 MeV ions are almost straight as shown in Fig. 2.4(c), because the elastic collisions are almost negligible and the ion interact predominantly with atomic electrons.

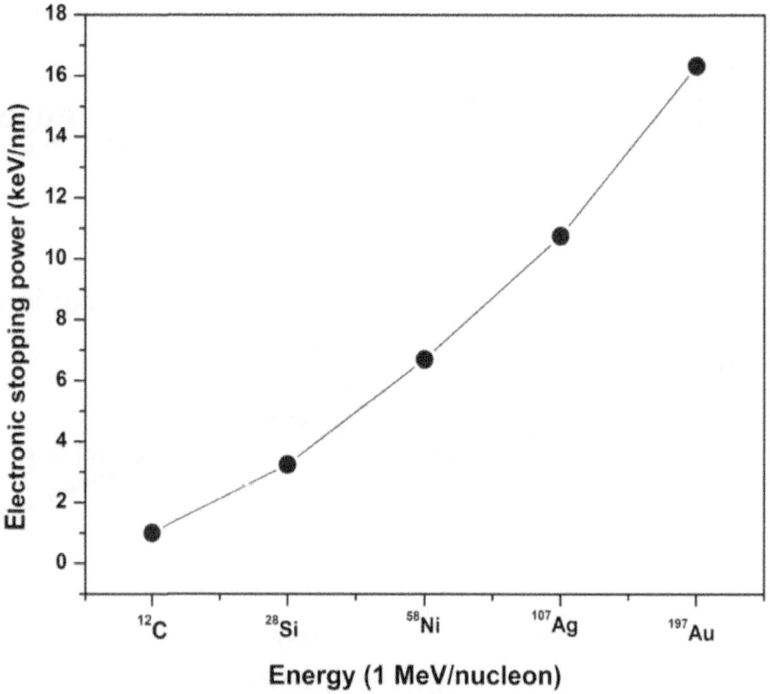

Figure 2.3: The values of S_e are shown for different ions of 1 MeV/nucleon to get an idea of large possible variation in S_e by changing ions of same velocity.

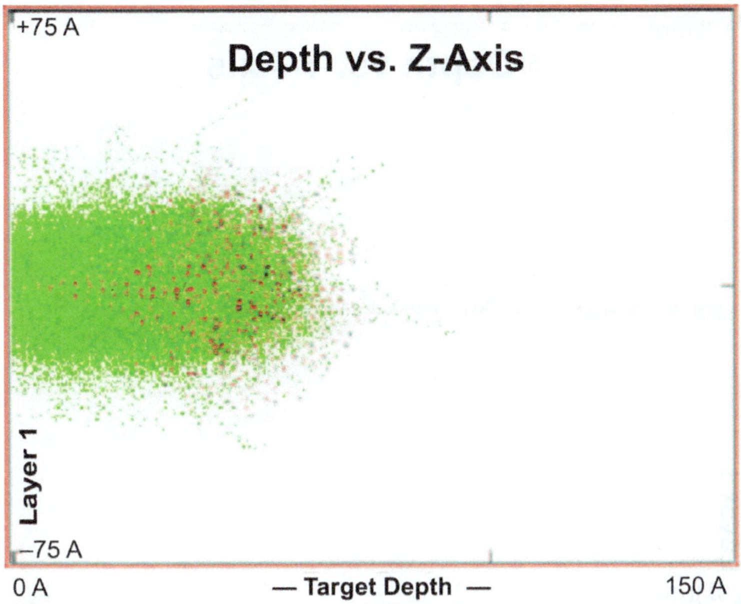

Figure 2.4(a): Trajectories produced by 1 keV Xe ions (red in colour) in Si along with the recoils (represented by green colour).

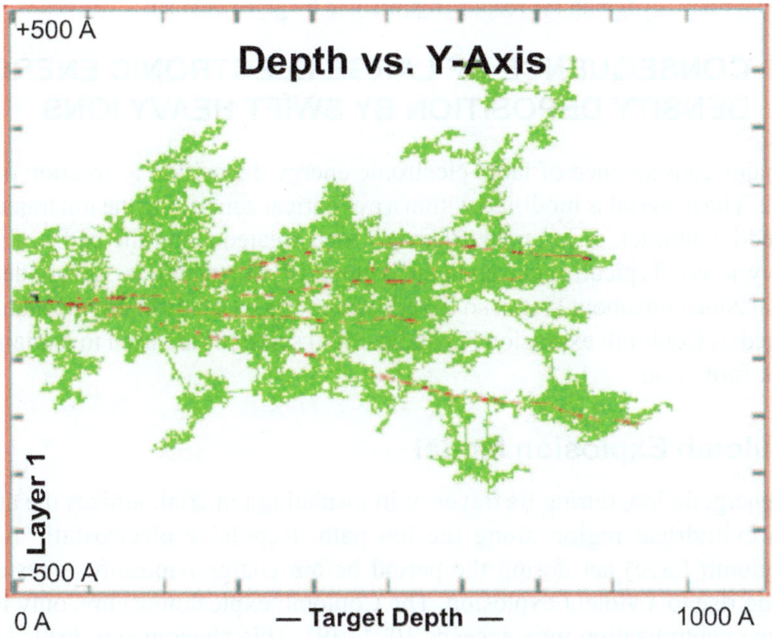

Figure 2.4(b): Trajectories produced by 100 keV Xe ions (red in colour) in Si along with the recoils (represented by green colour). The green colour dots show the cascade recoils.

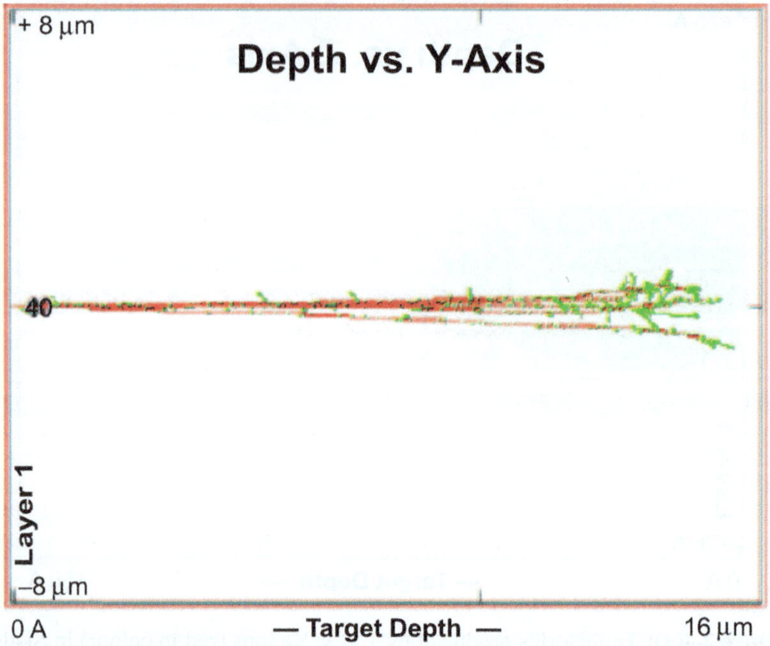

Figure 2.4(c): Trajectories produced by 1 MeV Xe ions (red in colour) in Si along with the recoils (represented by green colour).

2.3 CONSEQUENCE OF LARGE ELECTRONIC ENERGY DENSITY DEPOSITION BY SWIFT HEAVY IONS

A major consequence of large electronic energy deposition is creation of ion track. The material is modified within a cylindrical zone along the ion trajectory called latent track or ion track (damage zone created along the paths of swift heavy ions). Typical track diameter is upto 10 nm, which is surrounded by radial zone upto about ½ µm, so called 'penumbra'. Two different mechanisms named as Coulomb explosion [7] and thermal spike [8] are used to explain the track formation.

Coulomb Explosion Model

An energetic ion, during its traversal in insulating material, ionizes the atoms in a cylindrical region along the ion path. Repulsive electrostatic forces (Coulomb force) act during the period before charge neutrality is restored, giving rise to a violent explosion. The Coulomb explosion occurs, only if the charge neutralization time exceeds 10^{-14} s [9]. This phenomenon leads to the localized damage or change of the lattice within a narrow cylindrical zone in insulators and not in metals. Metals, typically, have plasmon frequency typically about 10^{14} Hz (obtained by the relation of plasmon frequency as $(ne^2/m\varepsilon_0)^{1/2}$, where n is the conduction electron density, e is the elementary charge, m is the electron mass, and ε_0 is the permittivity of free space). Therefore the electron return time to the ionized core is shorter than 10^{-14} s for metals and the ionized core is neutralized before the Coulomb explosion.

If two ions with charge of ne on each, are formed in material of dielectric constant ε, having average atomic spacing a_0, the force F between the charges can be expressed as

$$F = n^2e^2/\varepsilon\, a_0^2 \tag{2.2}$$

The local force per unit area (the electrostatic stress, σ_e) [7] can be written as

$$\sigma = n^2e^2/(\varepsilon a_0^4) \tag{2.3}$$

The mechanical tensile strength σ_M is approximately $E/10$, where E is Young's modulus of material. For Coulomb explosion to occur, $\sigma_e > \sigma_m$. In other words, the condition for the Coulomb explosion to occur is $\frac{n^2e^2}{\varepsilon a_0^4} > \frac{E}{10}$.

It is clear from the above equation that track formation is easily possible in materials of low mechanical strength, low dielectric constant, and small inter-atomic spacing. Fleischer et al. [7] demonstrated that plastics are more sensitive to radiation than glasses, as seen in experiments. The Coulomb explosion model could not predict the experimentally observed ion track in metals, for which the thermal spike model was invoked. Although the applicability of thermal spike model was shown in insulators also especially in several oxides [10, 11].

Thermal Spike Model

The thermal spike model considers the material as a two-component medium: (a) the electrons, described by the quasi-free electron gas theory and (b) the atomic lattice with known thermal properties. The projectile ion gives its energy to the electron gas, and then energy is transferred to the lattice via electron-phonon coupling in the hot electron system during the characteristic time 10^{-13} s [8]. It is assumed that when the temperature exceeds the melting point of the material, a transient molten state is formed in a localized zone of few nanometres existing for a short duration in typically ps time scale. The rapid thermal quenching of the hot cylindrical region results in an amorphous track along the ion path in a short span of time ($t \geq 10^{-11}$ s), with the diameter of typically upto 10 nm. The narrow cylinder of material that was rapidly heated to high temperature transiently and then rapidly quenched by thermal conduction, results in modified material within a cylindrical region.

Consequent to the energy $A(r)$ deposition in the electronic system, the temperature rise in electronic and lattice subsystems is governed by coupled non-linear differential equations in cylindrical geometry [8] as given below:

$$\rho C_e(T_e)dT_e/dt = d/dr\,[K_e(T_e)\,dT_e/dr] + [K_e(T_e)/r]\,dT_e/dr - g(T_e - T) + A(r) \quad (2.4)$$

$$\rho C(T)dT/dt = d/dr\,[K(T)\,dT/dr] + [K(T)/r]\,dT/dr + g\,(T_e - T) \quad (2.5)$$

where C_e, C, K_e and K are the specific heats and thermal conductivities for the electronic and atomic systems respectively, ρ is the density of the material, r is the radius of cylindrical geometry with heavy ion path as the axis, g is the electron-phonon coupling constant, and T_e and T are the temperatures for the electronic and atomic systems, respectively. The electron-phonon coupling factor g depends on the velocity of sound s in the material, thermal conductivity K and the temperature T of atomic system according to following relation [8].

$$g = \pi^4\,(K_B n_e s)^2/18K_e T \quad (2.6)$$

where K_B is the Boltzmann constant and n_e is the electronic number density.

It has been observed experimentally [11] that only continuous tracks are formed if the energy deposition along the ion path exceeds a threshold value of S_e, which depends on the physical properties of the target material as given in thermal spike equations.

Thermal spike model predicted that the damage zones or the ion track diameters for two ions having same S_e but different velocity can be different as was observed in experiments [12]. The slower moving ion creates larger diameter than the faster moving ion. This is also known as Velocity effect [13] and is one of the important aspects to understand the affects of the SHI on materials.

There are other versions [14, 15] of thermal spike model different from the formulation given above.

Pressure Spike

Some materials undergo phase transition under swift heavy ion irradiation, similar to the phase transition expected under the high pressure. It gives a clue that the ion beam can induce a pressure spike. It can be seen that when the material expands in the ion track during transient melt phase, it exerts a pressure to the surrounding annular region and thus the temperature spike results in a pressure spike. Similar argument can be given in the case of Coulomb explosion that the ions getting exploded (due to electrostatic repulsion force) may exert a pressure to the surrounding region. Schematically such a possibility of pressure spike region around the ion track is shown in Fig. 2.5. Different research groups have considered pressure spike in explaining the ion beam induced results and simulations [16-18].

The above three models are normally the basis to understand the modifications of materials by SHI. For metals, exclusively thermal spike model is applicable whereas for the other materials, combination of models may be required for explaining the modification of materials by SHI. In other words, there could be competing processes of Coulomb spike, thermal spike and pressure spike, responsible for the modification in the materials.

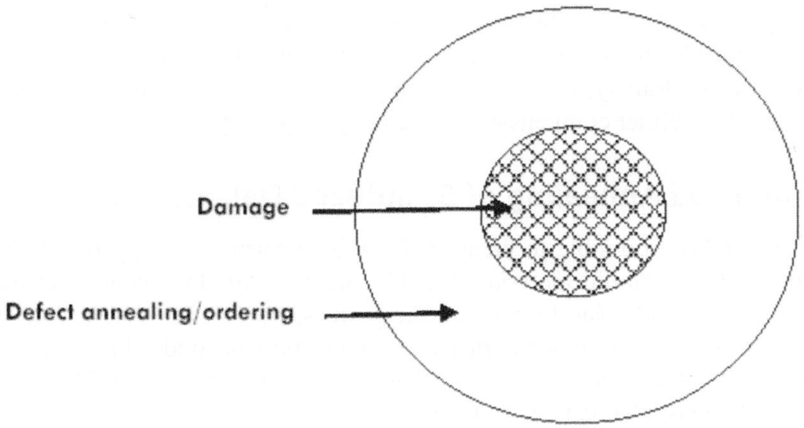

Figure 2.5: Schematic showing the annular region surrounding the track core, experiencing a pressure arising from expansion of material due to temperature spike or Coulomb explosion, as an effect of impact of swift heavy ion.

Hammering Effect/Plastic Deformation and Viscoelastic Model

Metallic glasses and amorphous solids under SHI irradiation undergo an increase in dimension perpendicular to the beam axis whereas dimension parallel to beam axis shrinks. In this anisotropic effect of SHI, sputtering effects are negligible and the volume is conserved. This being analogous to the effect produced by hammer is called as hammering effect, also known as plastic deformation [19]. Experiments performed on various crystalline metals and alloys show that the dimensional changes occur in amorphous structure but not in crystalline structures. The results are explained on the basis that the passage of SHI gives rise to mechanical stresses [20]. Trinkaus proposed a viscoelastic model to understand the origin of plastic flow in the form of anisotropic growth of amorphous solid under swift heavy ion irradiation. The anisotropic growth is attributed to the relaxation of shear stresses coupled to thermal expansion in a nanometric cylindrical zone (during temperature spike) and freezing of the associated strain upon cooling. Such strain zones created by individual ions accumulate and hence the anisotropic growth increases with fluence [21, 22].

2.4 COOPERATIVE EFFECTS OF NUCLEAR AND ELECTRONIC ENERGY LOSSES

Normally one considers the electronic and nuclear energy loss induced materials modifications as independent processes, but the effects produced by the two modes of energy loss need not be independent of each other. In fact, the materials modification induced by ion beam can have contribution from both modes of energy losses. It is important to realize that there can be a cooperative effect of nuclear and electronic energy losses of energetic ions in materials modifications. There can be four types of cooperation between S_e and S_n induced effects. These can be either compensatory type or synergetic type.

S_e Induced Annealing of S_n Induced Defects

It has not been possible to create defects by electronic energy loss of SHI in crystalline metallic materials like Cu, Ag and Au. The damage in these materials is totally due to nuclear energy loss, S_n. In some targets like Fe, continuous amorphized latent tracks are created only beyond a threshold value (~4 keV/A) of S_e. Below this value, S_e induces partial annealing of the defects created by elastic collisions (S_n) [23].

Ion induced amorphization and recrystallization of solid matrices by energetic ions is an excellent example of the combined role of S_e and S_n. Klauminzer and Meier [24] studied U_3Si_2, an important material used in nuclear reactor. Ion irradiation induced amorphization of this material leads to swelling and growth, which needs to be avoided. It was found that for lower S_e (~8 keV/nm), the full amorphization achieved was solely due to elastic collisions and S_e had negligible influence. Increasing S_e to ~20 keV/nm led to partial annealing of the S_n induced amorphization with a crystalline fraction, f_c ~0.7. A stationary state between S_n induced amorphization and S_e induced recrystallization is thus achieved. The latter process led to a crystalline fraction f_c ~0.95 when S_e was increased to 32 keV/nm. The understanding of the result is based on a competition between direct amorphization by elastic collision and recrystallization due to a thermal spike.

S_n Induced Modification during the Time Scale of S_e Induced Thermal Spike

It is found that in some situation the two modes of energy loss, S_e and S_n re-enforce the effect of each other. There seems to be a synergetic effect that leads to an extent of materials modification that cannot be accounted for by a simple addition of the modifications separately induced by S_e and S_n, but by a combined effect of both. Such complex synergetic effects of electronic and nuclear energy loss demand certain experimental conditions, and the study of sputtering provides a good tool to investigate this aspect.

To examine the combined role of S_e and S_n induced sputtering, effects of 200 MeV Ag and 100 MeV Au ion irradiations on InP surface were studied [25]. Both the ions have same S_e in InP. However, they induce very different surface modifications. 100 Mev Au irradiation results into large size pits on the surface whereas 200 MeV Ag ion irradiation does not affect the surface. Thus it is not a case of electronic sputtering. Can it be due to the high value of S_n of Au ions? The large pits that one sees in 100 MeV Au irradiation could be due to the combined effect of S_n and S_e. Although the velocity effect can also be invoked to explain this experimental result. Since the velocity of Au ions of 100 MeV is roughly half of the velocity of 200 MeV Ag ions, the damage expected by slower ions can be expected to be more as expected from the velocity effect.

Sputtering of metals such as Ti and Zr was studied with high energetic ion beam of 210 MeV I and 230 MeV Au beams where electronic energy loss dominates, and clear synergy between S_e and S_n was observed [26]. Sputtering was studied by collecting the sputtered particles on a catcher foil like high quality single crystalline Si and studying the amount collected by Rutherford Back Scattering (RBS) in channelling mode. The sputtering is likely to occur within the thermal spike period of a few picoseconds. During the short time the lattice stays above the vapourization temperature where nuclear stopping can result into much higher sputtering yield.

These studies suggest that, though small in magnitude, the effect of nuclear energy loss can be augmented by the presence of electronic excitations created by the high energy heavy ion.

S_e Induced Atomic Mobility Accelerated by S_n Induced Defects

This is another kind of synergetic effect of S_e and S_n. Unlike the previous case, the defect creation due to S_e and S_n in this case is decoupled in time. It was shown [27] that irradiation of Co/Ge bilayers by 100 MeV Au ions results into mixing at the interface leading to the formation of 4 nm thick Co-Ge alloy at room temperature. Changes occurring across the Co/Ge interface were studied by Rutherford Backscattering Spectrometry and cross-sectional TEM. Since the energy is such that $S_e < S_{eth}$, mixing in principle should not occur due to electronic energy loss process. Going by nuclear energy loss, one needs a fluence $\sim 10^{16}$ ions cm^{-2} to effect any mixing. But mixing was observed at a fluence of 10^{14} ions cm^{-2}. This implies that S_n alone cannot account for mixing, and there must be a synergetic effect of S_e and S_n. But the synergy here is different from the synergy of S_e and S_n in modifying the materials during thermal spike period as discussed above. To understand this unusual synergetic effect, one presumes that S_n builds up point defects with increasing irradiation fluence. At the fluence of 10^{14} ions/cm^2, there are enough point defects like vacancies and interstitials, which accelerate atomic diffusion at the interface during the thermal spike created by electronic energy loss.

Athermal Process

The above discussion pertains to thermodynamic equilibrium condition of electron and lattice degrees of freedom. In the wake of SHI, the electrons equilibrate to a maximum temperature in a time scale of a few femtoseconds. Subsequent transfer of energy from electrons to lattice heats up the lattice to a temperature of about a thousand Kelvin or above in a few picoseconds. In the background of these thermal processes, different athermal processes have been envisaged where materials modification does not involve increase in temperature, either transient or over long periods of time. The electronic excitation in one such process, for example, drives the electronic subsystem from a bonding to an anti-bonding state in a short time scale of a few femtoseconds [28]. The lattice becomes soft. In this state, the nuclear excitation S_n becomes much more effective in inducing enhanced atomic disorder. In another process, athermal annealing has been achieved in phosphorous ion implanted silicon [29]. In a study on YBCO it is observed that ions induce modification not only through creation of amorphized latent tracks along the ion path, but also through creation of atomic disorder in the oxygen sublattice in the Cu–O chains of YBCO by the secondary electrons. These electrons are emitted radially from the tracks during the passage of the SHI. It is found that it is a consequence of the inelastic interaction of the SHI induced low-energy secondary electrons with the YBCO lattice which reduces the critical temperature [30-32]. The secondary electrons can even lead to relaxation of surface stress [33, 34]. Although one cannot rule out the possibility of relaxation of stress by temperature rise in an annular region surrounding the ion track due to temperature spike.

Swift heavy ions can create a large density of defects, such as the ion track, and are capable of defect annealing as well, depending on the value of S_e. The schematic Fig. 2.6 depicts both these possibilities. One can say that at low values of S_e defect annealing is achieved whereas with increase in S_e,

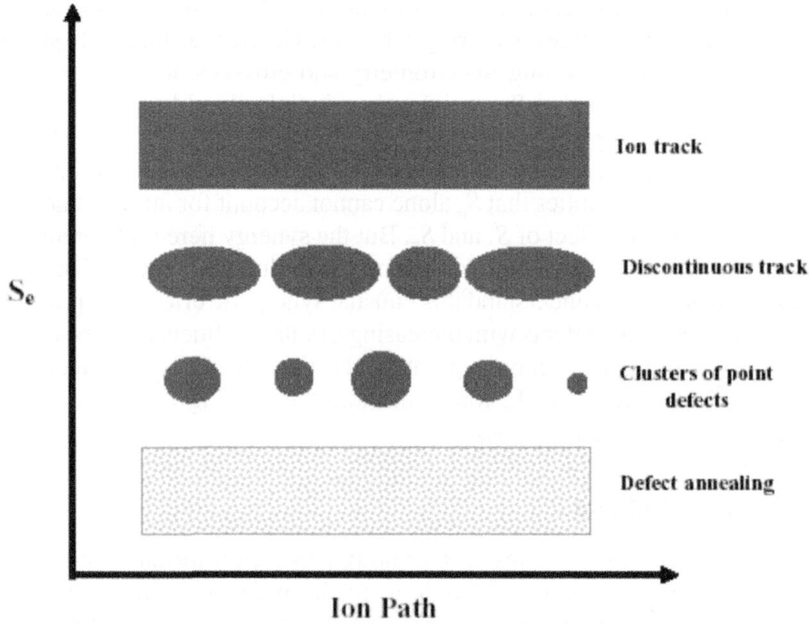

Figure 2.6: Schematic showing the possibility of defect annealing and defect creation by changing the values of S_e. Concept from Ref. 11 and Ref. 23.

defect creation starts dominating and first cluster of points are created along the ion path. With further increase in S_e, one gets discontinuous ion tracks and finally beyond a threshold of S_e, an ion track is formed. The schematic is only a guideline of possible effects of ion beam in materials. For example, there are materials, like most of the metals, where there is no formation of ion track. The ion track corresponds to the central core region where there is a transient melt phase and the surrounding annular region has lower temperature than central core region and can have either low density of defects or can cause defect annealing or ordering in the materials. It is thus also possible that an ion impact causes an ion track and an annular region around this undergoes defect annealing. Thus defect annealing and defect creation are competing process as a consequence of each ion impact. This is depicted in Fig. 2.7, where temperature

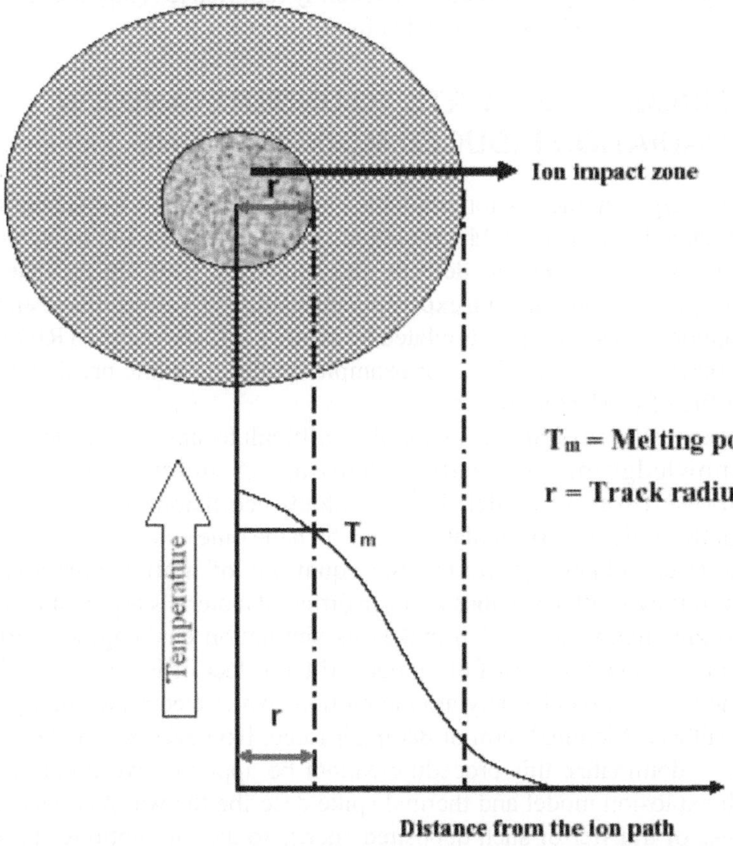

Figure 2.7: Centre core region of circle shows the damaged track core surrounded by less damaged zone or a zone where defect annealing/ordering also can take place due to favourable temperature. The central core region undergoes a melt zone for a short duration, whereas the temperature in surrounding region is below melting point. [Reprinted from N. Bajwa, Alka Ingale, D.K. Avasthi, Ravi Kumar, K. Dharamvir and V.K. Jindal, *Nucl. Instr. and Meth.*, **B212** (2003) 233 with permission from Elsevier.]

at different radial distance from ion path is shown. The central core region which melts transiently becomes ion track and the surrounding annular region has lower temperature where defect annealing or ordering can occur. Depending on the value of S_e, the diameter of the ion track and the surrounding annular region can vary.

Detailed experiments with different ion beams at high energy irradiation of fullerene thin films are explained with this picture, where the amorphization in central core region (ion track region) and the dimer formation in a surrounding annular region as ordering is considered [35, 36]. The annular region can also be under transient pressure spike due to expansion of material in track core during temperature spike. The coexistence of temperature and pressure in annular region results in dimer formation. The experiments on defect annealing in CNTs by high energy ion irradiation, recently reported [37, 38], are cases of defect annealing by SHI as discussed above.

2.5 SIMULATION EFFORTS TO UNDERSTAND ION IRRADIATION INDUCED MODIFICATIONS

In the low energy regime, the ion irradiation and its consequence on material is well simulated by TRIM/SRIM simulation codes, where the ion trajectories, energy loss S_e and S_n, recoils, defects, and sputtering rates are predicted in reasonably good agreement with experimental results. The effect of low energy ion irradiation in materials are simulated by another software called TRIDYNE [39]. It considers various effects for example sputtering while predicting the depth profile of implanted ions.

The consequence of SHI on materials is difficult to simulate due to lack of exact knowledge of involved mechanisms. A fundamentally more straightforward way to model multiple atomic collisions is provided by molecular dynamics (MD) simulations, in which the time evolution of a system of atoms is calculated by solving the equations of motion numerically. Modifications by swift heavy ions (SHI) in different materials is predominantly by electronic energy loss which makes its simulation challenging. Various simulation tools have successfully modelled the effect of lower energy ions where one needs to handle only atomic motions while electronic energy loss part is dealt as a frictional term or damping force. However, when electronic energy loss dominates, this procedure cannot be applied. Two models, viz. Coulomb explosion model and thermal spike describe the way to understand the process of transfer of such deposited energy to atomic motions but only atomistic simulations like molecular dynamics (MD) simulations and Monte Carlo simulations can provide detailed description of the modifications at the atomic level.

Atomistic simulations can, in principle, provide information about defect production, atomic density variation, electronic sputtering and motion across interfaces but they too have limitations as they do not give any information at electronic level. MD simulations have been used to understand electronic sputtering by energizing atoms around the ion path to mimic energy transfer due to electronic energy loss [40]. The same methodology has been used to study amorphization due to SHI irradiation in crystalline silicon [41]. The two-temperature thermal spike model can be coupled with MD simulations to simulate sputtering and to understand how it is influenced by details of the initial temporal and spatial energy distribution [42]. Similar approach has been used to model SHI induced modifications and for investigation of ion tracks [43]. In all these simulations which utilize thermal spike concept, the main issue is the electron phonon coupling constant as the result are very sensitive to its value. Better estimate of its value can be obtained by density functional theory [44]. Unified efforts involving various simulation techniques and accessibility of faster and powerful computers may gradually lead to better understanding of the modifications by SHI in different materials with minimal assumptions.

2.6 PERSPECTIVES OF ION-SOLID INTERACTION

Over the years, an important dimension of materials science research involving fusion of nanoparticles science with ion beam has emerged. This has opened up studies on physical phenomena in nanometric regime and femtosecond to picosecond time domain. Nanoparticles can be considered as intermediate states of matter with microscopic aspects, presenting analogies with nuclei, atoms, molecules, and macroscopic properties described by thermodynamic equilibrium in condensed matter. It is of fundamental interest to understand how mesoscopic system and properties evolve under high energy ion irradiation.

The issue of the stability of nanoparticles under ion-irradiation needs to be examined. This needs understanding of ion-matter interaction in the nano-scale range of the particle size. The response of nanoparticles and hence the modification induced in them due to ion irradiation, is expected to be governed by the confinement effects. This includes the consequence of the confinement of a high density of energy in a small volume, the energy that an energetic ion can deposit in a nanoparticle.

In nano-world ion induced defects, dislocations, strains etc. need to be examined in the light of the finite size of the nanoparticles as compared to the length scales associated with the irradiation induced defects. The thermal properties like the melting temperature, heat capacity, thermal conductivity etc. will be modified. There exists report that nanoparticles are radiation soft materials. Grain of SnO_2 nanoparticles for example fragment and even evaporate under 950 MeV Pb ion irradiation while micron size SnO_2 particles remain radiation resistance under similar irradiation condition [45]. On the contrary there are also reports that nanoparticles are radiation hard materials. One can damage bulk material by ion irradiation, but not nanoparticles. Shinde et al.

[46] demonstrated higher radiation resistance of strontium ferrite nanoparticles than their bulk counterparts to ion irradiation.

What happens upon ion/solid interaction, when the target size is reduced to nanometre scale? Collision cascades in nano dimension system, such as free-standing or embedded nanoparticles, will behave completely different from cascades in semi-infinite matter for which well-established computer codes exist since long. On reducing particles to nanoscale, surface effects will dominate and probably lead to enhanced sputtering. The increased phonon scattering at the particle boundaries should give rise to higher damage, but, on the other hand, the larger and close surface area of such particles should act as a sink for defects that will anneal easier. Diffusing species should readily be trapped at the particle surface so that one should expect a gettering effect of the particle environment.

2.7 SUMMARY

The field of ion-solid interaction, a former spin-off from nuclear science, is now a well established science with good foundation and useful applications emerging. However, due to the relatively short life time of the ion-solid interaction, many details are still unknown. Therefore, ion-solid interaction is still a challenging scientific field that offers a lot of promising applications, and justifies especially the invitation to interested scientists from other fields to contribute to this discipline with their specific expertise. Modern technologies require new materials with exquisite properties. Ion beam based methods to engineer the materials properties has the potential to create such new materials. However, the interaction of impurities, ion beam created defects and the control of these processes on the nanoscale requires improved experimental and theoretical approaches. For example: (1) the control of the size distribution of nanocrystals formed in implanted material, (2) the investigation of the dynamics of site-specific implantations etc. The ion-solid interaction can create unique solids which cannot be created by other means.

REFERENCES

1. D. Bromley (Ed.), Treatise on Heavy Ion Science. Plenum Press, New York (1984).
2. J.F. Ziegler, J.P. Biersack and U. Littmark, The Stopping and Range of Ions in Solids, The Stopping and Ranges of Ions in Matter, *In:* J.F. Ziegler (Ed.), Vol. 1, Pergamon (1985).
3. http:www.srim.org/
4. J. Lindhard, M. Scharff and H.E. Schiott, *Mat. Fys. Medd. Dan.*, **33** (1963) 14.
5. H. Beethe, *Ann. Phys.*, **5** (1930) 325.
6. F. Bloch, *Z. Phys.*, **81** (1933) 363.
7. R.L. Fleischer, P.B. Price and R.M. Walker, *J. Appl. Phys.*, **36** (1965) 3645.

8. Z.G. Wang, C. Dufour, E. Paumier and M. Toulemonde, *J. Phys. Condens. Matter*, **6** (1994) 6733; M. Toulemonde, C. Dufour and E. Paumier, *Phys. Rev.*, **B 46** (1992) 14362.
9. G. Schiwietz, E. Luderer, G. Xiao and P.L. Grande, *Nucl. Instr. and Meth.*, **B 175-177** (2001) 1.
10. M. Toulemonde, S. Bouffard and F. Studer, *Nucl. Instr. and Meth.*, **B 91** (1994) 108.
11. M. Toulemonde, *Nucl. Instr. and Meth.*, **B 156** (1999) 1.
12. A. Meftah, F. Brisard, J.M. Costantini, M. Hage-Ali, A.P. Stoquert, F. Studer and M. Toulemonde, *Phys. Rev.*, **B 48** (1993) 920.
13. Z.G. Wang, Ch. Dufour, B. Cabeau, J. Dural, G. Fuchs, E. Paumier, F. Pawlak and M. Toulemonde, *Nucl. Instr. and Meth.*, **B 107** (1996) 175.
14. G. Szenes, *Phys. Rev.*, **B 51** (1995) 8026.
15. N. Medvedev, O. Osmani, B. Rethfeld and M. Schleberger, *Nucl. Instr. and Meth.*, **B 268** (2010) 3160.
16. R.E. Johnson, B.U.R. Sundqvist, A. Hedin and D. Fenyo, *Phys. Rev.*, **B 40** (1989) 49.
17. M.M. Jakas, *Nucl. Instr. and Meth.*, **B 193** (2002) 727.
18. Z.Y. Chen, *Nucl. Instr. and Meth.*, **B 207** (2003) 415.
19. S. Klamaunzer, *Phys. Rev. Lett.*, **51** (1983) 1987.
20. M.D. Hou, *Phys. Rev.*, **B 41** (1990) 1144.
21. H. Trinkaus and A.I. Ryazanov, *Phys. Rev. Lett.*, **74** (1995) 5072.
22. H. Trinkaus, *Nucl. Instr. and Meth.*, **B 146** (1998) 204.
23. Z.G. Wang, C. Dufour, E. Paumier and M. Toulemond, *Nucl. Instr. and Meth.*, **B 115** (1996) 577.
24. S. Klauminzer and P. Meier, 20th Inter. Conf. on Atomic Collisions in Solids (ICACS 20), Puri (2003).
25. J.P. Singh, R. Singh, N.C. Mishra, V. Ganesan and D. Kanjilal, *Nucl. Instr. and Meth.*, **B 179** (2001) 37.
26. H.D. Mieskes, W. Assamann, F. Gruner, H. Kucal, Z.G. Wang and M. Toulemonde, *Phys. Rev.*, **B 67** (2003) 155414.
27. T. Som, B. Satpati, P.V. Satyam, D. Kabiraj, Ajay Gupta and N.C. Mishra, *J. Appl. Phys.*, **96** (2004) 7141.
28. A. Rousse, C. Rischel, S. Fourmaux, I. Uschman, S. Sebban, G. Grillon, Ph. Balcou, E. Forster, J. Geindre, P. Audebert, J.C. Gauthier and D. Hulin, *Nature*, **410** (2001) 65.
29. J. Grun, R.P. Fischer, M. Peckerar, C.L. Felix, B.C. Covington, W.J. Desisto, D.W. Donnelly, A. Ting and C.K. Mehta, *Appl. Phys. Lett.*, **77** (2000) 1997.
30. D. Behera, T. Mohanty, S.K. Dash, T. Banerjee, D. Kanjilal and N.C. Mishra, *Radiation Measurements*, **36** (2003) 125.
31. R. Biswal, J. John, D. Behera, P. Mallick, Sandeep Kumar, D. Kanjilal, T. Mohanty, P. Raychaudhuri and N.C. Mishra, *Supercond. Sci. Technol.*, **21** (2008) 085016.
32. Y. Zhu, Z.X. Cai, R.C. Budhani, M. Suenaga and D.O. Welch, *Phy. Rev.*, **B 48** (1993) 6436.
33. T. Narushima, A.N. Itakura, T. Kawabe and M. Kitajima, *Appl. Phys. Lett.*, **79** (2001) 605.

34. T. Narushima, M. Kitajima and K. Miki, *J. Phys. Condens. Matter*, **16** (2004) 193.
35. Navdeep Bajwa, K. Dharamvir, V.K. Jindal, Alka Ingale, D.K. Avasthi, Ravi Kumar and A. Tripathi, *J. Appl. Phys.*, **94** (2003) 326.
36. Navdeep Bajwa, Alka Ingale, D.K. Avasthi, Ravi Kumar, A. Tripathi, Keya Dharamvir and V.K. Jindal, *J. Appl. Phys.*, **104** (2008) 054306.
37. Amit Kumar, D.K. Avasthi, A. Tripathi, D. Kabiraj, F. Singh and J.C. Pivin, *J. Appl. Phys.*, **101** (2007) 014308.
38. Kiran Jeet, V.K. Jindal, L.M. Bharadwaj, D.K. Avasthi and Keya Dharamvir, *J. Appl. Phys.*, **108** (2010) 034302.
39. W. Moeller, W. Eckstein and J.P. Biersack, *Computer Phys. Communication*, **51** (1988) 355.
40. E.M. Bringa, R.E. Johnson and M. Jakas, *Phys. Rev.*, **B 60** (1999) 15107.
41. Y. Sasajima, T. Akabane, T. Nakazawa and A. Iwase, *Nucl. Instr. and Meth.*, **B 264** (2007) 259.
42. S. Mookerjee, M. Beuve, S.A. Khan, M. Toulemonde and A. Roy, *Phys. Rev.*, **B 78** (2008) 45435.
43. P. Kluth, C.S. Schnohr, O.H. Pakarinen, F. Djurabekova, D.J. Sprouster, R. Giulian, M.C. Ridgway, A.P. Byrne, C. Trautmann, D.J. Cookson, K. Nordlund and M. Toulemonde, *Phys. Rev. Lett.*, **101** (2008) 175503.
44. E. Akcöltekin, S. Akcöltekin, O. Osmani, A. Duvenbeck, H. Lebius and M. Schleberger, *New Journal of Physics*, **10** (2008) 053007.
45. A. Berthelot, S. Hemon, F. Gourbilleau, E. Dufour and E. Paumier, *Nucl. Instr. and Meth.*, **B 146** (1998) 437.
46. S.R. Shinde, A. Bhagwat, S.I. Patil, S.B. Ogale, G.K. Mehta, S.K. Date and G.J. Marst, *Magn. Mater.*, **186** (1998) 342.

3

Ion Beam Analysis

3.1 INTRODUCTION

The Ion Beam Analysis (IBA) techniques are powerful tools to investigate the composition of a material in a non-destructive way. Ion beams impinging on a material induce several processes simultaneously which lead to emission of secondary radiation and particles as illustrated in Fig. 3.1. These processes have provided various analytical techniques [1] which have been classified based on the process (scattering/photon emission or reaction) and the mode of detection of the reaction products:

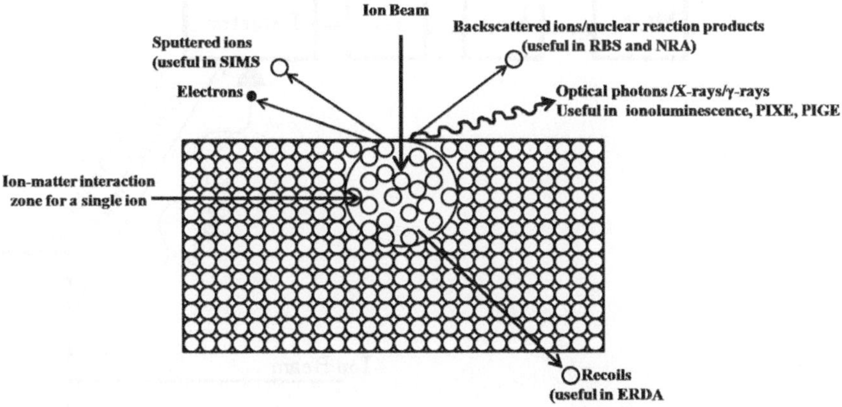

Figure 3.1: Various processes produced by ion beam interaction with materials which provide different techniques for analyzing the materials properties. (a) Photons/X-rays/ γ rays in ionoluminescence, PIXE and PIGE respectively. (b) Back scattered ions/ nuclear reaction products in RBS and NRA respectively. (c) Sputtered ions in SIMS (d) Recoils, in ERDA.

- PIXE and PIGE: Proton induced X-ray emission and Proton induced γ emission provide the elemental compositions.
- IL: Ionoluminescence, also known as Ion Beam Induced Luminescence, provides information about the ionic state of atoms by the study of the emitted photons in UV, Visible and IR region.
- RBS: Rutherford Back scattering Spectrometry provides depth profiles of elements. It is sensitive to heavy elements in light matrix.
- Ion-Channelling: Provides information on degree of crystallinity and related aspects from single-crystalline materials or epitaxial layers.
- ERDA: Elastic Recoil Detection Analysis is sensitive to light elements in a heavy matrix.
- LEIS and MEIS: Low Energy Ion Scattering and Medium Energy Ion Scattering, basically use the principle of RBS and ERDA in a few keV and ~100 keV region respectively.
- NRA: Nuclear Reaction Analysis provides isotope specific elemental profiling.
- CPAA: Charged Particle Activation Analysis provides isotopic light elements determination at sub ppb level.
- AMS: Accelerator Mass Spectroscopy provides an ultra sensitive technique to determine ratio of isotopes of an element, from a small quantity (~1 mg) sample.

Each technique provides unique information and, therefore, is utilized for specific applications. The sample to be analyzed is used as a target for a beam of accelerated ions. The interactions of the ion beam with the atoms (or the

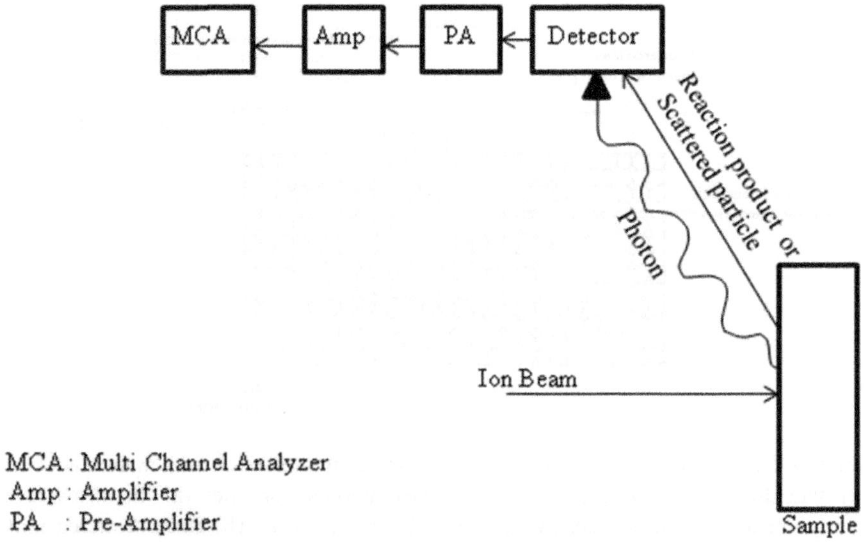

Figure 3.2: The instrumentation normally used to detect and analyze the signals from the events generated by the ion beam interaction with the material.

nuclei) of the target material induce from it the emission of secondary radiation (X-rays, γ-rays, particles), having an energy characteristic of the emitting atom or nucleus. Suitable nuclear detectors and signal processing electronic modules are then used to analyze the emitted radiation to quantify the presence of the different elements in the material as shown in Fig. 3.2.

In general, analysis done using ion beams are: Quantitative, Multi-elemental (Nuclear Reaction Analysis excluded), sensitive down to trace level, and non-destructive. Besides the techniques mentioned above there are several other areas where ion induced emissions provide excellent ways to obtain physical information. For example, the emitted electrons from the surface of the target irradiated with ions could be secondary electrons or back scattered electrons which are used in imaging techniques in Scanning Electron Microscopes (SEM). The emitted Auger electron can be used for on-line measurements providing crucial information on ion matter interaction [2-4]. Sputtered secondary ions from the surface produced by the primary ions are used in Secondary Ion Mass Spectroscopy (SIMS) to get elemental, isotopic or molecular compositions of the target.

3.2 PROTON INDUCED X-RAY EMISSION (PIXE)

It was first shown by Johansson et al. [5] that the analysis of X-ray emission following proton bombardment using silicon detectors constitutes a powerful multi-elemental analytic method. When a material is bombarded with energetic charged particles, the Coulomb interaction between these projectiles and the atomic electrons in the material results in ionisation of the atoms. Subsequent de-excitation of the ionised atoms leads to the emission of characteristic X-rays. Depending on the sub-shell from which the electron has been removed from the atom, as a result of the ionisation, K, L, M etc. X-rays are emitted. The energies of the emitted X-rays, detected by a suitable X-ray detector of the requisite resolution, give information on the various elements present in the material (Qualitative Analysis). The intensity of the X-rays gives information on the concentration of the elements (Quantitative Analysis).

Compared to electron based X-ray analytical techniques such as energy dispersive spectroscopy (EDS), PIXE offers much higher sensitivities. It is due to the fact that the background X-rays (Bremsstrahlung radiation) is three order of magnitude smaller in PIXE as compared to EDS.

The PIXE spectrum consists of the characteristic peaks of the elements present in the sample superposed on a continuous background, mainly arising from the Bremsstrahlung generated by the electrons ejected from these elements as a result of the Coulomb interaction of the incident projectiles. The PIXE spectrum is analyzed to extract the area of each peak after suitable backgroumd subtraction. The cross section for X-ray production for a given ion beam energy is a smoothly varying function of the atomic number of the element. The charactersitic X-ray yield is given by:

$$N_x = N_p\,(m\,N_o/A)\,\sigma_x\,\varepsilon\,T \tag{3.1}$$

where N_p is the number of incident protons, m is the mass of the element, A is its mass number, N_o is the Avogadro number, σ_x is the X-ray production cross section, ε is the detection efficiency and T is a factor to account for X-ray absorptions in the material and detector window.

The experimental arrangement consists of a vacuum chamber in which the sample to be analyzed (Target) is bombarded with the proton beam, a Si(Li) X-ray detector for detecting the emitted X-rays from the target, a data acquisition and analysis system and a Faraday Cup for charge measurement. Some suitable absorber may be some times inserted in front of the detector to absorb unwanted low energy background radiation.

Sample preparation is of utmost importance in PIXE. Ideal condition is thin self supporting target which can be directly bombarded with the proton beam inside the chamber. But normally some suitable backing is required. The backing should be thin so that the proton beam suffers minimum energy loss in it and should not have any impurities. Commonly used backings are Mylar, Kapton, carbon, formvar, nucleopore filter paper etc. Sample preparation is relatively easy for liquid samples and those which can be brought into liquid form (like water soluble samples). A drop of the solution allowed to dry up on a backing such as formvar, nucleopore filter paper etc. forms a good target.

Equation 3.1 for the number of X-rays detected can be written as

$$N_x(Z, E_p) = K(Z, E_p)\,Q_p\,m \tag{3.2}$$

where Q_p is the total charge collected at the Faraday Cup, N_x is the number of X-rays detected and K is a calibration constant, to be determined using a number of standard samples (chemical compounds dissolved in distilled de-ionised water and thin targets made out of these or other certified standard samples). Knowing N_x and K and by measuring Q_p the mass m of the element can be estimated.

For any analytical technique, it is necessary to have an idea of the minimum detection limit, i.e. the lowest concentration of an element which will give a detectable X-ray count rate over and above the background. The usual criterion is that $N_x = 3\sigma$, where σ is the standard deviation of the background counts in the expected peak region for a particular element. Typically the detection limit is about 0.1 ppm for $Z \sim 30$ for K X-rays. It depends on the X-ray energy.

It is possible to do measurements at atmospheric pressure by bringing out the proton beam from the high vacuum environment of the accelerator into the ambient of the laboratory. This technology makes it possible to measure a valuable artifact or precious material at atmospheric pressure outside the confines of an evacuated chamber without sampling. Problem in running in this configuration is that low energy X-rays from lighter elements are attenuated in air. However, we can purge the area between the sample and the detector with helium to minimize this effect.

Applications of the PIXE technique in areas such as Environmental pollution, Forensic science, Geochemical prospecting, Archaeological studies, etc. are described in Appendix 3.

3.3 PARTICLE INDUCED γ-RAY EMISSION (PIGE)

The particle induced γ-ray emission (PIGE) method is complimentary to the PIXE method. It provides additional ways to analyze the elemental compositions in materials by analyzing the energies of the γ-rays emitted by the materials irradiated by a few MeV ions. Usually proton beams are used to induce γ emission and Ge(Li) detector is used for γ-ray detection. It is very sensitive for the light elements and provides a technique to measure low concentrations. Simultaneous measurements of X-rays and γ-rays from the samples using PIXE and PIGE emissions have been used to measure inclusions in various ore deposits and other materials science studies [6, 7].

3.4 IONOLUMINESCENCE (IL)

When ion beam is incident on certain materials, it emits photons in UV, visible or IR region. The states in bandgap arising due to impurity atoms or the defect states, are populated due to interaction of incident ion and material. The decay of these populated states give photons in different energy regimes from UV to IR. The emitted light is analyzed by suitable detection system such as monochromator, detector, etc., which provides a spectrum of the intensities of different wavelengths. Such a study is referred to as ionoluminescence (IL) and has been extensively used for imaging and analysis of elements and chemical compounds [8, 9]. It is a powerful tool to determine ionic state of impurity atoms [10-12]. IL spectra provide signatures of the defects inside the material, enabling differentiation of natural specimens from imitations. Since the ion beam especially SHI's can modify the material, resulting in change of ionoluminescence, the on-line studies with ion fluence provide valuable information on modification induced by SHI. The ionoluminescence and PIXE are complimentary techniques. Their combination provides a powerful analytical tool, particularly in geochemistry [8].

3.5 RUTHERFORD BACKSCATTERING SPECTROMETRY (RBS)

Energetic ions of typically 1-3 MeV energy, incident on the sample, are elastically scattered from the atoms in the sample. The number of scattered ions and their energy is measured. These data provide information on the composition of the sample, the distribution of those components, and the thickness of the sample. The incident ion species, normally used, are positively charged ^{4}He atoms [13]. RBS performed with heavy ions is referred to as Heavy Ion Back Scattering (HIBS).

In an elastic collision the energy of scattered particle depends on the mass and the energy of incident particle, mass of the target atom and the angle of scattering. The measured scattered particle energy reveals the information about

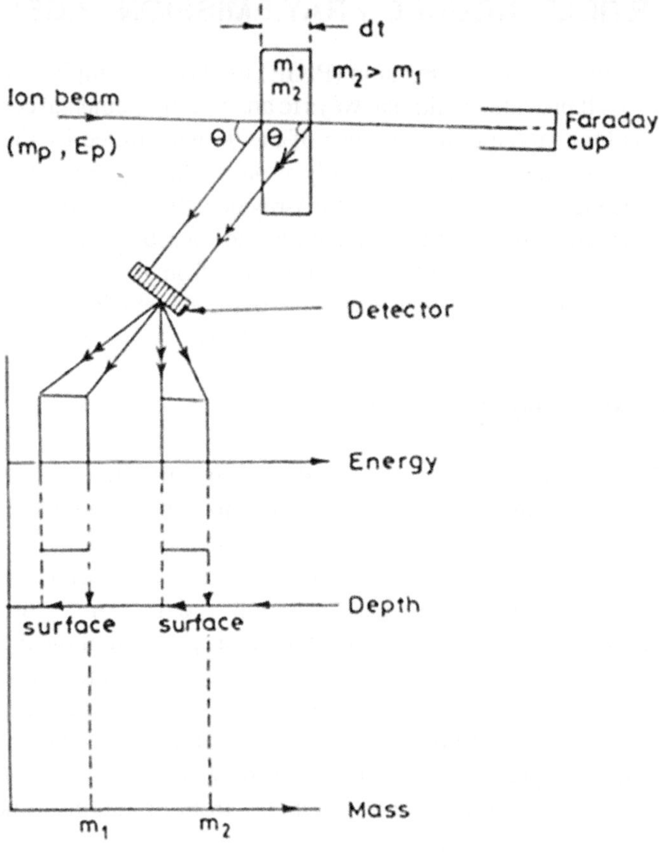

Figure 3.3: The ions of energy E_p and mass m_p are incident on a sample having two elements of masses m_1 and m_2. The back scattered particles having energies governed by kinematics, are detected in a detector at an angle θ. The energies of the back scattered particles provide information about mass of the elements. The ions scattered from the surface (shown by single arrow) have higher energy than those scattered from inside (shown by double arrow), give information of the depth profile as well as the film thickness as shown here. [Reprinted from D.K. Avasthi, *Physics Education,* Apr–June (1995) 28.]

the different elements in the material. The back angle for detector is chosen so that the scattered particle energy separation from different masses (dE/dm) is larger and the kinematic broadening ($dE/d\theta$) is small. The variation of energy of the scattered particle due to scattering from different inner layers of the sample enables one to reconstruct the depth profile, i.e. the information of concentration versus depth of an element. The schematic in Fig. 3.3 explains the basic principle of depth profiling by RBS. Consider ion beam of energy (E_p) striking a sample having two elements A and B. The backscattered ions at an angle θ from the elements at the surface will have the energies E_A and E_B,

$$E_A = K_A E_p \tag{3.3}$$

$$E_B = K_B E_p \tag{3.4}$$

where K is kinematic factor

$$K = [\{m_p \cos\theta + (m_t^2 - m_p^2 \sin^2\theta)^{1/2}\}/(m_p + m_t)]^2 \tag{3.5}$$

and m_p and m_t are the mass in amu of the projectile and the target respectively.

The energy loss of the backscattered ions is dependent on two processes: scattering events with nuclei of the sample and from electrons in the sample. The first process is dependent on the scattering cross-section of the nucleus and thus on its mass and atomic number. For a given measurement angle, nuclei of two different elements scatter incident ions to different degrees and with different energies, producing separate peaks on a plot of counts $N(E)$ versus energy E. The energies of scattered particle can be converted to a mass scale as shown in Fig. 3.3. These peaks are characteristic of the elements contained in the material, providing a means of analyzing the composition of a sample. Relative concentrations can be determined by measuring the peak heights.

The quantitative estimates of elements is made using the number of particles detected in a given solid angle Ω

$$Y = N_p N_t (d\sigma/d\Omega)\,\Omega \tag{3.6}$$

N_p is the number of incident particles, N_t is the number of target atoms per cm^2, and $(d\sigma/d\Omega)$ is the Rutherford scattering cross-section. The experimental determination of Y provides the value of N_t.

The second energy loss process, the stopping power of the sample electrons, does not result in large discrete losses such as those produced by nuclear collisions. Instead it creates a gradual energy loss dependent on the electron density and the distance traversed in the sample. This energy loss lowers the measured energy of ions which backscatter from nuclei inside the sample in a continuous manner dependent on the depth of the nuclei. The result is that instead of the sharp backscattered peaks the observed peaks on the $N(E)$ versus E plot trail off gradually towards lower energy as the ions pass through the depth occupied by that element. Elements which only appear at some depth inside the sample have their peak positions shifted by some amount which represents the distance an ion had to traverse to reach those nuclei.

When the scattering occurs inside of target, the backscattered particle energy E_A is smaller due to (i) energy loss of incoming ion in the sample and (ii) energy loss of scattered particle in the sample before reaching the detector. Thus reduced energy provides the information about the depth at which the scattering occurs and it becomes the basis for depth scale as shown in Fig. 3.3.

Depth profiling can be determined from the following equation [14]:

$$\Delta E = \Delta t\,[K\,(dE/dx)_{Ep} + 1/\cos\theta\,(dE/dx)_{kEp}] \tag{3.7}$$

where $(dE/dx)_{Ep}$ and $(dE/dx)_{kEp}$ represent rate of energy loss for incident ions and scattered ions respectively, ΔE is the energy width of the peak in RBS spectrum, E_p is the projectile energy, Δt is the film thickness, K is the kinematic factor, and θ is the scattering angle.

This kind of measurement determines the elemental composition; the chemical structure of the sample cannot be determined from the $N(E)$ versus E profile of RBS spectrum. However, it is possible to learn about the crystallinity of the sample through RBS channelling, discussed in the next section.

In practice, RBS is an analytical tool that uses elastic scattering of 0.1- a few MeV charged particles to analyze the surface and the few micrometres of solids. A typical RBS system consists of an accelerator and scattering chamber with sample manipulators and particle detectors. The scattered particles are detected by surface barrier detector. The signal from the detector is processed by nuclear electronics and multi-channel analyzer (MCA), which can be integrated with the computer. The data evaluation is accomplished using standard procedures and RUMP computer code [15]. The advantage of the low-energy RBS technique lies in its capability of quantitative analysis of major and minor constituents lying in the first 0.5 to 2.0 micrometres of a material. Depending on the sample structure and composition, the detection limits vary from 10^{11} to 10^{15} at. cm^{-2} for heavy and light elements respectively. The detection limits for some light elements may further be improved by using resonant scattering. The depth distribution of constituents can be reconstructed with a depth resolution of ~10 nm. The RBS technique is non-destructive since the erosion and the radiation degradation of the sample material by the particle impact is negligible. It is routinely used for quantifying impurity concentration, film thickness measurement, diffusion studies, and compositional analysis of novel phases. Novel phases synthesized by MeV ion beam mixing technique can be investigated by RBS. A few examples of the applications of RBS are given in Appendix 4.

3.6 RBS CHANNELLING

Ion channelling in a crystal is due to the repulsive force between the incident ions and the atoms in the lattice. Ions are deflected by this force causing a shadow cone behind the atoms. If the ion beam is aligned parallel to a major crystallographic direction, all atoms except the topmost ones are located in the shadow cone of other atoms and the ions travel mainly in the open channels between the atomic rows. This phenomenon is called RBS channelling. In such a condition the backscattering yield or x-ray production is drastically reduced and these events provide a monitoring signal to achieve channelling condition. RBS channelling is a powerful tool to provide information on (i) the content of crystallization, (ii) location of dopant atom, (iii) determination of strain in super lattices, etc. [16-20].

3.7 ELASTIC RECOIL DETECTION ANALYSIS (ERDA)

Elastic Recoil Detection is an ion beam technique to obtain elemental concentration depth profiles [21, 22]. An ion beam is directed at the sample to be depth profiled and as in RBS there is elastic interaction with the atoms of the sample. The incident energetic ions used typically have MeV energies, enough to kick out (recoil) the atoms being struck. The technique depends on putting in an appropriate detector to detect these recoiled atoms. The great advantage in ERDA is that all the atoms of the sample can be recoiled if a heavy incident beam is used, so a complete analysis of the sample is immediately available.

ERDA provides excellent technique for depth profiling light elements. The use of large area position sensitive telescope detectors makes it a highly sensitive technique with which ion beam induced modifications such as interface mixing, electronic sputtering etc. can be performed during irradiation itself, with its availability as an online facility.

The analysis of hydrogen is important in many areas of materials science. Hydrogen also has important effects on the chemical, physical and electrical properties of many materials. It is the lightest element in nature. Its detection and quantitative analysis is difficult by the conventional methods utilized for other elements. The technique of elastic recoil detection analysis (ERDA) using 1–2 MeV He^+ beam has been developed to quantitatively and simultaneously analyze hydrogen and its isotopes in solids. For example, such a facility using the 2 MeV Van-de-Graaff accelerator, facilitates H and D analysis in a material up to a depth of ~1 μm with a detection sensitivity of 0.1 at.% and depth resolution of about 100 Å.

Elastic recoil detection analysis (ERDA) is a powerful method to investigate elemental concentration of surfaces. Since each recoiling species carries the specific information about its own depth distribution in the sample, concentration depth profiles for all elements can be measured nondestructively at once and without interferences. This is especially well suited to quantitatively analyze light elements in heavy substrates. The time needed for a complete measurement of one sample is only minutes.

Mathematical Formulation of ERDA

When an elastic collision occurs with the incident particle of mass m_p and energy E_p with the atoms of mass m_r, the atoms in the sample recoil in forward direction. The recoil energy E_r of atom at an angle ϕ with respect to the beam direction can be expressed as

$$E_r = E_p\, 4\, m_p m_r \cos^2 \phi/(m_p + m_r)^2 \qquad (3.8)$$

If the projectile mass is greater than that of the target atom ($m_p > m_r$), maximum angle Θ of recoil is given by the relation

$$\Theta = \text{arc sin}\,(m_r/m_p) \qquad (3.9)$$

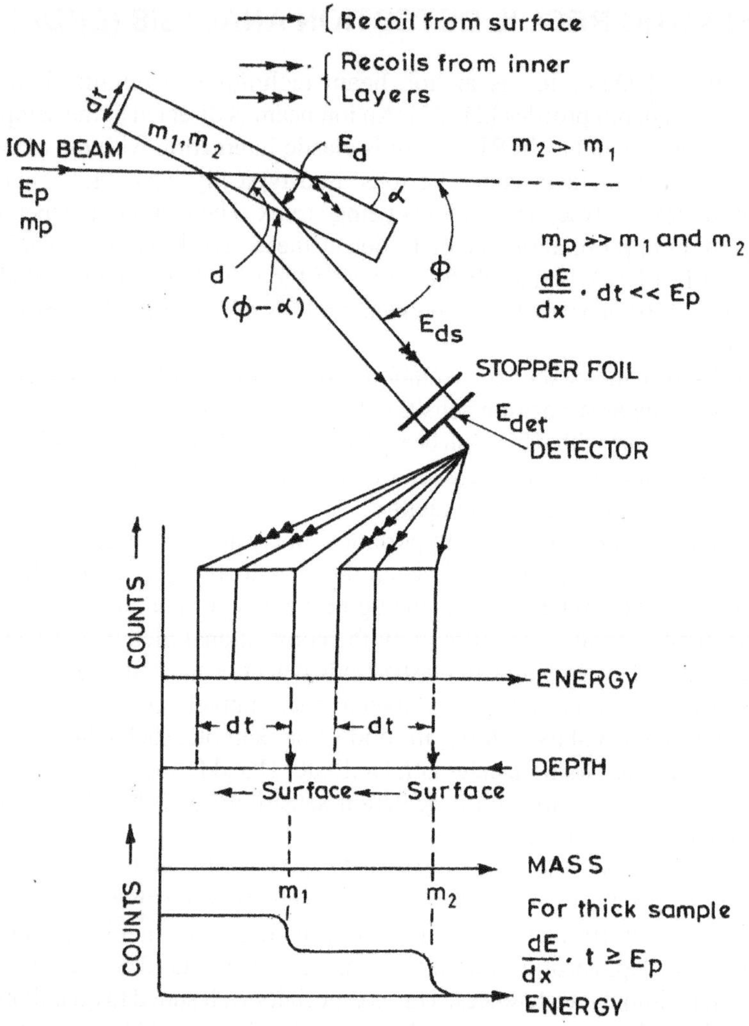

Figure 3.4: The ions are incident on a sample (tilted at an angle α), having two elements of masses m_1 and m_2. Unwanted ions and heavier recoils are stopped in a stopper foil. The recoils having different energies provide information about the mass of the elements. The recoils from the surface (shown by single arrow) have higher energy than those originated from inside (shown by double and triple arrows) and give information of the depth profile as well as film thickness. [Reprinted from D.K. Avasthi and W. Assmann, *Current Science,* **80** (2001) 1532 with permission from Indian Academy of Science.]

The scattered projectiles are thus blocked by putting the detector beyond the maximum scattering angle. If the sample has elements with mass more than projectile, the scattered projectiles are stopped in a stopper foil of appropriate thickness in front of the detector. Recoils of light elements are not stopped in the stopper foil due to smaller energy loss. The schematic of an ERDA set up is shown in Fig. 3.4.

The concentration of sample atoms N_r (in atoms/cm^2) is given by relation

$$N_r = Y \sin \alpha / (N_p \, \Omega \, d\sigma/d\Omega) \tag{3.10}$$

where Y is the number of recoils in a solid angle Ω, N_p is the number of incident ions, α is the tilt angle of the sample as seen in Fig. 3.4, and $d\sigma/d\Omega$ is the differential Rutherford recoil cross section given (at an angle ϕ) by the following relation.

$$d\sigma/d\Omega = [Z_p Z_r e^2 (m_p + m_r)]^2 / (4 E_p^{\,2} m_r^{\,2} \cos^3 \phi) \tag{3.11}$$

where Z_p and Z_r are the atomic numbers of the projectile ion and the recoiling ion from the sample respectively.

The recoil energy originating from a depth d, depends on (i) the kinematics governed by Eq.(3.8), (ii) the energy loss of the incoming ion in the sample material, (iii) the energy loss of the recoil in the sample and (iv) the energy loss of the recoil in the stopper foil as shown in the schematic in Fig. 3.4. The energy of a recoil generated at depth d is given by

$$E_d = k\,[E_p - (d/\sin \alpha)\,(dE/dx)_{in}] \tag{3.12}$$

where $(dE/dx)_{in}$ is the energy loss of incident ion in the sample material.

The recoil energy as detected by the detector is given by

$$E_{det} = E_d - (d/\sin(\phi - \alpha)\,(dE/dx)_{out} - \Delta E_{foil} \tag{3.13}$$

where $(dE/dx)_{out}$ is the energy loss of recoil ions in the sample, and ΔE_{foil} is the energy loss of recoil in the stopper foil.

The recoil energy can be converted to depth scale using equations (3.12) and (3.13). Recoil energy gives the information about the mass of the elements in the sample. The identification of the elements from the recoil energy is feasible if the difference in recoil energies of the elements is larger than the energy resolution of the set up. Heavier recoils have higher energy and appear as separate group as shown in Fig. 3.4.

Most common geometry used in ERDA is the reflection geometry where the sample is tilted at an angle and the detector is kept at an angle greater than the sample tilt angle. ERDA can be used in transmission geometry for self supporting thin film samples. The detector resolution becomes critical if the sample has elements with neighbouring masses, the recoil energies overlap and it becomes difficult to distinguish them. In such a situation, various types of the particle identification techniques are utilized for discrimination of different elements. These techniques are: Time of Flight (TOF) Spectrometer [23, 24], $\Delta E - E$ detector telescope [25-27], Magnetic Spectrograph [28], Bragg Curve Spectrometer (BCS) [29] etc.

The TOF technique is normally preferable with low energy ion beams so that the recoils have large flight time and the percentage error in flight time is small. The magnetic spectrometer provides excellent resolution but are expensive. The use of detector telescopes is an ideal choice for identifying neighbouring mass elements with reasonably good depth resolution. The detector telescope can be designed and fabricated according to the need.

Telescope arrangement [30] consists of two detectors; first one is thin detector, where a small part of the energy (ΔE) is deposited and the rest of the energy $E_{rest} = E - \Delta E$ is deposited in the second detector. The energy lost, ΔE in the first detector, depends on the product MZ^2 where Z is atomic number and M is the mass of recoil. Thus the energy lost in ΔE detector by the recoils of different elements having almost identical energies is different and can be used to identify the atomic number of recoil.

On-Line Monitoring with ERDA

Normally in ion beam techniques the ion dose required for analysis is small and there is no significant change in the sample. However in case of heavy ion ERDA, where it is known that SHI can induce changes in material under investigation, there are chances of modification of material during analysis. This however is a blessing in disguise as it provides a handle on the possibility of engineering the properties of materials [31-33]. A large area position sensitive gaseous telescope detector (discussed in Appendix 4) is an ideal system for such studies as large area enables analysis with small dose and position sensitive feature allows to retain depth resolution.

Elastic recoil detection analysis (ERDA) with SHI provides a unique possibility of on-line monitoring of compositional changes produced in the material by SHI irradiation. The probing ions provide information about the modifications produced by the preceding ions since the probe action takes place faster (in 10^{-17} s) in comparison to the time taken for material modification

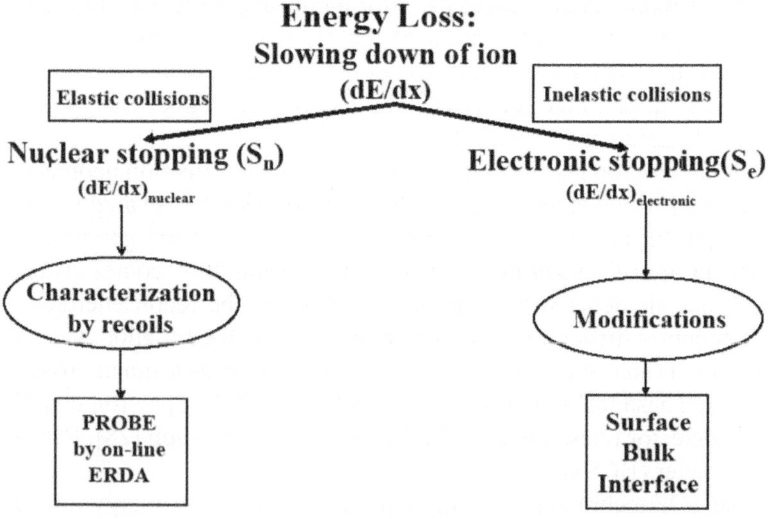

Figure 3.5: The recoils produced due to elastic collisions are used in ERDA for on-line analysis which provides the information of material modification (in surface bulk and interface) caused by electronic excitation. [Reprinted from D.K. Avasthi, S. Ghosh, S.K. Srivastava and W. Assmann, *Nucl. Instr. and Meth.*, **B 219-220** (2004) 206 with permission from Elsevier.]

($\sim 10^{-11}$ s) by already impinged ions [34]. Thus fluence dependence of SHI induced compositional changes in materials can be carried out using on-line ERDA. The incident ions produce the modification in material due to large electronic energy deposition (arising due to inelastic collisions) whereas the recoils produced by elastic collisions are used for characterization by ERDA. Schematic for this online monitoring of modification of material is shown in Fig. 3.5. A few examples of applications of ERDA are given in Appendix 4.

3.8 MEDIUM ENERGY ION SCATTERING (MEIS)

Medium energy (50–250 keV) beams of light ions (usually H^+ or He^+) are used to probe the surface and near-surface region. The beam is incident on a sample target along a known direction and the energy and angle of the scattered ions are measured simultaneously using an electrostatic analyzer with two-dimensional detector. Simple 'billiard-ball' mechanics determines the atomic mass of components of the target, whilst depth sensitivity arises due to the more subtle energy losses that occur as an ion passes through a solid. Structure determination is via the variation in scattered ion intensity that arises due to shadowing and blocking of the ions by atoms in the target along specific crystallographic directions. The ability to characterize materials in terms of their composition and structure with virtually monolayer resolution makes MEIS an ideal technique for the characterization of surfaces, thin films and nanostructures. It is sensitive to both the composition and structure of materials, with a depth resolution down to single atomic layer spacing [35-37].

3.9 LOW ENERGY ION SCATTERING (LEIS)

LEIS is in principle same as MEIS and RBS but a few keV energies are used to analyze the surface region. Special equipment is used for the detection of the positions, velocities and energies of the ions that have interacted with the surface. It allows determinations of relative positions of atoms in a surface lattice and elemental identity of those atoms [38-41]. LEIS is one of a very few surface-sensitive techniques capable of directly observing hydrogen atoms as recoil events [42].

3.10 NUCLEAR REACTION ANALYSIS (NRA)

NRA is a technique to obtain concentration versus depth distribution in thin films. The basic principle is as follows: An incoming ion enters the film and loses energy while penetrating the film. In the low energy regime, the ion beam interacts with the specimen only by means of their electric charges. The Coulomb barrier keeps the ion from reaching the nucleus. At higher energies the incoming ion can undergo a nuclear reaction with the nucleus of the atoms of the film causing a change in the nucleus. The emitted radiation and the reaction products can be detected with appropriate detectors to provide the desired information. When the ion energy used is not high the nuclear reaction takes place with only low-Z elements in the specimen. Because of this ability to pick out low-Z elements in high-Z matrices, analysis via nuclear reaction is well suited for the detection of light elements which is a major advantage of NRA [14].

When the incident particle succeeds in penetrating the Coulomb barrier of the target nucleus there are two types of nuclear reactions possible: (1) Direct reaction producing the products or (2) Compound Nucleus Reaction in which the particle is absorbed by the target nucleus A forming a compound nucleus B in an excited state which then loses its internal energy by emitting γ rays or break up giving the reaction products.

$$a + A \geq B^* \geq B + \gamma \tag{3.14}$$

$$a + A \geq B^* \geq C + d \tag{3.15}$$

The detection of the light particle d and the measurement of its energy enables to generate a depth profile of an element. If the nuclear reaction leaves the nucleus C in an excited state, its prompt γ decay can also be used for NRA.

As an example, consider a reaction used to profile hydrogen.

$$^{15}\text{N} + {}^{1}\text{H} \geq {}^{12}\text{C} + \alpha + \gamma\,(4.965\text{ MeV}) \tag{3.16}$$

This reaction has a resonance at 6.383 MeV energy. The emitted γ is characteristic of the reaction and the number of γ rays detected is proportional to the concentration at the respective depth of hydrogen in the sample. To obtain the depth information the stopping power of the projectile ion in the sample has to be known. The H concentration profile [43] of the sample is obtained by scanning the ^{15}N incident beam on the sample.

NRA can also be used non-resonantly. For example, deuterium can be profiled with He beam.

$$^{3}\text{He} + \text{D} \geq \alpha + p + 18.353\text{ MeV} \tag{3.17}$$

For a fixed energy of the incident ^{3}He particle the energy of the emitted proton depends on the depth of the deuterium atom in the sample.

3.11 CHARGED PARTICLE ACTIVATION ANALYSIS (CPAA)

Charged particle activation analysis technique uses high current light ions (^{1}H, ^{2}H, ^{3}He and ^{4}He) to induce nuclear reaction in light elements (B, Be, C, F etc.) in a sample [44, 45]. The decay of induced radioactivity is recorded to get concentration of traces (upto sub ppb level) of the light elements. The technique requires choosing a nuclear reaction depending on the light element, whose concentration is to be determined.

The difference between NRA and CPAA is that NRA uses prompt decay events of nuclear reaction, whereas CPAA uses delayed events having large lifetime. It has been a popular technique for in situ tribological investigation to test the abrasion rate of surfaces. The material surface to be tested is bombarded with light ions to produce activity. The loss in activity is monitored during operation of wearing/tearing of surfaces under process, to determine the abrasion rate.

3.12 ACCELERATOR MASS SPECTROMETRY (AMS)

Accelerator Mass Spectrometry is an ultra sensitive technique for detection and counting of individual nuclei [46, 47]. It is typically used for determining the ratio of the abundant to rare isotopes in any sample. It can determine extremely small ratio of two isotopes of an element for estimating the age of the sample (for example a rock). The process involved in a Mass Spectrometer is to take the sample, ionize it to produce ions, which are then separated according to their mass-to-charge ratio by the spectrometer and counted by appropriate detection system. Accelerator Mass Spectrometer (AMS) is a combination of mass spectrometer and a particle accelerator to provide a

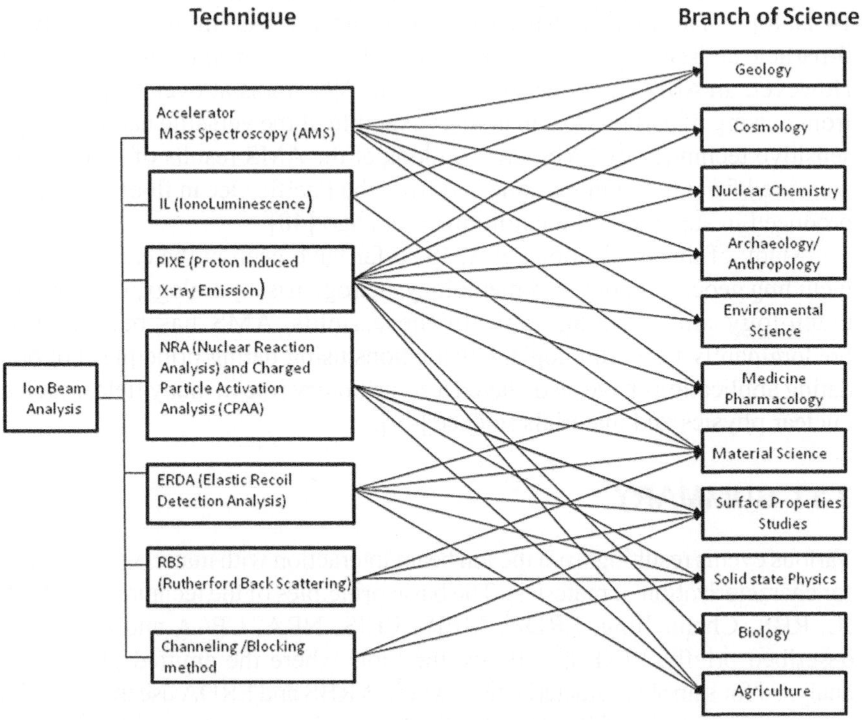

Figure 3.6: The uses of different ion beam characterization tools in different branches of science.

sensitive method to measure the natural abundances of very rare radioactive isotopes. The advantage of counting the radioactive atoms themselves rather than their decay products is well illustrated by radiocarbon dating, which requires the measurement of the number of ^{14}C atoms in a sample. In Accelerator Mass Spectrometry (AMS) the particle accelerator in conjunction with ion sources, analyzing magnets separates out interferences and detectors count single atoms in the presence of 1×10^{15} stable atoms. AMS can use as little as 1–2 milligrams sample, and under special circumstances samples as small as 50–100 μgm. This ultra sensitive technique finds application in various interdisciplinary branches of science as given in schematic in Fig. 3.6.

The ion source produces a beam of ions from a few milligrams of solid material. The element extracted (for example, a rock, rain water, a meteorite) is used in the ion source. Negative ions produced on the surface of the sample are extracted from the ion source and sent down the evacuated beam line towards the injector magnet. It bends the negative ion beam by 90° to select the mass of interest.

AMS is an efficient tool for detection of a number of cosmogenic long-lived radionuclides, such as ^{3}H, ^{10}Be, ^{14}C, ^{26}Al, ^{36}Cl, ^{41}Ca, ^{55}Mn, ^{59}Ni, ^{129}I etc. There are important applications of AMS in archaeological and geophysical studies, for example, in radiocarbon dating by the counting of the rare carbon-14 isotope. The long half-life of 5730 years for ^{14}C implies that only 15 β-particle emissions per minute are observed from 1 g of contemporary carbon. However, an AMS can be used to count the ^{14}C atoms at over 15 per second from one mg of carbon. Just to give an example of the versatile nature of super sensitive technique of AMS, one can look at the AMS results of ^{60}Fe isotope measured from the samples collected from the Pacific Ocean floor, which was produced in supernova three millions years ago [48].

Many of the applications of AMS so far have been in the earth sciences, including geochronology, sedimentology, petrogensis, hydrology, volcanology, glaciology and oceanography. In these fields, AMS has been applied predominantly for chronological evaluations using dating principles but non-dating applications have also played a contributory role in many fields, such as nuclear physics and materials science [49].

3.13 SUMMARY

Various events resulting from the ion beam interaction with materials are utilized for characterization of materials. The basic principles of the techniques—PIXE, IL, RBS, Channelling, ERDA, MEIS, LEIS, NRA, CPAA and AMS—are described briefly. PIXE and IL are the tools where the emitted photons are analyzed for sample characterization, whereas RBS and ERDA use the scattering events for analysis of the sample. Backscattering events provide the depth profiling of high Z elements and the recoils provide the depth profile of low Z elements. Ion channelling is a powerful tool to quantify crystallinity of the sample, determine the location of a foreign atom in a host crystal, determine the strain in superlattice, etc. NRA and CPAA are the techniques based on nuclear reactions where the prompt and delayed events, respectively, are used. In the supersensitive technique AMS, the sample itself is used in ion source of accelerator with state-of-art detector to discriminate the isotopes. There are a large number of applications of these techniques in different branches of science as shown in Fig. 3.6. A glimpse of some examples of the application of PIXE, RBS and ERDA is given in the appendices. The swift heavy ions provide a possibility of on-line monitoring of modification of materials, using a large area position sensitive gaseous telescope detector.

REFERENCES

1. Handbook of Modern Ion Beam Materials Analysis. Editors M.A. Nastasi, J.R. Tesmer, Material Research Society (2009), ISBN: 1558992545.
2. G. Schiweitz, P. Grande, B. Skogvall, J.P. Biersack, R. Köhrbrück, K. Sommer, A. Schmoldt, P. Goppelt, I. Kadar, S. Ricz and U. Stettner, *Phys. Rev. Lett.*, **69** (1992) 628.
3. G. Xiao, G. Schiwietz, P.L. Grande, N. Stolterfoht, A. Schmoldt, M. Grether, R. Köhrbrück, A. Spieler and U. Stettner, *Phys. Rev. Lett.*, **79** (1997) 1821.
4. G. Schiwietz, G. Xiao, P.L. Grande, E. Luderer, R. Pazirandeh and U. Stettner, *Euro Phys. Letts.*, **47** (1999) 384.
5. T.B. Johansson, R. Akselsson and S.A.E. Johansson, *Nucl. Instr. and Meth.*, **B 84** (1970) 141.
6. M. Volfinger, *Journal of Radioanalytical and Nuclear Chemistry*, **253** (2002) 413.
7. A. Ene, I.V. Popescu and T. Badica, *Journal of Optoelectronics and Advanced Materials*, **8** (2006) 222.
8. K.G. Malmqvist, M. Elfman, G. Remond and C. Yang, *Nucl. Instr. and Meth.*, **B 109/110** (1996) 227.
9. C. Yang, N.P-O. Homman, L. Johansson and K.G. Malmqvist, *Nucl. Instr. and Meth.*, **B 85** (1994) 808.
10. H. Nagabhushana, B. Umesh, B.M. Nagabhushana, B.N. Lakshminarasappa, Fouran Singh and R.P.S. Chakradhar, *Phil. Mag.*, **89** (2009) 995.
11. H. Nagabhushana, S.C. Prashantha, B.N. Lakshminarasappa and Fouran Singh, *Journal of Luminescence*, **128** (2008) 7.
12. H. Nagabhushana, S.C. Prashantha, B.M. Nagabhushana, B.N. Lakshminarasappa, Fouran Singh and R.P.S. Chakradhar, *Solid State Communications*, **147** (2008) 377.
13. W.K. Chu, J.W. Mayer and M.-A. Nicolet, Rutherford Backscattering Spectrometry, Academic Press. Orlando, FL (1978).
14. Leonard C. Feldman and James W. Mayer, Fundamentals of Surface and Thin Film Analysis, North Holland (1986).
15. L.R. Doolittle, *Nucl. Instr. and Meth.*, **B 9** (1985) 344.
16. D.S. Gemmel, *Rev. Mod. Phys.*, **46** (1974) 129.

17. J.W. Mayer, L. Eriksson and J.A. Davies, Ion Implantation in Semiconductors, Academic Press, New York (1970).
18. P.K. Sahoo, S. Gupta, A. Pradhan and V.N. Kulkarni, *Nucl. Instr. and Meth.*, **B 216** (2004) 313.
19. A.P. Pathak, S.V.S. Nageswara Rao, A.M. Siddiqui, G.B.V.S. Lakshmi, S.K. Srivastava, S. Ghosh, D. Bhattacharya, D.K. Avasthi, D.K. Goswami, P. Satyam, B.N. Dev and A. Turos, *Nucl. Instr. and Meth.*, **B 193** (2002) 319.
20. S.V.S.N. Rao, Anand P. Pathak, Azher M. Siddiqui, D.K. Avasthi, Claudiu Muntele, Daryush Ila, B.N. Dev, R. Muralidharan, F. Eichhorn, R. Groetzschel and A. Turos, *Nucl. Instr. and Meth.*, **B 212** (2003) 442.
21. J.L. 'Ecuyer, *Nucl. Instr. and Meth.*, **B 149** (1978) 271.
22. J.L. 'Ecuyer, C. Brassard, C. Cardinal, J. Chabbal, L. DescMnes, J.P. Labrie, B. Terreault, J.G. Martel and R.St.-Jacques, *Journal of Appl Phys.*, **47** (1976) 381.
23. H.A. Rijken, S.S. Klien and M.J.A. de Voigt, *Nucl. Instr. and Meth.*, **B 64** (1992) 395.
24. J.P. Thomas, M. Fallavier, D. Ramdane, N. Chevarier and A. Chevarier, *Nucl. Instr. and Meth.*, **218** (1983) 125.
25. R. Yu and T. Gustafsson, *Surface Sci.*, **177** (1986) L987.
26. D.K. Avasthi, D. Kabiraj, A. Bhagwat, G.K. Mehta, V.D. Vankar and S.B. Ogale, *Nucl. Instr. and Meth.*, **B 93** (1994) 480.
27. D.K. Avasthi, S.K. Hui, E.T. Subramaniyam and B.R. Mehta, *Vacuum*, **47** (1996) 1061.
28. G. Dollinger, C.M. Frey, A. Bergmaier and T. Fastermann, *Nucl. Instr. and Meth.*, **B 136-138** (1998) 603.
29. A. Tripathi, S. Mandal, D.O. Kataria, D.K. Avasthi and S.K. Datta, *Nucl. Instr. and Meth.*, **B 129** (1997) 423.
30. D. Bromley (Ed.), Treatise on Heavy Ion Science, Plenum Press, New York (1984).
31. D.K. Avasthi, W. Assmann, H. Nolte, H.D. Mieskes, H. Huber, E.T. Subramaniyam, A. Tripathi and S. Ghosh, *Nucl. Instr. and Meth.*, **B 156** (1999) 143.
32. D.K. Avasthi, W. Assmann, H. Nolte, H.D. Mieskes, S. Ghosh and N.C. Mishra, *Nucl. Instr. and Meth.*, **B 166-167** (2000) 345.
33. S.A. Khan, M. Kumar and D.K. Avasthi, *Nucl. Instr. and Meth.*, **B 266** (2008) 1912.
34. D.K. Avasthi, S. Ghosh, S.K. Srivastava and W. Assmann, *Nucl. Instr. and Meth.*, **B 219-220** (2004) 206.
35. M. Copel, *IBM Journal of Research & Development*, **44** (2000) 571.
36. E.P. Gusev, M. Copel, E. Cartier, I.J.R. Baumvol, C. Krug and M.A. Gribelyuk, *Appl. Phys. Lett.*, **76** (2000) 176.
37. T. Kobayashi, G. Dorenbos, S. Shimoda, M. Iwaki and M. Aono, *Nucl. Instr. and Meth.*, **B 118** (1996) 584.
38. M. Aono, *Nucl. Instr. and Meth.*, **B 2** (1984) 374.
39. Horst Niehus, Werner Heiland and Edmund Taglauer, *Surface Science Reports*, **17** (1993) 213.
40. H. Niehus, *Nucl. Instr. and Meth.*, **B 33** (1988) 876.
41. R.S. Williams and J.A. Yarmoff, *Nucl. Instr. and Meth.*, **B 218** (1983) 235.

42. A.V. Mijiritskii, U. Wahl, M.H. Langelaar, D.O. Boerma, *Nucl. Instr. and Meth.*, **B 136-138** (1998) 1097.
43. W.A. Lanford, *Solar Cells*, **2** (1980) 351.
44. K. Shikano, H. Yonezawa and T. Shigematsu, *Journal of Radioanalytical and Nuclear Chemistry*, **167** (1993) 81.
45. David C. Rididle and Emile A. Schweikert, *Anal. Chem.*, **46** (1974) 395.
46. A.E. Litherland, *Nucl. Instr. and Meth.*, **B 5** (1984) 100.
47. M. Suter, *Nucl. Instr. and Meth.*, **B 223-224** (2004) 139.
48. K. Knie, G. Korschinek, T. Faestermann, E.A. Dorfi, G. Rugel and A. Wallner, *Phys. Rev. Lett.*, **93** (2004) 171103.
49. David Fink, *Nucl. Instr. and Meth.*, **B 268** (2010) 1334.

4

Engineering of Materials by Swift Heavy Ion Beam Mixing

4.1 INTRODUCTION

Ion beam mixing (IBM) is a phenomenon, at interface between two layers, in which the atoms of one layer mingle with the atoms of the other elements under the influence of ion beam traversal through them. When a thin film deposited on a substrate is subjected to ion irradiation, intermixing of the atoms in the film with that of in the substrate takes place. Such atomic mixing at the interface forced by the energetic ion beam is referred to as ion beam mixing. It can be understood by the schematic given in Fig. 4.1. The two layers after ion exposure get mixed at the interface resulting in the mixed region of the properties different from that of the individual two layers. If one of the two layers is very thin, say about ten nm or smaller, it can result in complete mixing of the layer. At low energies, of the order of keV to a few hundreds of keV, irradiation causes a mixing at the interface due to elastic collision cascades,

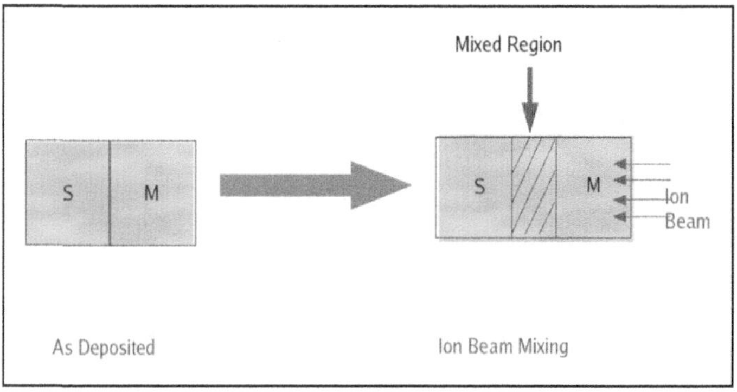

Figure 4.1: The schematic showing the ion beam mixing.

which is primarily a result of ballistic effects. The kinetic energies of ions are transferred to target atoms on collision. Ion energies are sufficiently high to break molecular bonds and initiate relocations within an atomic lattice. During this ballistic process, impinging ions displace atoms and electrons of the target material several lattice sites away, resulting in relocations there and interface mixing.

Ballistic ion beam mixing involves two basic processes, recoil mixing and cascade mixing, which happen simultaneously as a result of ion bombardment. In recoil mixing, atoms are relocated by single collision events. Recoil mixing is predominantly seen at large angles as a result of soft collisions, with the number of atoms undergoing recoil implantation varying linearly with ion dose. Recoil implantation, however, is not the dominant process in ion beam mixing. Most relocated atoms are part of a collisional cascade in which recoiled atoms participate, initiating a series of lower energy lattice displacements, which is referred to as cascade mixing. In addition to the cascade mixing, the elastic collisions produce considerable defects in both the layers. These defects assist the diffusion across the interfaces, referred to as radiation enhanced diffusion (RED). The energy lost by the ions in elastic collisions, called nuclear energy loss, is considered to be basically responsible for ion beam mixing. It has been shown [1, 2] that the number of atoms in the mixed region are proportional to the nuclear energy loss and the ion fluence. The rate of energy loss of the bombarding ions in the target material depends on its energy in a complex manner. The nuclear energy loss peaks at low ion energies and decreases monotonically for higher values.

Ion beam mixing finds applications in the generation of new phases, novel materials and enhancing the adhesion of the film to substrate. Low energy ion beams have been very widely used over the last three decades for meeting these objectives [1-10]. The ion energy and species are chosen in such a way that they impart a large amount of energy to the atoms at the interface of the thin film and the substrate by the process of elastic collisions. Normally, the fluence required for ion beam mixing with low energy ions is of the order of 10^{16} ions cm^{-2}.

The situation differs completely for ions of higher energy (several MeV per nucleon). Such high energy ions are referred to as swift heavy ions (SHI). Here the initiation of atomic transport by the incident ion is much less obvious. Stopping of the ion at these energies almost exclusively occurs by electronic excitation and ionization of the target atoms (electronic stopping), while nuclear stopping contributes $<1\%$ to the total energy loss. Initially it was believed that the energy deposited by the ions in the electronic system will not result in any atomic relocation causing any ion beam mixing process. It was observed in nineties that the electronic energy deposition beyond a certain threshold can cause the movement of atoms, which can lead to the mixing, at fluence two orders of magnitude smaller than that in low energy case. When such a high energy ion penetrates a solid it leaves behind a highly excited electronic system in a small cylindrical zone around its path, which results in the formation of ion track having material different from surrounding. Part of the energy deposited by incident ion is transferred to the lattice via two different mechanisms: Coulomb-explosion [11] or thermal spike [12, 13] caused by electron–phonon coupling, which are two models to explain the formation of ion track. The passage of high energy heavy ions through certain materials creates extensive defects along the ion path forming latent tracks. The materials prone to getting defects by SHI irradiation are referred to as S_e sensitive materials.

The first observation of mixing mediated by swift heavy ions was reported in Fe/Si system in 1993 by Dufour et al. [14]. The signature of mixing in SHI irradiated Ti/Si system and diamond-like carbon (DLC) film deposited on Si substrate were observed at the interfaces [15, 16]. These experiments provided evidences of the ion beam mixing mediated by the swift heavy ions passing through the material without any direct displacement of atoms through elastic collisions. It is now more or less established that the swift heavy ion beam mixing is due to the inter-diffusion during transient temperature spike.

Ion beam mixing is an effective tool for synthesizing silicides and germanides. Formation of silicides and germanides [17-21] has been of interest because of their applications. The formation of silicides due to the defect created (by ion beam) at metal/Si interfaces is of great interest owing to its use in interconnections and contacts in integrated circuit technology. Understanding of defect creation is fundamental for electronic device technology. Ion beam mixing has definite advantages over the mixing by thermal annealing process. The striking feature of ion beam mixing is that (i) it has spatial selectivity, (ii) it is a low temperature process and (iii) thermodynamically immiscible systems, in principle, can be mixed.

The interface mixing can be analyzed by Rutherford Backscattering Spectrometry (RBS) [22] and Elastic Recoil Detection Analysis (ERDA) [23]. However, since the ion beams can induce surface roughness, extra care needs to be taken while analyzing the samples by RBS. The RBS spectra of a mixed layer and a roughened (but unmixed) layer may appear similar. The roughness at the surface and the interface mixing produces similar effects in the RBS spectrum [24]. Thus the roughness of the irradiated samples should be taken into account while determining the extent of the mixed layer thickness.

Secondary Ion Mass Spectrometry (SIMS) provides excellent technique for depth profile of elements at interface. X-ray reflectivity and X-ray standing wave techniques are employed for quantifying mixed region. X-ray diffraction is used for phase identification, Mossbauer Spectroscopy is employed, if one of the elements in the system under investigation is Mossbauer active. The cross-sectional Transmission Electron Microscopy is a powerful characterization technique for direct viewing of the interfaces.

4.2 SHI INDUCED MIXING FOR MATERIAL ENGINEERING

Adhesion Improvement through IBM

Adhesion between the two surfaces is of scientific and technological interest. In most of the techniques the enhancement of adhesion is induced by the creation of an intermediate layer induced by solid state reaction between the components. The quality of the bonding then depends on the physiochemical characteristics of the intermediate layer. For example, in the case of metal-ceramic system, the bonding optimization needs high temperature and high pressures. Ion beam provides a way to induce the modification of the interface at room temperature and in non-thermodynamic–equilibrium conditions.

Adhesion strength between metals and polymers for metallization application can be improved by an order of magnitude by heavy ion irradiation. SHI induced mixing in Cu film on teflon polymer results in a better adhesion of the metal layer on polymer than the adhesion obtained by conventional means [25, 26]. Despite Cu being insensitive to S_e, a very strong mixing was observed as investigated by RBS. This shows that SHI induced mixing in metal/polymer system takes place even if the metal layer is S_e insensitive. The cause of mixing is believed to be due to the diffusion of Cu in the transiently molten track of polymer.

Metal Silicides and Germanides by IBM

Metal silicides are compounds formed by metal on silicon exhibiting metallic conduction behaviour. They have low metal-like resistivity and high temperature stability. Metal silicides, in general, are chemically more stable [27]. Some of them are very good candidates for direct band gap material like beta $FeSi_2$, wide band gap semiconductor like SiC etc. Low electrical resistivity of metal germanides make them potentially better candidates for use as contact and interconnect materials [28] than the well-known metal silicides. Ferromagnetic metal germanides (ferromagnetic metal doped semiconductors) have opened up exciting possibilities of their use as ferromagnetic semiconductors in magnetic devices [29]. Ferromagnetic semiconductor, as the name implies, has semiconducting properties and long-range ferromagnetic order. Thus, in brief, there is wide interest in mixing of metal/semiconductor systems.

Ion beam mixing provides a technique for producing metal silicides, particularly for metastable phases and homogeneous mixing at the interface. It has been used for long for the formation of metal silicides with low energy ions where elastic collisions (nuclear stopping) are predominant.

Atomic transport occurring across the interface of Cu/Ge and Co/Ge layers leading to the synthesis of Cu_3Ge and Co_2Ge phases, under MeV Kr and Ar ion irradiations, was investigated in detail [30].

It was found that in the case of Cu/Ge systems both Cu and Ge are mobile but Cu is the dominant moving species. On the other hand, in the case of the Co/Ge system only Ge atoms are mobile. The atomic movements during ion irradiation in these two systems was explained on the basis of distribution of defect density and effects of thermodynamic forces.

Attention is now towards materials modification and phase changes using ion beam mixing with high energy ions [31-33].

Mixing in Fe/Si System

Fe with Si makes different silicides depending on the ratio of Fe and Si contents and the temperature. $FeSi_2$ exists in two phases α and β. It is a very good candidate for direct band gap semiconductor. The effects of 200 MeV Ag ion irradiation and thermal annealing of Fe/Si multilayers with overall composition $Fe_{50}Si_{50}$ was investigated [34] with two multilayers of Fe layer thickness of 19 A (having amorphous structure) and 62 A (having crystalline structure). X-ray reflectivity studies show that although both the thermal annealing and heavy ion irradiation result in significant inter-diffusion between the Fe and Si layers, the final product of thermal annealing depends upon the initial structure of the Fe layer whereas in the case of SHI irradiation, the formed phases are similar and independent of initial structure.

The irradiation experiments on thin ^{57}Fe layers deposited on Si were performed [35] by varying the energy loss of the ions in the Fe layer between 32 and 47 keV/nm by taking different heavy ion beams and energies. The formation of iron silicide (α-$FeSi_2$), was evident from the Conversion Electron Mossbauer Spectroscopy (CEMS) measurement as shown in Fig. 4.2. The

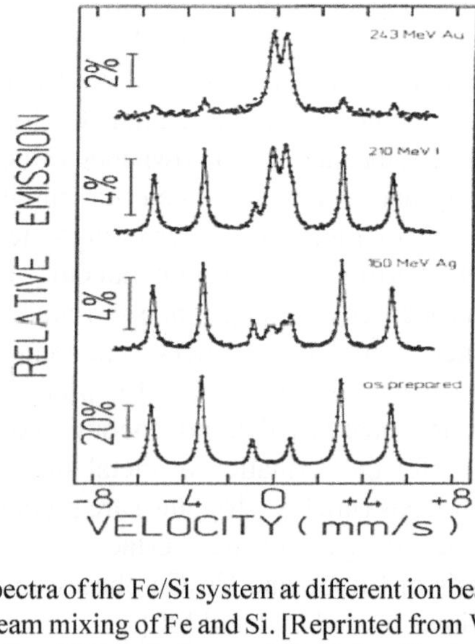

Figure 4.2: CEMS spectra of the Fe/Si system at different ion beam parameters, showing the signature of ion beam mixing of Fe and Si. [Reprinted from W. Assmann, M. Dobler, D.K. Avasthi, S. Kruijer, H.D. Mieskes and H. Nolte, *Nucl. Instr. and Meth.*, **B 146** (1998) 271 with permission from Elsevier.]

occurrence of this phase, which is thermodynamically stable above 960°C, was an indication of high temperature conditions along the ion tracks.

The irradiations using GeV ion beam (resulting in very high electronic energy deposition) of Fe/Si multilayers were performed [14] where Fe was in crystalline form and Si was in amorphous state. The formation of totally mixed layer at a fluence of 10^{13} ions cm^{-2} took place. The temperature spike was given as possible cause of mixing. The phase formation was not reported in this work. The amorphous nature of Si was considered to be responsible for mixing in Fe/Si [36]. It was shown, as an outcome of several experiments, that there is a S_e threshold below which no mixing occurs [37].

The studies of SHI induced mixing by on-line ERDA were initiated by Avasthi et al. [38-41] utilizing the capabilities of the ion beam in modification as well as characterization of materials. On-line studies performed on mixing in Fe/Si system induced by SHI provided quantitative information on the dependence of mixing on the electronic excitation. It was found that about 56 atoms of Fe and 85 atoms of Si mix with each other for each incident Au ion of 230 MeV and 33 atoms of Fe and 33 atoms of Si mix with each other for each incident I ion of 210 MeV. It was also shown that mixing in Fe/Si does not take place, in the case when oxygen concentration at the interface is high.

SHI induced mixing has been qualitatively suggested to be due to the temperature spike. Quantitative support to this hypothesis was given by Srivastava et al. [42]. The hypothesis that it is a consequence of a transient molten state diffusion was tested by a detailed study of 230 MeV Au ion induced mixing at an Fe/Si interface. The experimental diffusivities of Fe in Si, and vice versa, believed to be occurring during the temperature spike was estimated using experimental data and inputs of thermal spike model. The required ion track diameter and duration of the transient melt phase was estimated theoretically from the thermal spike model. A fair agreement of the experimental diffusivities to the liquid state diffusivities (reported in the literature) was the first comprehensive verification of the hypothesis that SHI induced mixing is a consequence of inter-diffusion during transient melt phase. The occurrence of SHI induced mixing between Fe and Si is unexpected since crystalline Si is known to be insensitive to SHI. It was suggested that Si also melts near the Fe/Si interface.

Mixing in Ti/Si, Ni/Si, Mn/Si, Mo/Si and W/Si

Nickel and titanium silicides are of interest for ohmic contact in semiconductors. Different phases of silicides are made in Ti/Si and Ni/Si systems depending on the temperature and the ratio of the contents of these. High electronic energy deposition by 350 MeV Au^{+} ions at different fluences for SHI induced mixing in Ti/Si was studied [43]. Rutherford backscattering spectra (RBS) and the X-ray reflectivity measurements indicated mixing at the interface, leading to the formation of titanium di-silicide ($TiSi_2$) phase which was evident in grazing incidence X-ray diffraction data.

The irradiation of Ni/Si and Ti/Si systems by 95 MeV Au ions at a fluence of 10^{13} ions cm^{-2} and subsequent annealing lead to the formation of silicides [44]. Thermal annealing of the SHI irradiated Ni/Si system provided the required energy to the atoms for diffusion across the interface. The defects produced by SHI irradiation enhances the interface diffusion during annealing. However, the mixing in Ni/Si is shown to take place, without annealing when irradiated by higher energy ions (350 MeV Au ions) [45].

Manganese silicide is considered to be low cost thermoelectric material. The mixing in Mn/Si by swift heavy ion irradiation results in formation of manganese silicide phase, confirmed by X-ray diffraction and photoelectron spectroscopy [46, 47]. Molybdenum silicide has high melting point and has excellent corrosion resistance. SHI induced mixing was investigated at the Mo/Si interface in single and multilayer samples. The extent of interfacial mixing was estimated through X-ray reflectivity measurements and the phases of the crystalline molybdenum silicides (Mo_3Si_5, h-$MoSi_2$, t-$MoSi_2$) formed in the mixed zone were identified by grazing incidence X-ray diffraction. An experimentally observed upper limit for the threshold of defect creation by electronic energy loss in Mo was established [48].

Tungsten silicide is used as contact material in microelectronics. The mixing in W/Si by 120 MeV Au ions results in formation of t-W_5Si_3 and t-WSi_2 phases [49], evidenced by glancing angle X-ray diffraction.

Comparison of Mixing in Co/c-Si and Co/a-Si

Cobalt silicide has low electrical resistivity and is a useful material for semiconductors and VLSI technology. Multilayer heterostructure of a-Si(50 nm)/Co(50 nm)/a-Si(50 nm) on Si(100), prepared by an e-beam evaporation technique under UHV conditions, were irradiated by 120 MeV Au ions with fluence of 10^{13} to 10^{14} ions cm^{-2} [50]. SIMS spectra of the pristine and irradiated samples at different fluences shown in Fig. 4.3 indicate significant mixing at the interface. The investigations by X-ray diffraction confirmed the formation of different phases of cobalt silicides, Co_2Si, CoSi and $CoSi_2$. Fig. 4.4 gives the X-ray diffraction confirming the silicide phase, without annealing [33]. This suggests that SHI irradiation can be used for making silicide buried layer at room temperature without thermal annealing. The cross sectional TEM showed that deposited Si was in a-Si form at the interface, which play an important role in maximizing the effect of SHI irradiation. It is to be noted that the irradiation of Co on Si did not yield significant mixing as discussed in the next paragraph. The metal layer on a-Si results in higher mixing than that on the Si substrate simply because the a-Si layer is sensitive to SHI irradiation whereas Si substrate, being crystalline, is not sensitive to SHI irradiation.

In another study [51], thin films of Co deposited on Si were irradiated by 120 MeV Au beam irradiation at a fluence of 10^{13} ions cm^{-2} and subsequently subjected to thermal annealing to induce inter-diffusion at the interface. Complete intermixing of Co and Si in the interfacial region to form CoSi was

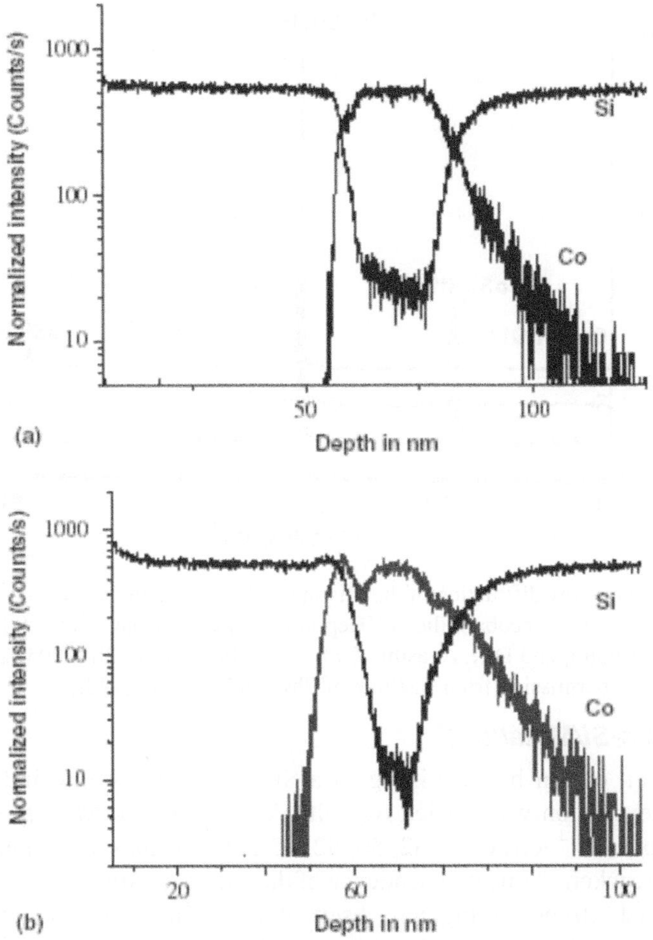

Figure 4.3: The SIMS spectra of the pristine as well as ion irradiated a-Si/Co/a-Si system, indicating the mixing of Co and a-Si. [Reprinted from B.R. Chakraborty, S.K. Halder, N. Karar, D. Kabiraj and D.K. Avasthi, *J. Phys. D: Appl. Phys.*, **38** (2005) 2836 with permission from Institute of Physics Publishing Ltd.]

achieved following thermal annealing (at 400°C) of the SHI irradiated CoSi system. The radiation enhanced diffusion mechanism explains the mixing in SHI irradiated and subsequently annealed samples. Thermally assisted (ion generated) defect migration across the CoSi interface enhanced defect mediated atomic mobility, leading to the formation of cobalt silicides, in accordance with the thermodynamically favoured route. The post-SHI irradiation annealing temperature, required for the formation of crystalline phases (400°C), is lower than that reported for low energy ion beam mixing cases where post-irradiation annealing temperature in excess of 700°C is required for the formation of phase. Due to lower processing temperatures, SHI irradiation and subsequent annealing is a promising silicidation technique in solid state technology.

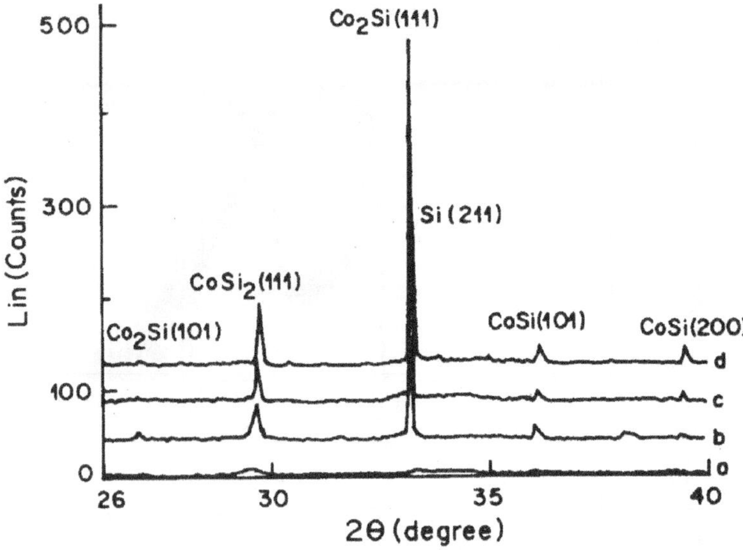

Figure 4.4: The X-ray diffraction of the pristine and ion irradiated a-Si/Co/a-Si layers, showing the formation of cobalt silicide. [Reprinted from B.R. Chakraborty, S.K. Halder, N. Karar, D. Kabiraj and D.K. Avasthi, *J. Phys. D: Appl. Phys.*, **38** (2005) 2836 with permission from Institute of Physics Publishing Ltd.]

Mixing in a-Si/Metal/a-Si

Investigation of ion beam mixing in a-Si/Metal/a-Si system induced by electronic excitation was carried out with Au ions of 120 MeV at different fluences up to 10^{14} ions cm^{-2} [32, 50, 52, 53]. Here metals Ti, V, Fe, Mn, Ni and Nb were taken. A linear dependence of the atomic mixing with the fluence was observed. Strong mixing was observed at the fluence of 10^{14} ions cm^{-2}, which was explained by the inter-diffusion at interface during transient melt phase. The contribution of nuclear energy loss to the observed mixing was shown to be insignificant on the basis of the reported results of mixing at low energy ions.

The use of a-Si layer for silicide formation has two advantages. The deposition of a-Si on Si followed by deposition of metal layer without breaking the vacuum, makes the Si component free of oxygen. This was possible by using four pocket electron gun. Another advantage is that the a-Si layer is more sensitive to S_e than the crystalline Si substrate and therefore stronger mixing is expected in case of a-Si. A study [32, 50, 52, 53] was made on the mixing in metal/a-Si, with different metal layers of Ti, V, Fe, Mn, Ni and Nb and it was shown [52] that the content of mixing is dependent on the electron phonon coupling of the metal component as shown in Fig. 4.5.

A detailed study of mixing was done on a-Si/Fe/c-Si trilayers with 350 MeV Au^{26+} ions at fluences of $(0.6–11.3)\times10^{14}$ ions cm^{-2}. Si/Fe/Si trilayers were made with 12 nm amorphous Si and 45 nm polycrystalline Fe films

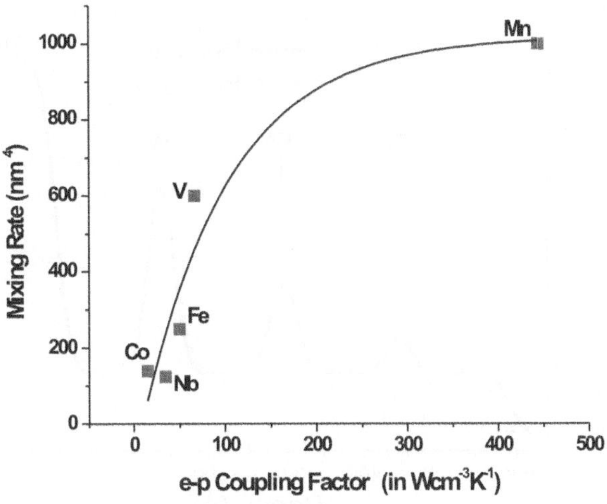

Figure 4.5: The dependence of the mixing content in the metal/a-Si systems on the electron-phonon coupling of the metal layer. [From Diva, Ph.D. thesis, Agra University]

deposited on Si(100) wafers and irradiated. The mixing rate at the upper a-Si/Fe interface was found to be three times as high as that at the lower Fe/c-Si interface. Ion irradiation at a moderate fluence ($6.7 \times 10^{14}/cm^2$) induced a magnetic anisotropy in the sample, which was magnetically isotropic after deposition. At the highest fluence, full interface mixing occurred and the magnetic anisotropy almost disappeared [54].

The mixing efficiencies of different metals with a-Si were investigated using the X-ray standing wave technique combined with X-ray reflectivity which can provide concentration profile of a marker layer with sub-nanometer accuracy [55]. These techniques have been used to compare mixing efficiencies of three different metals, namely, Ti, Fe and W with Si, by embedding thin marker layers of the three metals in a single Si layer, which is then irradiated with 100 MeV Au ions to a fluence of 1×10^{13} ions cm^{-2}. Combined fitting of reflectivity and fluorescence data yields the electron density profile, which essentially reflects the concentration profiles of the three metallic marker layers (Fig. 4.6) [56, 57]. Perusal of Fig. 4.6 shows that irradiation results in significant intermixing of the metal layers with Si. However, there is large variation in the efficiency of mixing of different metals. While Ti exhibits a strong mixing with Si, Fe exhibits relatively less mixing.

Intermixing of W layer with Si is rather small. This is in accordance with the mixing efficiencies predicted by the thermal spike model. Fe layer exhibits intermixing even though the electronic energy loss S_e in Fe layer is below the threshold value of 30 keV/nm for bulk Fe. However, the observed intermixing can be understood in terms of a reduction in the threshold S_e value in thin film as compared to that in bulk metal. In the case of thin film, a decrease in the electron mobility due to scattering from the grain boundaries and surface/

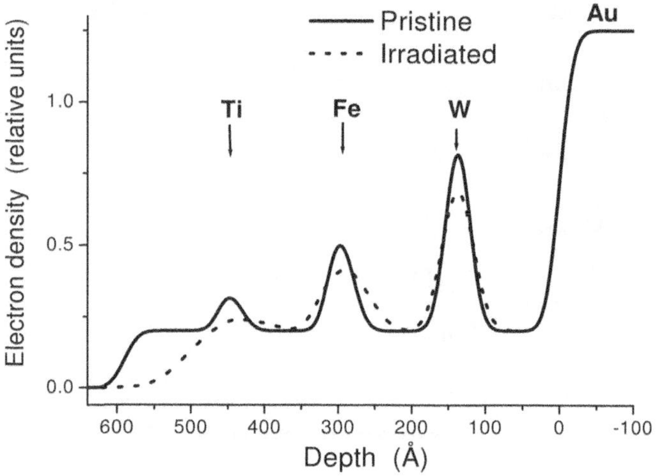

Figure 4.6: Electron density profiles in pristine and irradiated samples as obtained from simultaneous fitting of X-ray reflectivity and X-ray fluorescence data. The positions of the Ti, Fe, and W layers are marked by arrows. [Reprinted from Ajay Gupta, Carlo Meneghini, Amit Saraiya, Giovanni Principi and D.K. Avasthi, *Nucl. Instr. and Meth.*, **B 212** (2003) 458 with permission from Elsevier.]

interfaces and a possible modification in the electron-phonon interaction can result in a decrease in threshold S_e value as compared to the bulk [58]. By moving the antinodes of X-ray standing waves to different depths in the multilayer, XAFS measurements can be done as a function of depth [57, 59]. In this way it is found that, after an irradiation fluence of 1×10^{13} ions cm^{-2} in the above system, in the centre of the layer an equiatomic FeSi phase is formed, while away from the centre $FeSi_2$ phase is observed. Deep inside the Si layer, isolated Fe atoms in Si are observed. On the other hand, XAFS study alone would have suggested the formation of $FeSi_2$ phase only. Similar studies done in W/Si system irradiated with 350 MeV Au ions also show formation of different phases as a function of distance from the W layer [60]. These studies suggest the importance of doing depth resolved measurements in understanding the irradiation induced mixing process.

Synthesis of Germanides

Nickel germanides are of interest from microelectronics point of view. The formation of Ni_2Ge alloy by 100 MeV ion irradiation of Ni/Ge bilayer structure at different temperatures and fluences was investigated [61]. High resolution cross-sectional transmission electron microscopy was performed for thickness determination and structure determination which provided strong evidence of the formation of Ni_2Ge. Figure 4.7 shows cross section TEM images at the interface along with high resolution image which confirms the Ni_2Ge phase.

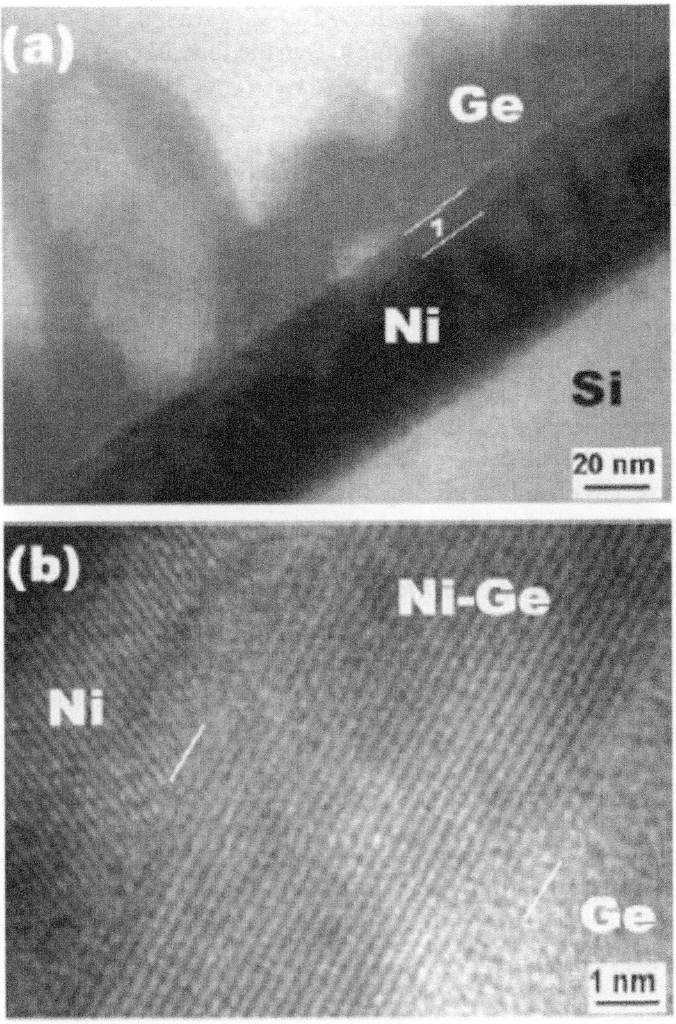

Figure 4.7: The cross sectional TEM image showing the formation of Ni_2Ge phase after irradiation of Ni/Ge system. [Reprinted from T. Som, B. Satpati, P.V. Satyam, D. Kabiraj, P. Ayyub, S. Ghosh, Ajay Gupta, B.N. Dev and D.K. Avasthi, *Nucl. Instr. and Meth.*, **B 212** (2003) 206 with permission from Elsevier.]

Mixing in Co/Ge and Au/Ge Systems

Cobalt germanides are of interest for contact materials. The interface modification in Co/Ge bilayer system using 100 MeV Au ion irradiation was studied at different temperatures and fluences [62]. The irradiation at low temperature did not show any appreciable interface mixing. Formation of a thin (3–4 nm) Co-Ge alloy layer was observed after room temperature irradiation to a fluence of 1×10^{14} ions cm^{-2}. This mixing was attributed to synergetic effects of nuclear and electronic energy loss processes in the bilayer system.

The irradiation of bilayer Au/Gu by 120 MeV Au ions at a fluence of 10^{13} ions cm^{-2} and subsequent annealing results in the formation of $Au_{0.6}Ge_{0.4}$ alloy [63].

Mixing in Cu/Ge System

Copper germanides is considered to be an advanced contact material for MOSFETs. The ion beam mixing in Cu/Ge bilayer, which were irradiated by 120 and 140 MeV Au ions and Cu/Ge multilayer by 120 MeV Au ions was studied [64, 65]. The pristine as well as irradiated Cu/Ge multilayer samples were annealed at 200 °C temperature. The concentration versus depth profiles at the interface were determined by Rutherford backscattering of the pristine and irradiated samples. The irradiated annealed samples showed considerable mixing as compared to pristine annealed due to radiation effected diffusion. On the basis of several experiments with different ions and energies it was concluded that the IBM increases with S_e and fluence. Irradiation at room temperature gave higher mixing than the irradiation at liquid nitrogen temperature.

Formation of Special Semiconductors by Mixing

Mixing in Si/C System

SiC has different phases of interest depending on the phase diagram. It is mainly of interest from the point of view of large band gap semiconductor with good stability at higher temperature. It is therefore known as wide band semiconductor for applications at high temperatures.

The irradiation of multilayer of Si and C by 100 MeV Au ions results in formation of SiC [66]. This was confirmed by glancing angle XRD of the pristine and irradiated systems. The formation of SiC is also reported [67] by the irradiation of Si and fullerene (C_{60}) multilayer at fluence of 10^{13} ions cm^{-2}. Figure 4.8 shows the X-ray diffraction of irradiated a-Si fullerene multilayer, indicating the formation of SiC.

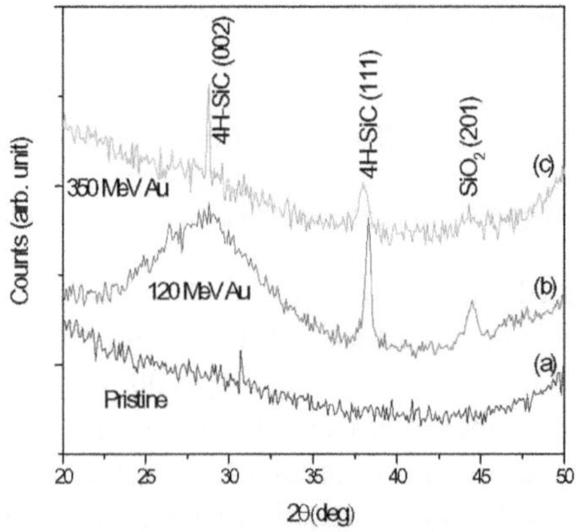

Figure 4.8: The X-ray diffraction of the pristine and ion irradiated a-Si/C_{60} multilayers, showing the formation of silicon carbide. [Reprinted from S.K. Srivastava and D.K. Avasthi, *Defence Science Journal*, **59** (2009) 425.]

Mixing in In/Se Layer

Indium selenide is a potential candidate for photo voltaic applications. The irradiation of In/Se bilayer by 40 MeV Si ions and subsequent annealing leads to the formation of the InSe phase [68]. The temperature required for the formation of InSe phase (for the irradiated sample) is much lower than that required in the annealing of pristine In/Se pristine layer. It is similar to the observation for mixing as in the Co/Si case [50].

Mixing in In/Sb

Indium antemonide is a narrow bandgap semiconductor. In/Sb is integrated with Si by a process of swift heavy ion beam mixing. The samples of In/Sb deposited on the Si substrate were irradiated using 100 MeV Au ions having fluences from 1×10^{12} to 6×10^{13} ions cm^{-2}. Phase formation due to ion beam mixing was detected using high-resolution X-ray diffraction measurements. X-ray photoelectron spectroscopy measurements indicated that both In and Sb were embedded in the Si substrate with an irradiation fluence of 3×10^{13} ions cm^{-2}. Formation of InSb phase was observed in the irradiated sample, at a fluence of 1×10^{13} ions cm^{-2} and higher, without any post-irradiation annealing [69].

Mixing in Al/Sb

Aluminium antimonide is a promising semiconductor material for high temperature applications. It was synthesized by mixing of Al/Sb by 160 MeV Ag ions at a fluence of 2×10^{13} ions cm^{-2}. The phase was revealed by X-ray diffraction [70].

SHI Induced Mixing in Metal/Metal Interface

Swift heavy ions can induce significant modifications in metallic systems above a threshold value of the electronic energy loss S_e, which varies considerably from metal to metal. Studies have shown that, in thin films and multilayers, swift heavy-ion-induced effects are much more pronounced and occur at substantially lower values of $(dE/dx)_e$ as compared to the bulk metals. The observed effects include defect annihilation or creation, intermixing at the interfaces, modification of the surface and interface roughnesses, and latent track formation. Many of the features of electronically induced modifications in metals can be understood in terms of the thermal spike model. However, there exists experimental evidence to suggest that, similar to low-energy ion irradiation, in the case of swift heavy ion irradiation also, thermodynamic parameters like heat of mixing can play an important role in determining the nature of the induced modifications. In thin films and multilayers, the magnitude of electronic energy loss $(dE/dx)_e$ and the ion fluence can be used as parameters for selectively modifying the surface/interface roughness or interdiffusion at the interfaces. Thus, swift heavy ions can be used to manipulate the interface-sensitive properties of the multilayers [71].

Comparison of Mixing in Ti/Fe and Ti/Au

The two systems were chosen such that in one system (Ti/Fe), both Ti and Fe are sensitive to S_e, while in other system (Ti/Au), only Ti is S_e sensitive and the other Au is S_e insensitive. It was shown that mixing is higher when both the layers are S_e sensitive [72].

Mixing in Ni/Ti

The alloys of Ni and Ti are of interest as shape memory alloys. The electronic excitations induced ion beam mixing in metallic bilayers, Ni/Ti bilayers, irradiated at 80 K with GeV Ta ions was investigated [73, 74]. A very strong inter-diffusion across the interface indicating the ion beam mixing was observed in Ni/Ti system as a result of dense electronic excitations.

Mixing in Magnetic Multilayers

Mixing in Fe/Ni Multilayer

Alloys of Fe and Ni are of interest for magnetic properties. The Fe mixed with Ni results in different phases of interest from magnetic properties point of view. Effects of 120 MeV Au ion irradiation on a Fe/Ni multilayer were analyzed in detail [75]. The characterization of pristine and irradiated films were made through X-ray diffraction, X-ray reflectivity, magneto-optic Kerr effect and four probe resistivity measurements. The formation of $FeNi_3$ and FeNi alloy phases, evidenced by XRD, were observed due to SHI induced mixing. The saturation magnetization of the film increased after irradiation because of the high magnetic moment of the mixed phase. The measured resistivities of the pristine and irradiated multilayers were found to be in agreement with the theoretically predicted values, considering the contributions from interfaces and grain boundaries. It is suggested that the mixing takes place due to an inter diffusion during the transient melt phase predicted by the thermal spike model.

Mixing in Fe/Zr

Alloy of Fe and Zr is a magnetic material. SHI induced mixing and phase change for Fe/Zr bilayer was studied [76]. A complete mixing was observed at a fluence of 10^{13} ions cm^{-2} of high energy U ions. The electronic energy depositions in Zr and Fe by ion beam were 4.5 and 6.9 keV/nm respectively. Mixing takes place since both Fe and Zr are S_e sensitive and the ions deposited very high electronic energy loss in the bilayer, which is more than the S_e threshold for both the layers.

Mixing in Fe/Cr Multilayer

The Fe/Cr multilayer is of interest from the giant magnetoresistance (GMR) point of view. The irradiation of this multilayer by 200 MeV Ag ions resulted in mixing at the interface, revealed by X-ray reflectivity measurements. It resulted in decrease of GMR [77].

Mixing in Fe/Tb Multilayer

The transitional metal-rare earth (TM-RE) multilayers (such as Fe/Tb) have interesting property of perpendicular magnetic anisotropy (PMA) and is considered as potential candidate for recording media. The swift heavy ion irradiation causes significant changes at the interfaces and influence the magnetic properties. In experiments of irradiation of multilayers by 80 MeV Si and 150 MeV Ag ions, the modifications were studied [78]. It was noticed that irradiation with 80 MeV Si ions causes interface roughening whereas 150 MeV Ag ions results in interface mixing as revealed by X-ray reflectivity and Mossbauer spectroscopy. The PMA decreased with ion irradiation in both cases. The decrease in PMA was more in case of Ag than that of Si ions.

The mixing in Tb/Fe amorphous multilayers by SHI irradiation was explained by thermal spike model [79, 80]. There was partial recrystallization of amorphous Fe layer in SHI irradiated multilayer [81]. Threshold of mixing was established between 5 and 15 keV/nm. SHI irradiated FeTb/Co multilayer showed an enhancement in the PMA, alongwith the mixing in Fe-Co at a low fluence of 7×10^{12} ions cm^{-2} [82].

Asymmetry of Mixing at the Interfaces

X-ray waveguide structures have been used for precise depth profiling of thin marker layers. The system to be studied forms the cavity of a planar waveguide which consists of two layers of a high-Z element like Au or Pt [83]. It is found that the mixing induced by 100 MeV Au ions in a Si/Fe/Si trilayer is not symmetric at the two interfaces; Si-on-Fe interface gets mixed more strongly as compared to the Fe-on-Si interface. This asymmetry of mixing is also observed in a number of metal-semiconductor as well as metal-metal systems like Ti/Si and Fe/Ti [84, 85]. Conversion electron Mossbauer spectroscopy measurements show that the origin of this asymmetry is a difference in the structure of the two interfaces.

4.3 INTERFACE MODIFICATION IN THERMO-DYNAMICALLY IMMISCIBLE SYSTEMS

Interface Modification of Pt/C and Ni/C

Pt/C and Ni/C multilayers are of interest for optical applications as X-ray mirror. Two bilayer systems were chosen in such a way that (i) one S_e sensitive layer is common and (ii) one system is miscible while the other is immiscible. It was seen [86] that the mixing occurs in the Ni/C multilayer and the demixing occurs in the Pt/C multilayer under 100 MeV ion irradiation. In the present case Ni/C is miscible whereas Pt/C is immiscible. In both the multilayers, C is S_e sensitive. The study suggested that the interface of the immiscible system gets smoothened or sharpened, whereas the interface in the miscible system mixes as a result of swift heavy ion irradiation.

Study on Mixing in Fe/Bi and Fe/Au Interfaces

Fe/Bi and Fe/Au multilayers are of interest from magnetic properties point of view. Therefore the influence of SHI at these interfaces attracts attention. It was shown [87] that the Fe/Bi interface, having positive heat of mixing, remains immiscible despite large electronic energy deposition. Normally a strong mixing at interface is expected if both the layers at the interface are S_e sensitive. In the present case, both the Fe and Bi are S_e sensitive, still no mixing is observed due to the fact that the heat of mixing is positive and thermodynamical concepts prevail, resulting in no inter-diffusion at interface. Thus the Fe/Bi due to being immiscible system does not mix under swift heavy ion irradiation and shows the similar behaviour as was seen [86] in Pt/C (an immiscible system). Similarly the interface smoothening has been observed [88] as a result of irradiation by SHI of Fe/Au system. This observation is a bit in contrast, with what is observed in low energy ion beam mixing. It has been seen [89] that the ion beam mixing can occur in thermodynamically immiscible systems at low energies and high fluences, due to recoils and cascade collisions at the interface region.

4.4 METAL/INSULATOR, SEMICONDUCTOR/INSULATOR AND INSULATOR/INSULATOR SYSTEMS

It has been shown that the SHI induced mixing of Ni thin film on polymer results in mixing of Ni in polymer [90]. A thin film of Ni on polytetrafluoro-ethane was irradiated with 120 MeV Au ions. The mixing in the system was revealed by RBS analysis and was explained as a consequence of reaction taking place in molten ion track of polymer and hot zone around the ion path in Ni metal layer during transient temperature spike. The evolution of several gases (CF, CF_3, C_2F_3 etc.) from the bottom polymer layer was monitored by on-line quadrupole mass analyzer. The reaction of F with metal layer was revealed by electron spectroscopy for chemical analysis. Further investigation showed the formation of Ni nanostructures embedded in carboneous matrix [91].

Apart from the above cases of metal/semiconductor and metal/metal interfaces, there have been several experiments [92-98] on insulating layer (such as oxides, fluorides etc.) on Si or insulator substrate like glass, quartz, alumina etc., where the key conclusion has been that the beam is a consequence of interdiffusion across the transiently melt ion track, which is quite expected because the ion track formation is favourable in insulating media. The observed mixing did not have any dependence on Young's modulus of the systems [95], which suggests that there is no correlation of ion beam mixing with Coulomb explosion. The observed threshold of mixing had a clear correlation with the product of melting temperature and specific heat, therefore it could be explained by the thermal spike model proposed by Szene [12].

Mixing in the system CuO/glass was shown by on-line ERDA, utilizing the capabilities of the ion beam in terms of modification as well as characterization [38, 41].

4.5 CONCLUSIONS AND FUTURE PROSPECTS

A large number of compoud phases such as silicides, germanides, carbides, metal alloys and chalcogenides could be synthesized by swift heavy ion beam mixing.

One can make the following conclusions on the basis of a large number of experiments performed so far. These could provide guidelines for planning further explorations on IBM with SHI.

(i) The two layers should be thermodynamically miscible, i.e. the heat of mixing should be negative, to get SHI induced mixing. In the immiscible systems, demixing has been observed which means the interface becomes sharp. This could be utilized for engineering the interface, when one needs to make the interface sharper, like in X-ray mirror applications.

(ii) SHI induced mixing has a threshold of electronic energy loss, beyond which it occurs.The IBM increases with ion fluence.

(iii) One of the two layers must be S_e sensitive in order to have SHI induced mixing, and the mixing rate is quite large in the cases where both the layers are S_e sensitive.

(iv) Even if one of the layers is S_e sensitive, the SHI induced mixing takes place if the melting point of the S_e insensitive layer is smaller than the melting point of S_e sensitive layer.

(v) The SHI induced mixing can be explained as inter-diffusion at the interface during transient melt state.

(vi) The experiments on ion beam mixing in several metal/a-Si system indicates that the content of mixed layer is dependent on the electron-phonon coupling of the metal layer. It indirectly supports the hypothesis of thermal spike model in explaining the SHI induced mixing.

(vii) SHI induced mixing in several metal/semiconductor [14, 32, 33, 39, 42, 44, 48, 50, 53, 65, 67] and metal/metal [72, 73, 75] systems is well explained as inter-diffusion across the interface during transient melt phase.

(viii) Although Coulomb explosion [11] is considered to be most suitable model for track formation in insulators, it fails [95] to explain the mixing in insulator thin film on insulating substrate. The concept of thermal spike explain the mixing in such cases also.

(ix) SHI induced mixing has not been observed in thermodynamically immiscible system, in contrast to observations with the low energy induced ion beam mixing in such systems. In the immiscible systems, the SHI irradiation induces smoothening of the interface, which is of importance in optical and magnetic multilayers. This is an area which requires further experiments and investigations.

(x) Irradiation followed by annealing leads to the formation of mixed phase at lower temperature than that obtained in annealing of the pristine samples. This is because of the enhanced diffusivity due to defects in the SHI irradiated sample.

(xi) Metal/insulator system with immiscible combination under SHI ion irradiation is shown to produce metal-insulator nanocomposite (metal particles embedded in insulator). Even the low energy ion irradiation can be very effective for this purpose. This is an important futuristic area of investigation where the kinetics for the growth of nanostructures can be studied and investigated with low energy as well as SHI.

(xii) For the mixing in the metal/a-Si, Fig. 4.5 could be a guideline to predict the content of mixing and the film thickness can be chosen to achieve complete mixing.

REFERENCES

1. M.M. Mitan, D.P. Pivin, T.L. Alford and J.W. Mayer, *Thin Solid Films*, **411** (2002) 219.
2. René Pretorius, Christiaan C. Theron, André C. Vantomme and James W. Mayer, *Critical Reviews in Solid State and Materials Sciences*, **24** (1999) 1.
3. M. Nastasi and J.W. Mayer, *Materials Science and Engineering: R: Reports*, **12** (1994) 1.
4. S. Matteson, B.M. Paine and M.-A. Nicolet, *Nucl. Instr. and Meth.*, **182-183** (1981) 53.
5. Tamou Yoshitaka, Li Jian, Stephen W. Russell and James W. Mayer, *Nucl. Instr. and Meth.*, **B 64** (1992) 130.
6. J.-P. Hirvonen, M.A. Elve, J.W. Mayer and H.H. Johnson, *Materials Science and Engineering*, **90** (1987) 13.
7. L.S. Hung and J.W. Mayer, *Nucl. Instr. and Meth.*, **B 7-8** (1985) 676.
8. A.H. Hamdi and M-A. Nicolet, *Thin Solid Films*, **119** (1984) 357.
9. S. Dhar, P. Schaaf, K.P. Lieb, N. Bibic, M. Milosavljevic, T. Sajavaara, J. Keinonen and A. Traverse, *Nucl. Instr. and Meth.*, **B 205** (2003) 741.
10. Y.V. Kudryavtsev, Y.P. Lee, J. Dubowik, B. Szymañski, J.Y. Rhee, *Phys. Rev.*, **B 65**, (2002) 104417.
11. R.L. Fleischer, P.B. Price and R.M. Walker, *J. Appl. Phys.*, **36** (1965) 3645.
12. G. Szenes, *Phys. Rev.*, **B 51** (1995) 8026.
13. Z.G. Wang, C. Dufour, E. Paumier and M. Toulemomde, *J. Phys: Condensed Matter*, **6** (1994) 6733.
14. C. Dufour, Ph. Bauer, G. Marchal, J. Grilhe, C. Jaouen, J. Pacaud and J.C. Jousset, *Euro. Phys. Lett.*, **21** (1993) 671.
15. Nita Dilawar, S. Sah, B.R. Mehta, V.D. Vankar, D.K. Avasthi and G.K. Mehta, *Vacuum*, **47** (1996) 1269.

16. D.K. Avasthi et al. Proc. of DAE Symp. on Solid State Physics, 1996.
17. M. Mäenpää, L.S. Hung, M.G. Grimaldi, I. Suni, J.W. Mayer, M.-A. Nicolet and S.S. Lau, *Thin Solid Films*, **82** (1981) 347.
18. L.S. Hung, Q.Z. Hong and J.W. Mayer, *Nucl. Instr. and Meth.*, **B 37-38** (1989) 414.
19. J.O. Olowolafe, E.G. Colgan, C.J. Palmstr and J.W. Mayer, *Thin Solid Films*, **138** (1986) 245.
20. K. Tao, C.A. Hewett, S.S. Lau, Ch. Buchal and D.B. Poker, *Appl. Phys. Lett.*, **50** (1987) 1343.
21. R.S. Averback, L.J. Thompson, J.J. Moyle and M. Schalit, *J. Appl. Phys.*, **53** (1982) 122.
22. W.K. Chu, J.W. Mayer and M.A. Nicolet, Backscatt. Spectrometry, Academic Press, NY 1978.
23. J.L'Ecuyer, C. Brassard, C. Cardinal and B. Terreault, *Nucl. Instr. and Meth.*, **149** (1978) 271.
24. J. Sun, W. Bolse, K.P. Lieb and A. Traverse, *Mater. Sci. Eng.*, **A 196** (1995) 229.
25. L. Wang, N. Angert, D. Ruck, C. Trautmann and J. Vetter, *Nucl. Instr. and Meth.*, **B 83** (1993) 503.
26. D.M. Ruck, *Nucl. Instr. and Meth.*, **B 166-167** (2000) 602.
27. M. Ekman and V. Ozolins, *Phys. Rev.*, **B 57** (1998) 4419.
28. S. Dhar, T. Som and V.N. Kulkarni, *J. Appl. Phys.*, **83** (1998) 2363.
29. S.A. Wolf, D.D. Awschalom, R.A. Buhrman, J.M. Daughton, S. von Moinar, M.L. Roukes, A.Y. Chtchelkanova and D.M. Treger, *Science* **294** (2001) 1488.
30. Sankar Dhar and V.N. Kulkarni, *Thin Solid Films*, **333** (1998) 20.
31. S. Chaudhary, S. Biswas, A. Gupta, D.K. Avasthi, D. Bhattacharyya, S. Teichert and D.K. Sarkar, *Nucl. Instr. and Meth.*, **B 217** (2004) 589.
32. K. Diva, D. Kabiraj, B.R. Chakraborty, S.M. Shivaprasad and D.K. Avasthi, *Nucl. Instr. and Meth. Phys. Res.*, **B 222** (2004) 169.
33. B.R. Chakraborty, S.K. Halder, N. Karar, D. Kabiraj and D.K. Avasthi, *J. Phys. D: Appl. Phys.*, **38** (2005) 2836.
34. Pratima Dhuri, Ajay Gupta, S.M. Chaudhari, D.M. Phase and D.K. Avasthi, *Nucl. Instr. and Meth.*, **B 156** (1999) 148.
35. W. Assmann, M. Dobler, D.K. Avasthi, S. Kruijer, H.D. Mieskes and H. Nolte, *Nucl. Instr. and Meth.*, **B 146** (1998) 271.
36. S.K. Sinha, D.C. Kothari, A.K. Balmuragan, A.K. Tyagi and D. Kanjilal, *Surface and Coating Techn.*, **158-159** (2002) 214.
37. Ph. Bauer, C. Dufour, C. Jaouen, G. Marchal, J. Pacaud, J. Grilhe and J.C. Jousset, *J. Appl. Phys.*, **81** (1997) 116.
38. D.K. Avasthi, W. Assmann, H. Nolte, H.D. Mieskes, S. Ghosh and N.C. Mishra, *Nucl. Instr. and Meth.*, **B 166-167** (2000) 345.
39. D.K. Avasthi, S. Ghosh, S.K. Srivastava and W. Assmann, *Nucl. Instr. and Meth.*, **B 219-220** (2004) 206.
40. D.K. Avasthi, W. Assmann, H. Nolte, H.D. Mieskes, H. Huber, E.T. Subramaniyam, A. Tripathi and S. Ghosh, *Nucl. Instr. and Meth.*, **B 156** (1999) 143.
41. D.K. Avasthi, W. Assmann, A. Tripathi, S.K. Srivastava, S. Ghosh, F. Gruner and M. Toulemonde, *Phy. Rev.*, **B 68** (2003) 153106.
42. S.K. Srivastava, D.K. Avasthi, W. Assmann, Z.G. Wang, H. Kucal, E. Jacquet, H.D. Carstanjen and M. Toulemonde, *Phys. Rev.*, **B 71** (2005) 193405.

43. V. Sisodia, D. Kabiraj, W. Bolse and I.P. Jain, *Appl. Surf. Sci.*, **252** (2006) 4016.
44. V. Sisodia, R.K. Jain, D. Bhattacharya, D. Kabiraj and I.P. Jain, *Radiation Measurements*, **36** (2003) 657.
45. K. Zhang, K.P. Lieb, V. Milinovic, M. Uhrmacher and S. Klaumunzer, *Nucl. Instr. and Meth.*, **B 249** (2006) 167.
46. P.C. Srivastava, M.K. Srivastava and P.S. Pandey, *Appl. Surf. Sci.*, **254** (2008) 5116.
47. M.K. Srivastava, T. Shripathi, M.M. Phase and P.C. Srivastava, *Appl. Surf. Sci.*, **256** (2010) 1664.
48. D. Bhattacharya, G. Principi, A. Gupta and D.K. Avasthi, *Nucl. Instr. and Meth.*, **B 244** (2006) 198.
49. Garima Agarwal, Vaibhav Kulshrestha, Pratibha Sharma and I.P. Jain, *J. Colloid and Inter. Sci.*, **351** (2010) 570.
50. B.R. Chakraborty, D. Kabiraj, K. Diva, J.C. Pivin and D.K. Avasthi, *Nucl. Instr. and Meth.*, **B 244** (2006) 209.
51. D. Bhattacharya, S.K. Srivastava, P.K. Sahoo, T. Som, V.N. Kulkarni and D.K. Avasthi, *Surface and Coating Technology*, **158-159** (2002) 59.
52. K. Diva, Ph.D. Thesis, Agra University, India.
53. K. Diva, B.R. Chakraborty, R.S. Chauhan, J.C. Pivin and D.K. Avasthi, *J. Phys. D: Appl. Phys.,* **41** (2008) 185305.
54. K. Zhang, K.P. Leib, V. Melinovic and P.K. Sahoo, *J. Appl. Phys.*, **100** (2006) 053501.
55. S.K. Ghose, B.N. Dev and Ajay Gupta, *Phys. Rev.*, **B 64** (2001) 233403.
56. Ajay Gupta, Carlo Meneghini, Amit Saraiya, Giovanni Principi and D.K. Avasthi, *Nucl. Instr. and Meth.*, **B 212** (2003) 458.
57. Ajay Gupta, Parasmani Rajput and Carlo Meneghini, *Phys. Rev.*, **B 76** (2007) 195401.
58. Ajay Gupta and D.K. Awasthi, *Phys. Rev.*, **B 64** (2001) 155407.
59. Parasmani Rajput, Ajay Gupta, Carlo Meneghini, D.K. Avasthi, Nora Darowski, I. Zizak and A. Erko, *Hyperfine Interactions*, **185** (2008) 689.
60. A. Gupta, N. Darowski, I. Zizak, C. Meneghini, G. Schumacher and A. Erko, *Spectrochimica Acta*, **B 62** (2007) 622.
61. T. Som, B. Satpati, P.V. Satyam, D. Kabiraj, P. Ayyub, S. Ghosh, Ajay Gupta, B.N. Dev and D.K. Avasthi, *Nucl. Instr. and Meth.*, **B 212** (2003) 206.
62. T. Som, B. Satpati, and P.V. Satyam, Ajay Gupta and N.C. Mishra, *J. Appl. Phys.*, **96** (2004) 7141.
63. T. Som, B. Satpati, P.V. Satyam, P. Ayyub and D. Kabiraj, *Nucl. Instr. and Meth.*, **B 212** (2003) 151.
64. Sarvesh Kumar, R.S. Chauhan, R.P. Singh, D. Kabiraj, P.K. Sahoo, C. Rumbolz, S.K. Srivastava, W. Bolse and D.K. Avasthi, *Nucl. Instr. and Meth.*, **B 212** (2003) 242.
65. S. Kumar, R.S. Chauhan, D.C. Agarwal, M. Kumar, A. Tripathi, W. Bolse and D.K. Avasthi, *Nucl. Instr. and Meth.*, **B 266** (2008) 1759.
66. K. Asokan, S.K. Srivastava, D. Kabiraj, S. Mookerjee, D.K. Avasthi, J.C. Jan, J.W. Chiou, W.F. Pong, L.C. Ting and F.Z. Chien, *Nucl. Instr. and Meth.*, **B 193** (2002) 324.
67. S.K. Srivastava, D. Kabiraj, B. Schattat, H.D. Carstanjen and D.K. Avasthi, *Nucl. Instr. and Meth.,* **B 219** (2004) 815 and S.K. Srivastava and D.K. Avasthi, *Defence Science Journal*, **59** (2009) 425.

68. R. Sreekumar, P.M. Ratheesh Kumar, C. Sudha Kartha, K.P. Vijayakumar, D. Kabiraj, S.A. Khan and D.K. Avasthi, *Nucl. Instr. and Meth.*, **B 244** (2006) 190.
69. R. Sreekumar, P.M. Ratheesh Kumar, C. Sudha Karth, K.P. Vijayakumar, D. Kabiraj, S.A. Khan, D.K. Avasthi, Y. Kashiwaba and T. Abe, *Semiconductor Science and Technology*, **21** (2006) 382.
70. R.K. Mangal, M.K. Singh, Y.K. Vijay and D.K. Avasthi, *Bull. Mater. Sci.*, **29** (2006) 653.
71. Ajay Gupta, *Vacuum*, **58** (2000) 16.
72. Sarvesh Kumar, R.S. Chauhan, S.A. Khan, W. Bolse and D.K. Avasthi, *Nucl. Instr. and Meth.*, **B 244** (2006) 194.
73. Z.G. Wang, C. Dufour, S. Euphrasie and M. Toulemonde, *Nucl. Instr. and Meth.*, **B 209** (2003) 194.
74. R. Leguay, A. Dunlop, F. Dunstetter, N. Lorenzelli, A. Braslau, F. Bridou, J. Corno, B. Pardo, J. Chevallier, C. Colliex, A. Menelle, J.L. Rouvière and L. Thomé, *Nucl. Instr. and Meth.*, **B 122** (1997) 481.
75. S.K. Srivastava, Ravi Kumar, A. Gupta, R.S. Patel, A.K. Majumdar and D.K. Avasthi, *Nucl. Instr. and Meth.*, **B 243** (2006) 304.
76. C. Jaouen, A. Michel, J. Pacaud, C. Dufour, Ph. Bauer and B. Gervais, *Nucl. Instr. and Meth.*, **B 148** (1999) 176.
77. Amitesh Paul, A. Gupta, S.M. Chaudhari, D.M. Phase, S. Ghose and D.K. Avasthi, *Nucl. Instr. and Meth.*, **B 156** (1999) 158.
78. A. Gupta, Amitesh Paul, R. Gupta, D.K. Avasthi and G. Principi, *J. Phys. Cond. Matt*, **10** (1998) 9669.
79. F. Richomme, A. Fnidiki, J. Teillet and M. Toulemonde, *Nucl. Instr. and Meth.*, **B 107** (1996) 374.
80. F. Richomme, J. Teillet, A. Fnidiki and M. Toulemonde, *Nucl. Instr. and Meth.*, **B 122** (1997) 507.
81. J. Juraszek, A. Fnidiki, J. Teillet, F. Richomme, M. Toulemonde and W. Keune, *Nucl. Instr. and Meth.*, **B 146** (1998) 244.
82. J. Juraszek, A. Grenier, J. Teillet, E. Cadel, N. Tiercelin, I. Monnet and M. Toulemonde, *Nucl. Instr. and Meth.*, **B 267** (2009) 912.
83. Ajay Gupta, Parasmani Rajput, Amit Saraiya, V.R. Reddy, Sigrid Bernstorff and H. Amenitsch, *Phys. Rev.*, **B 72** (2005) 075436.
84. Parasmani Rajput, Ajay Gupta, V.R. Reddy, Vasant Sathe and D.K. Avasthi, *J. Phys.: Cond. Matt.*, **19** (2007) 036221.
85. Parasmani Rajput, Ajay Gupta and D.K. Avasthi, *Nucl. Instr. and Meth.*, **B 266** (2008) 1680.
86. Ajay Gupta, Suneel Pandita, D.K. Avasthi, G.S. Lodha and R.V. Nandedkar, *Nucl. Instr. and Meth.*, **B 146** (1998) 265.
87. A. Gupta, R.S. Chauhan, D.C. Agarwal, S. Kumar, S.A. Khan, A. Tripathi, D. Kabiraj, S. Mohapatra, T. Som and D.K. Avasthi, *J. Phys. D: Appl. Phys.*, **41** (2008) 215306.
88. C. Rumbolz, W. Bolse, S. Kumar, R.S. Chauhan, D. Kabiraj and D.K. Avasthi, *Nucl. Instr. and Meth.*, **B 245** (2006) 145.

89. Amrithapandian, B.K. Panigrahi, S. Rajgopalan, A. Gupta, K.G.M. Nair, A.K. Tyagi and A. Narayanasamy, *Phys. Rev.*, **B 69** (2004) 165411.
90. Jai Prakash, A. Tripathi, S.A. Khan, J.C. Pivin, F. Singh, Jalaj Tripathi, Sarvesh Gupta and D.K. Avasthi, *Vacuum*, **84** (2010) 1275.
91. Jai Prakash, A. Tripathi, J.C. Pivin, Jalaj Tripathi, A.K. Chawla, R. Chandra, S.S. Kim, K. Asokan and D.K. Avasthi, *Adv. Mat. Lett.* (2011), DOI: 10.5185/amlett.2010.(2187).
92. B. Schattat, W. Bolse, S. Klaumünzer, F. Harbsmeier and A. Jasenek, *Nucl. Instr. and Meth.*, **B 191** (2002) 577.
93. Wolfgang Bolse and Beate Schattat, *Nucl. Instr. and Meth.*, **B 190** (2002) 173.
94. Wolfgang Bolse and Beate Schattat, *Nucl. Instr. and Meth.*, **B 209** (2003) 32.
95. B. Schattat and W. Bolse, *Nucl. Instr. and Meth.*, **B 225** (2004) 105.
96. B. Schattat, W. Bolse, A. Elsanousi and T. Renz, *Nucl. Instr. and Meth.*, **B 230** (2005) 240.
97. R. Nakatani, R. Taniguchi, Y. Chimi, N. Ishikawa, M. Fukuzumi, Y. Kato, H. Tsuchida, N. Matsunami and A. Iwase, *Nucl. Instr. and Meth.*, **B 230** (2005) 234.
98. H. Paulus, T. Bolse and W. Bolse, *Nucl. Instr. and Meth.*, **B 245** (2006) 117.

5

SHI for Synthesis and Modifications of Nanostructured Materials

5.1 INTRODUCTION

Nanostructured materials may be defined as those materials whose structural elements—clusters, crystallites or molecules—have dimensions in 1 to 100 nm range. Their importance for applications is due to remarkable variations in their fundamental electrical, optical and magnetic properties with their size. Nanostructures are considered as the building blocks of future technologies.

The most common and economical approach of generating nanostructured materials presently appears to be the chemical route. However, in most of the applications, one requires thin films in suitable matrix, which is not achievable by chemical route. There are other routes of creating nanostructures like vapour phase condensation, thermal co-deposition, pulsed laser deposition, and various methods related to the use of ion beams. Energetic ion beam is emerging as a powerful technique for generating different nanostructures because it provides remarkable ways to control the structures which can be approached in various ways due to the availability of variety of ion beams with diverse and controllable characteristics.

Energetic ion is an excellent tool for the creation of nanostructures on the surface or in the near-surface region of a solid. In this process the energy of the ion is transferred to the solid almost instantaneously into a highly localized volume of nm dimensions. The deposited energy density by the ions can exceed the binding energy of the solid resulting in a nanometer scale zone of non-equilibrium conditions. Because of the energy dissipation into the cold surrounding, this highly excited zone rapidly "cools" down within some tens to hundreds of picoseconds. Such extremely short and intense solid state excitations, localized in a nano-scale volume, cannot be achieved by any other technique, providing unique ways for engineering the properties of materials. A single high energy heavy ion can be used to generate non-equilibrium phases with novel structural and physical properties in nm-dimensions, which can be used as the basis for functional nanostructures. Moreover, single-ion irradiation is not the only possibility to generate nano-scale structures utilizing energetic heavy ions. Increasing the ion fluence and the number of modified zones, the interaction and finally the overlap of these zones give rise to surface instabilities and self-organization effects.

Ion beams can circumvent many difficulties because of the continuous developments and unique possibilities in ion beam technologies. Ions can now be focussed on to spots of a few nanometers in diameter and can be scanned very fast over the structures in programmed fashion. Furthermore, bombardment with energetic ions can cause chemical reactions locally or the sputtering of atoms from one part of the structure into another part and remove or mix the atoms in pre-defined areas.

Ion irradiation provides controlled ways to create or modify nanostructures. Ions with low energies (the region where the energy loss is dominantly via elastic collisions) induce a collision cascade in a controlled range of depths and high energy ions (where the energy loss is dominantly due to electronic process) perturb the structure of the target in narrow channels (with diameters of the order of 10 nm) in a region upto a few microns. SHIs have a unique feature of depositing a large energy per unit length, which can produce an amorphous zone or a modified region along the ion path in a cylindrical region of diameter of upto 10 nm, referred to as ion track. Transformations in the material under ion impingement may result from the disordering correlated to the atomic displacements or from local changes of compositions, because of the diffusion of atoms during the cascade (radiation induced) or of residual defects (radiation enhanced), when the latter are mobile at the temperature of the experiments. The topology of the target surface is also affected by these various types of events as well as by the ejection of atoms (sputtering).

The role of ions in nanostructuring of materials can be divided in three energy regimes: (i) very low energies, upto a few keV, suitable for synthesis of thin films by sputtering, (ii) energies from tens of keV to a few MeV for implantation related processes and (iii) the swift heavy ions, energies >1 MeV/nucleon, where electronic energy loss is dominant. Focus here is on the nanostructuring with swift heavy ions.

Ion matter interaction leads to microstructructural changes in materials which can provide unique ways to engineer the properties of materials directed towards specific application but in order to achieve this objective it is necessary to understand the ion-matter interaction in nanoscale regime.

Nanoscale Materials under Irradiation

The irradiation-induced generation of defects in nanosystems is different from bulk materials because of the limited size of the system in one or more directions. Many ions traverse these tiny objects without inelastic interaction so that little energy is deposited. This means that the probability for energy loss of an impinging particle is low in nano targets and decreases with increasing ion energy. Correspondingly, the amount of damage in a nanoscale object is low when the particle energy is high, provided that displacement cascades do not play a significant role.

The reduced dimensionality, on the other hand, may give rise to a different temperature profile after the impact so that the local temperature may exceed the melting temperature of the material. This is particularly important for ion irradiation of zero-dimensional objects. For example, just 30 eV energy transferred to an isolated fullerene (C_{60}) molecule, which is an amount of energy that can easily be transferred by a single ion impact, will raise the 'temperature' in the fullerene to about 2,000 K. Furthermore, the large fraction of surface atoms in nanosystems results in a high sputtering yield (removal of atoms from the system). The finite size of the system also affects the electronic structure and thus the mechanisms of conversion of electronic excitations to kinetic energy of the atoms.

The response of some structures to the ion beam can be such that the restructuring of the systems is governed by principles of self-organization. Then, the desired modification of the nanosystems can be achieved even without the need to focus the beam.

Irradiation-induced Self-organization Phenomena

In 'self-organization' a system develops some kind of structuring by itself, that is, without selective intervention from outside. Surface instabilities and the resulting self-organization processes play an important role in nanotechnology since they allow for large-array nano-structuring. Such instabilities in thin film systems can be triggered by energetic ion bombardment and the subsequent self-assembly of the surface can be controlled by fine-tuning of the irradiation conditions. The role of the ion in such processes is two fold. If the instability is latently present in the virgin sample but self-assembly cannot take place because of kinetic barriers, the ion impact may just supply the necessary atomic mobility. On the other hand, the surface may become instable due to the ion beam induced material modifications and further irradiation then results in its reorganization.

Nanomaterials are far from thermal equilibrium under intense ion irradiation. Defects are continuously generated and heat is released from the particles. Because only a minor fraction of the transferred energy is stored in persistent defects (most defects anneal immediately), there is a considerable energy flux through the particles and the conditions for self-organization in dissipative systems may be fulfilled. The transformation of graphite to diamond under ion irradiation is an example of such a process where a system is driven to a state (diamond) that is energetically less favourable but has a higher degree of order. Ion irradiation of a graphite–diamond interface at high temperature can lead to the growth of diamond at the expense of graphite. Conversely, heating without irradiation leads to the relaxation towards equilibrium—that is, to the transformation of diamond to graphite [1]. This interesting ion beam induced effect is not nanostructuring but that of change in phase of the material.

Ordered nanostructures can be fabricated with ion irradiation by self-organization originating from an interplay between disorder, sputtering and self-assembly due to irradiation-stimulated diffusion [2]. A delicate balance between different processes can result in the formation of nanoclusters or quantum dots with unique electronic or magnetic properties and in the creation of metallic nanoparticles in dielectric matrices.

In the field of nanostructured materials, the energetic ions provide ways for the controlled synthesis and modification. The important issues in nano materials are: (i) the control of the size of particles, (ii) the size distribution, (iii) creation of special structures such as alignment of particles, (iv) shape transformation of nanoparticles from spherical to prolate and oblate and (v) alignment of nanoparticles in a row like a nanostring. These will result in interesting properties such as tuning of surface plasmon resonance, tilting the plane of magnetization in film of magnetic nanoparticles embedded in matrix, etc. and many more yet to be explored and understood.

The processes involve the reorganization of a metastable system under the influence of ion irradiation. Metastable states resulting from the deposition of a film on the substrate with lattice mismatch and nonwetting characteristics can undergo phase segregation and formation of nanostructures of the film materials on the substrate. Silver films on Si surface, for example, leads to the formation of nano size islands of silver under ion irradiation [3].

A metastable state can often be realised in systems prepared under conditions which induce nonstoichiometry in the system. If the system possesses sufficient kinetic stability in ambient temperature and pressure, it continues to remain in a metastable state. Irradiation by high-energy ion beams can excite the system in highly localized regions of a few nm dimension and can lead to the transition of the metastable state to a stabler one in confined volume. Such a process can lead to the formation of nanoparticles along the ion track. Si nanocrystals for example can be produced in the nanometric size ion tracks of SHI irradiated Si.

5.2 SYNTHESIS OF NANOSTRUCTURED MATERIALS UNDER ELECTRONIC EXCITATION

Ion beams offer possibilities for synthesizing nanoparticles under extreme physical conditions and in highly localized regions, involving processes far from equilibrium. Traditional approach with ion beams is ion implantation followed by annealing leading to supersaturated solid solution and subsequent nanoparticle formation. This has led to formation of exotic phases such as Si nanocrystals in SiO_x matrix. One can combine low energy ion implantation with the highly localized excitation capability of SHI. Basic idea is that the diffusion coefficient of atom is widely different in defected and defect-free medium.

Nanostructures by Phase Separation

SHI irradiation leads to the formation of semiconductor nanostructures (Si and Ge) embedded in their respective sub-oxides (SiO_x and GeO_x) [4-7]. This is due to the ion beam induced separation of semiconductor phase from oxide phase of the semiconductor suboxide matrix. The process of phase separation takes place in the molten zone formed due to high temperature rise as a result of passage of the swift heavy ion, most likely by spinodal decomposition as discussed later.

Semiconductor nanocrystallites embedded in thin films can be synthesized with the help of swift heavy ions. In an experiment with SiO_x films deposited on Si substrates with 100 MeV Ni^{4+} ions irradiation, and fluence between 10^{12} and 10^{14} ions per cm^2, amorphous SiO_x films did not show any photoluminescence (PL) before irradiation. However, after irradiation the films exhibited clear peaks in the PL correlated with the luminescence from Si nanoclusters embedded in the matrix of SiO_x [8].

SHI irradiation of oxygen deficient germanium oxide (GeO_x with x <2) leads to the formation of Ge nanocrystals. X-ray diffraction, micro-Raman studies of the pristine and irradiated films show the formation of Ge nanocrystallites. Figure 5.1 shows the X-ray diffraction of the pristine and irradiated GeO_x films indicating the formation and growth of Ge nanoparticles, evident by the evolution and decrease in fwhm of Ge peak with fluence. Formation of Ge nanocrystallites can be explained on the basis of phenomenon

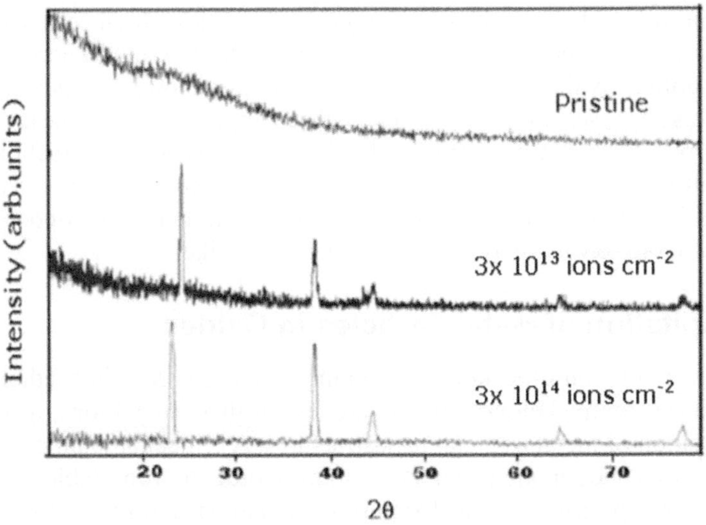

Figure 5.1: The X-ray diffraction patterns of the pristine and irradiated GeO_x films indicating the formation and growth of Ge nanoparticles, revealed by the evolution and decrease in fwhm of Ge peak with fluence. [Reprinted from Shyama Rath, D. Kabiraj, D.K. Avasthi, A. Tripathi, K.P. Jain, Manoj Kumar, H.S. Mavi and A.K. Shukla, *Nucl. Instr. and Meth.*, **B 263** (2007) 419 with permission from Elsevier.]

of ion beam induced phase separation [7, 9, 10]. The possibility of preferential sputtering of oxygen, which can lead to precipitation of Ge, is ruled out by on-line monitoring of Ge and O recoils. The observed sputtering is stoichiometric sputtering [11].

In similar manner the SHI irradiation of InO_x films lead to the formation of In nanoparticles [12]. The films irradiated with 120 MeV Ag ions of varying fluences indicate a phase separation of the In and In_2O_3 comprising largely of nanometric crystallites of 35 to 45 nm size. This temperature spike leads to a high temperature-induced phase separation in suboxides. The observed phase separation during irradiation is understood to proceed in a large number of small steps, in which oxygen-rich regions in films gradually become more oxygen-rich and oxygen-poor region become more poor. The net result is therefore a phase separation of In_xO_y into In and In_2O_3. Stoichiometry of the film monitored by on-line ERDA measurements at different fluences shows that the O/In ratio decreases with increasing fluence, indicating that there is a preferential loss of oxygen in the film [12].

In all these experiments of SHI irradiation with suboxide of Si, Ge and In, leading to the formation of respective nanocrystals in matrix, the precipitation and phase separation is induced by electronic energy deposition. It is interesting to note that if one performs the SHI irradiation of the perfect oxides such as SiO_2, GeO_2 and In_2O_3, the formation of nanoparticles does not take place. The phase separation leading to nanocrystal formation is due to spinodal decomposition, which is a mechanism by which a solution of two or more components can separate into distinct regions (or phases) with distinctly different chemical compositions and physical properties. This mechanism differs from classical nucleation in that phase separation due to spinodal decomposition occurs uniformly throughout the material—not just at discrete nucleation sites [13]. This phase separation due to spinodal decomposition is not due to transport of macroscopic matter by diffusion, but it is due to the local fluctuations in the concentration of the constituent. In the spinodal decomposition, oxygen-rich regions in the material become more oxygen-rich and oxygen-poor regions become more oxygen-poor, during temperature spike.

Precipitation of Metal Particles in Oxides

The ion irradiation of samples containing metal ions M^{n+} in triethoxysilane gel, induces an oxidation-reduction process of these metal ions, resulting in precipitation of metal as metal nanostructures. A critical amount of energy is required to reduce enough number of atoms in order to form stable nuclei. The rate of precipitation or the formation of nanostructures is found to be proportional to the total electronic energy deposition ($S_e \times \varphi$), where φ is fluence [14].

In other metastable systems such as silicate glasses containing dissolved Ag, a single ion impact forms a stable nucleus even with He ions of a few MeV having an electronic stopping power of the order of 30 eV/A [15]. Figure 5.2

shows the UV-visible spectra of the pristine and irradiated films of silicate glass containing a few % of Ag. The pristine film does not show any absorbance, indicating the presence of Ag as solid solution, whereas the irradiated films show absorbance at ~3 eV region, increasing with fluence. The absorbance is due to SPR, indicative of the formation of Ag nanoparticles. Since absorbance increases with fluence, the precipitation kinetics is simply a linear function of the transferred energy ($S_e \times \varphi$) per unit length. The interesting characteristics of the particles formed in all these systems is their narrow size distribution, compared to that obtained by annealing treatments at high temperatures of same materials.

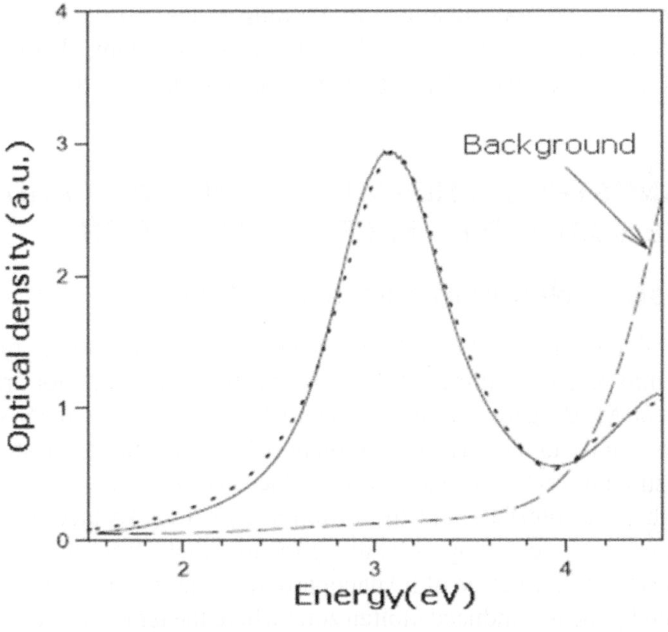

Figure 5.2: The UV-Visible spectra of the pristine and irradiated films of silicate glass containing a 3.5% of Ag. The absorbance in the irradiated films is due to SPR, indicative of the formation of Ag nanoparticles. Dotted line in absorbance peak is the fitted curve by Mie theory. [Reprinted from J.C. Pivin, G. Roger, M.A. Garcia, Fouran Singh and D.K. Avasthi, *Nucl. Instr. and Meth.*, **B 215** (2004) 373 with permission from Elsevier.]

Reduction Process Leading to Nanostructural Transformations

Copper oxide has two stoichiometric forms: cuprous oxide (Cu_2O) and cupric oxide (CuO) having cubic and monoclinic crystal structures respectively. Phase purity and stability of Cu_2O is important for its application in solar cell and other devices. SHI provides a way to achieve the desired transformation. Balamurgan et al. investigated the effect of large electronic excitation (24 keV/nm) generated by the 120 MeV Ag ions on copper oxide nanoparticles using X-ray diffraction, Raman spectroscopy and photoluminescence [16]. The XRD revealed that the pristine films were having both the phases CuO and Cu_2O and the irradiation induced a reduction process resulting in phase change from CuO to Cu_2O. The size of nanocrystallites of Cu_2O can be tuned by ion fluence [16]. The phase transition is primarily due to atomic rearrangements caused by dense electronic excitations. The crystalline quality of the films improved with irradiation.

Annealing the glass containing copper oxide above a certain temperature (855K) shows the existence of nanoparticles both of copper and copper oxide [17]. Below this temperature, there is no signature of nanoparticles. The SHI irradiation of these samples show a clear signature of nanoparticles of copper only and with a narrow size distribution. Nanoparticles formation occurs only in the first 40% of ion stopping range, corresponding to the region where electronic energy loss is higher than 5 keV/nm, clearly showing the role of electronic energy deposition in nucleation and growth of the particles [17].

5.3 NANOSTRUCTURES WITHIN ION TRACK AND AT THE SURFACE BY SELF-ORGANIZATION

Synthesis of Nanoparticles in Thin Films

Kumar et al. demonstrated the SHI induced conversion of indium oxide (IO) thin film into nanocrsytalline IO thin film by 100 MeV Ag^{8+} ion irradiation [18]. The 100 MeV Ag ions produce latent tracks in IO but not in the Si substrate. This discontinuity in the track formation results in a compressive stress at the IO/Si substrate interface. The ion-beam-induced compressive stress at the IO/ silicon substrate interface and the difference in their surface energies are responsible for the dewetting of IO along the ion tracks from the Si substrate, which leads to the formation of IO nanoparticles. The size of the nanoparticles is restricted by the ion-induced molten zone where the temperature rises above the melting point of the IO film. This method can be efficiently used for the synthesis of nanoparticles of a variety of materials on different substrates with a judicious choice of incident ion energy and fluence.

Ion tracks can be used to confine the growth of nanoparticles resulting in narrow size distribution. The ion implantation followed by annealing is a standard technique to synthesize embedded nanoparticles in a matrix. In the ion implantation process, the depth and amount of nanoprecipitates can be controlled by varying the energy and fluence of the ion during implantation. Limitation of this method is that it always results in broad size distribution due to lateral straggling of implanted species. In the case of swift heavy ions nanoclusters grow inside the tracks. The boundary of latent tracks produced by SHI irradiation acts as a capping agent to restrict the growth of nanoparticles inside the track. The size of nanoparticles can be controlled by controlling the parameters of the MeV ion beams [19].

Thin film of SnO_2 transforms from amorphous to nanocrystalline form under the swift heavy ion irradiation [20]. The substrate plays important role in this process. Tin oxide thin films of thickness of 100 nm grown on both amorphous substrate (SiO_2 layer on Si, fused silica) and crystalline substrate (crystalline quartz and α-sapphire) by electron beam evaporation method, are amorphous in nature. The irradiation of these films by 100 MeV Ag ions leads to the transformation from amorphous to nanocrystalline form. During SHI irradiation both the competitive processes like amorphization and crystallization take place. The rate of crystallization is more in crystalline substrates than in the films grown on amorphous substrates.

Nanocrystalline TiO_2 structures, formed by irradiating polycrystalline thin films of TiO_2 with 100 MeV Au^{8+} ion, have nanoparticles nucleated along the ion trajectory with the shape and size of nanoparticles dependent on the irradiation fluence [21].

Surface Nanostructures by Self-organization

Thin oxide Fe_2O_3 and NiO films can be restructured on a sub-μm scale by triggering structural instabilities and subsequent self-assembly processes with high fluence SHI irradiation [22-25]. This way regular arrays of nano-structures at the surface and at the interface of the layer system can be achieved, which cover a large area.

SHI irradiation at small angles of incidence with surface at liquid nitrogen temperature results in restructuring of crystalline NiO films. Study of the nanopatterning of the surface of NiO thin film under the 100 MeV Ag ion bombardment at fluence of 6×10^{14} ion cm^{-2} showed that the continuous film starts to crack at low fluence normal to the beam direction and at high fluence the material between the cracks begins to shrink and self-organizes into the periodic lamella structures (Fig. 5.3) [26]. The periodic lamella formation also depends on the nature of substrate as shown in Fig. 5.3 for 100 MeV Ag ion irradiation at a fluence of 6×10^{14} ion cm^{-2}.

The self-organization of NiO thin film can be understood in the light of the ion matter interaction at high energy. Each incident ion produces a Gaussian-like temperature profile in the vicinity of ion path in the material [27-29]. High temperature causes the transient local melting of the lattice within a cylindrical zone. The energy rapidly dissipates into the cold surroundings resulting in the solidification. The melting and subsequent re-solidification of materials generates the uni-axial tensile stresses along the ion track [30, 31]. When the tensile stresses along the beam direction overcome the fracture strength of the NiO surface, it shows periodic cracking perpendicular to the beam direction due to Grinfeld instability [32, 33]. Grinfeld showed by the linear surface stability analysis that the competition between elastic strain energy and surface energy leads to the amplification of perturbation of specific periodicity on such stressed surface. The formation of cracks releases elastic stress energy

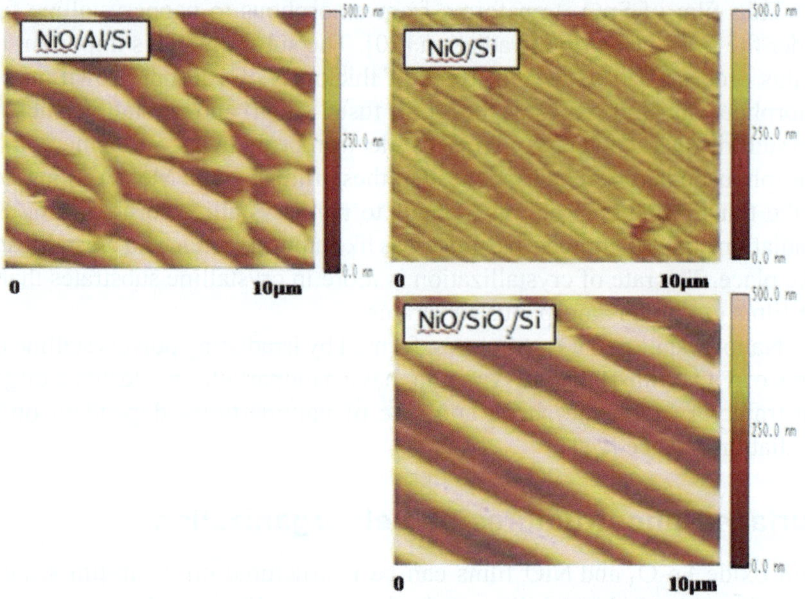

Figure 5.3: The AFM micrographs of the NiO thin film on different substrates irradiated by 100 MeV Ag ions at fluence of 6×10^{14} ions cm^{-2}. [Reprinted from R.S. Chauhan et al., *Vaccum* (Accepted) with permission from Elsevier.]

and consumes energy for the creation of the crack surface. For each stressed surface there is an optimum cracking distance, which minimizes the total free energy. The lamellae formation can be explained by the hammering effect [34-36]. When fluence increases, the material between cracks shrinks and develops the lamellae structure. The width of the lamellae decreases and the separation between them increases with the increase of fluence. The irradiated film on Al substrate has higher width and lack of periodicity of lamellae as the energy deposited by SHI in the film is easily transferred to Al substrate, since the resistivity of the Al is less than Si and SiO_2.

Online ERDA [37] measurements show that there is reduction in areal concentration of both nickel and oxygen with ion fluence due to the electronic sputtering. Sputtering takes place upto the fluence 1×10^{13} ions cm^{-2} fluence and beyond this fluence the surface cracks and reorganization of NiO surface starts. The stoichiometry (Ni/O) of the film increases from 0.74 to 0.91 up to 1×10^{13} ions cm^{-2} and beyond which it remains constant with fluence. When the stoichiometry approaches 0.9, the sputtering of the NiO film ceases. This implies that once plastic deformation starts, there is no loss of NiO material. It appears that initially the sputtering of NiO thin film takes place under the ion bombardment and whenever ions impart enough ($S_e \times \varphi$) into the film such that the tensile stress generated by ion beam exceeds the fracture strength of the NiO surface, the plastic deformation of film occurs. The energy deposited in the film is consumed in plastic deformation and sputtering ceases.

Self Affine Structures at ZnO Surface by SHI Irradiation

The formation of self affine nanostructures on the surface of evaporated ZnO thin film occurs under the heavy ion irradiation [39]. Surface roughness increases upto a critical fluence and decreases beyond that fluence. Density of the nanostructures increases from 8×10^9 ions cm^{-2} to 3×10^{10} ions cm^{-2} upto the fluence of 1×10^{12} ions cm^{-2} due to the reduction in the size of the nanostructures and after this critical fluence the density of nanostructures again decreases due to evolution of bigger structures as shown in Fig. 5.4. The reduction in size and increase in areal density of the nanostructures up to a specific fluence can be explained in terms of ion induced stress generation and change in surface energy due to total energy deposited by the ion beam. The values of roughness and growth parameters extracted from the power spectral density of AFM images, reveal that the surface diffusion is responsible for nanostructuring. The value of the roughness parameter determined from the power spectral density spectra is found to be around 0.9, which implies that the energy deposition by swift heavy ion provides energy to the surface atoms which causes the atomic mobility at the surface to be so high that it prevents

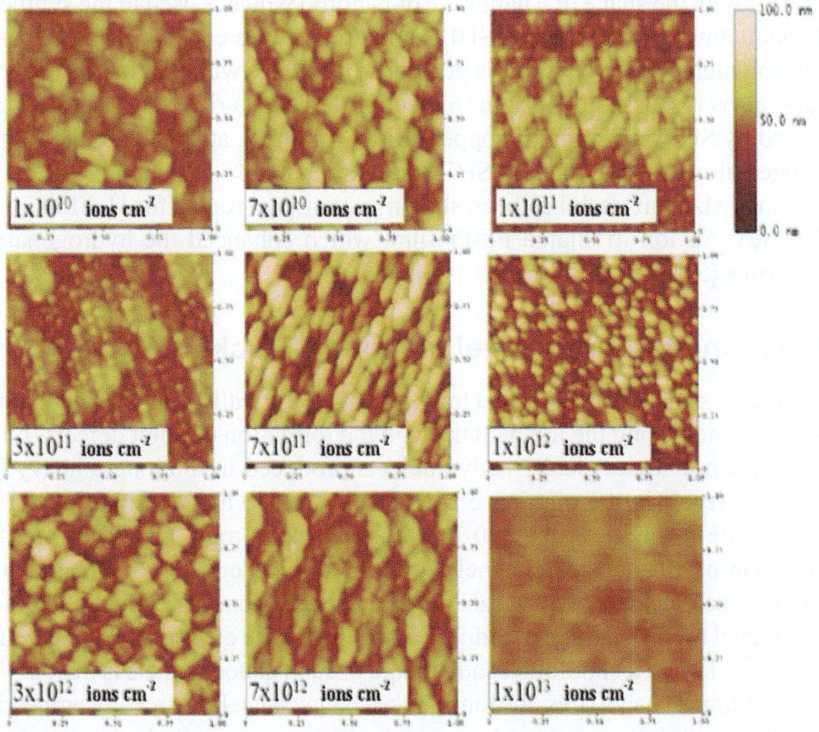

Figure 5.4: Surface nanostructures in ZnO films at different fluences of [100 MeV Ag ions [Reprinted with permission from D.C. Agarwal, R.S. Chauhan, D.K. Avasthi, S.A. Khan, D. Kabiraj and I. Sulania, *J. Appl. Phys.*, **104** (2008) 024304, copyright (2008) American Institute of Physics.]

the roughening of the surface under the swift ion irradiation. The results of online Elastic Recoil Detection (ERD) analysis show that there is reduction in areal concentration of both zinc and oxygen with ion fluence, which is a clear indication of sputtering under the ion bombardment. The sputtering in this energy regime is dominantly due to electronic energy loss. The total sputtered thickness is 0.05 nm at a fluence of 1×10^{13} ions cm^{-2}, which is much less than the sputtered thickness for low energy case where fluence is very high ($\sim 10^{17}$ ions cm^{-2}). Therefore, the role of sputtering yield in creation of surface nanostructures for irradiated ZnO thin film at normal incidence can be assumed to be negligible. The stoichiometry of the ZnO thin film remains almost same under the irradiation.

Creation of Nanowires by SHI

SHI can produce tracks in the insulators, semiconductors and metals beyond certain threshold of electronic energy loss, S_{eth}. This specific feature is utilized to achieve nanostructures and to modify the existing nanoparticles in specific shapes, discussed later in Section 5.4. Ion track is inherently a nanodimensional cylinder with the shape of a nanowire or nanorod which is used in the synthesis of conducting C nanowires by SHI irradiation of fullerene, DLC and Si based gels/polymers [40-49]. Synthesis of carbon nanowires with SHI in carboneaous materials is discussed in Chapter 6, Section 1. The Cu rich nanostructures are formed by SHI irradiation of copper nitride film [50], attributed to the loss of nitrogen along the ion track in SHI irradiation.

Nano size ion track has been shown as transport routes for H diffusion in 120 MeV Ag ion irradiated Pd-Pr film, which enhanced the hydrogenation properties [51].

Depinning of Fermi Level along Ion Tracks

Apart from structurally modified track region, electrically modified nano track can be formed along the ion path through the formation of point defects. It has been shown that electrically active defects produced in Si along a heavy ion track can result in the pinning of the Fermi level to a deep level [52]. This occurs within a region of <100 nm around an ion track and results in the formation of a nanosized channel with a modified Fermi level embedded in bulk Si. Such nanochannels might act as electronic nanowires due to the local Fermi level modification that induces a bending of the electronic bands around the ion track. The technological advantage of such nanochannels is that, unlike the "traditional" nanowires, the ion-induced nanochannels are readily embedded into the Si crystal, a semiconductor in contrast to an oxide, by conventional ion implantation. Although no evidence of ion track in crystalline Si exists, Vines et al. [52] show the evidence of single ion impact on crystalline Si by using scanning capacitance microscopy.

Etched Ion Tracks as Templates for Nanostructures

The modified zone in the ion tracks specially in polymers can be chemically etched to achieve nanopores of controlled diameters. These etched nano or submicron pores can then be used as templates to grow nanostructures of desired materials by electro-deposition. Etched ion tracks find variety of applications [53-55]. The resonant tunnelling behaviour was observed in deposition of p and n type material in micropores of polymer, created by etching of SHI irradiated polymer [56]. Au nano chain of Au nanoparticle aggregates in the etched ion track of silica have been synthesized giving interesting photoluminescence properties [57].

These submicron pores in polymer can be used for controlled drug delivery [58]. Etched ion track find application in filtering selective gases. The tracks in nanocomposite ($CoO.6ZnO.4Fe_2O_4$ particles embedded in polycarbonate) film show higher permselectivity [59]. The track etched porous membranes can be used as sensors when functionalized by grafting other molecules inside or as templates to grow nanocylinders or nanotubes of desired materials [60-64].

5.4 MODIFICATION OF THE METAL-DIELECTRIC NANOCOMPOSITE FILMS

Engineering the Shape and Size of Ag and Au Particles in Polymer Matrix

The shape of Au nanoparticles grown in teflon matrix (Au-teflon nanocomposite) by tandem deposition are not of spherical shape. It was observed that the shape of the particles becomes spherical and the size starts growing with fluence in 120 MeV irradiation of Au-teflon nanocomposite (deposited on TEM grid) [65]. Formation of spherical particles indicates the possibility of the existence of transient molten state as in the molten state the metal will have spherical shape like a liquid drop. The average size changed from about ~9 nm to ~14 nm after irradiation, which shows that the particles coalesce together to form bigger particles under SHI irradiation, like in Oswald ripening. Figure 5.5 gives the TEM images of Au polymer nanocomposite before and after irradiation with 120 MeV Au ions.

Different behaviour is observed in the case of Ag particles embedded in teflon matrix. The teflon films containing Ag nanoparticles deposited on TEM grids irradiated with 100 MeV Au ions at different fluences, show slight reduction of the particle size and loss of Ag in the films, due to electronic sputtering similar to the observation of surprisingly high sputtering yield in low dimension gold films under SHI irradiation [66], which is explained on the basis of large rise in temperature spike in low dimension system. In an experiment on Ag-teflon nanocomposite it is observed that SHI irradiation leads to reduction in size and size distribution of Ag nanoparticles [67].

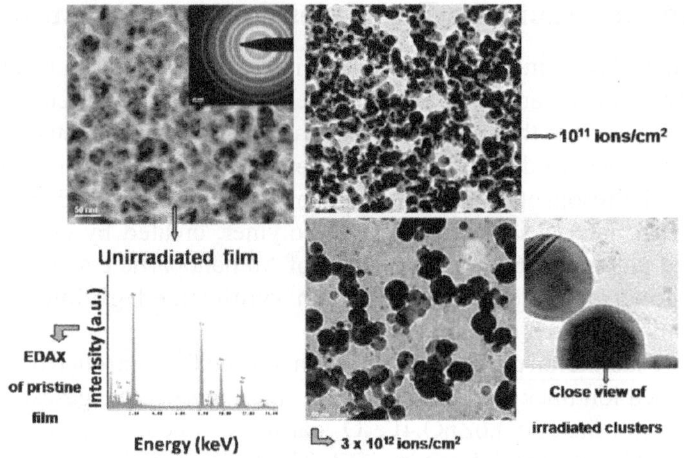

Figure 5.5: The TEM images show the pristine and irradiated Au polymer nanocomposite. The irradiation was with 100 MeV Au ions at a fluence of 10^{11} and 3×10^{12} ions cm^{-2}. EDAX shows presence of Au. [Reprinted from A. Biswas, D.K. Avasthi, D. Flink, J. Kanzow, U. Sch€urmann, S.J. Ding, O.C. Aktas, U. Saeed, V. Zaporojtchenko, F. Faupel, R. Gupta and N. Kumar, *Nucl. Instr. and Meth.*, **B 217** (2004) 39 with permission from Elsevier.]

Engineering the Shape and Size of Embedded Nanoparticles in Silica Matrix

The nanoparticles embedded in silica matrix are of interest due to their unique properties. The nanoparticle size, its shape, and interparticle separation are the key parameters deciding the properties of the nanocomposite thin film. These can be engineered with SHI irradiation. The most common matrix has been silica, where it is well established that SHI induced ion tracks in silica are responsible for modifications in embedded nanoparticles. Some of these experiments are briefly described here.

Au-silica Nanocomposite (NC) Thin Films

Reduction and growth of Au particle size with ion fluence

The Au-silica NC thin films with metal fraction of 5 at.%, irradiated with 100 MeV Ag ions show reduction in particle size with fluence [68]. The average interparticle separation in this case is ~14 nm. The metal fraction and average particle size are used for the estimation of average interparticle separation. The particle size decreased from 10.1 nm to 8.8 nm at a fluence of 10^{13} ions cm^{-2} as shown in Fig. 5.6. For the Au-silica NC thin films with metal fraction of about 10 at.%, (having an estimated inert particle separation ~2 nm), growth of Au particles was observed from about an average diameter of 4 nm to 9 nm under the irradiation by 90 MeV Ni ions up to a fluence of 10^{14} ions cm^{-2} [69]. In-situ XRD was performed to take the snap shots of the growth of the Au particles in silica matrix under SHI irradiation, as shown in Fig. 5.7.

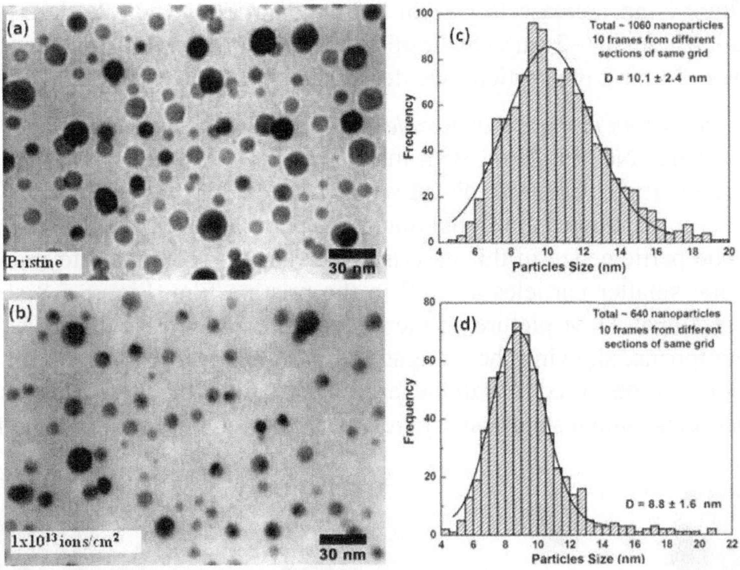

Figure 5.6: The TEM pictures show the pristine and irradiated of Au-silica nanocomposite thin films (with metal fraction of 5 at.%), irradiated with 100 MeV Ag ions. [Reprinted from D.K. Avasthi, Y.K. Mishra, F. Singh and J.P. Stoquert, *Nucl. Instr. and Meth.*, **B 268** (2010) 3027 with permission from Elsevier.]

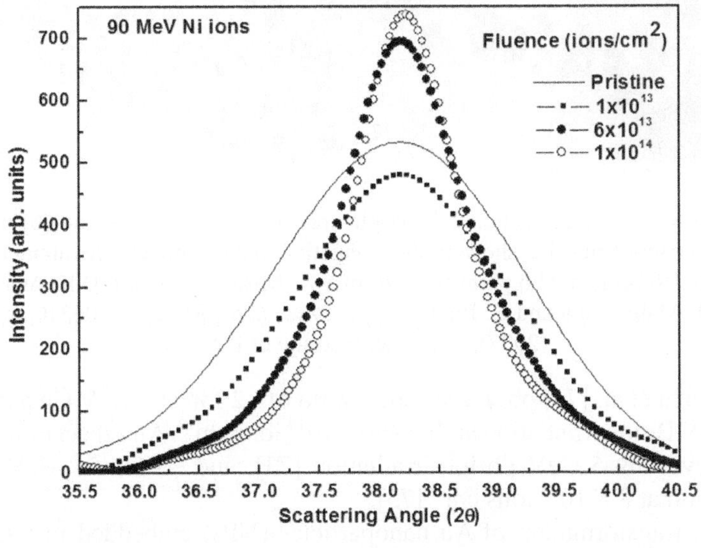

Figure 5.7: The X-ray diffraction patterns recorded by in-situ XRD was performed to take the snap shots of the growth of the Au particles in silica matrix under the irradiation by 90 MeV Ni ions at different fluences. [Reprinted with permission from Y.K. Mishra, D.K. Avasthi, P.K. Kulriya, F. Singh. D. Kabiraj, A. Tripathi, J.C. Pivin, I.S. Bayer and A. Biswas, *Appl. Phys. Lett.*, **90** (2007) 073110, copyright (2007) American Institute of Physics.]

There is an increase in size of Au nanoparticles when the interparticle separation is small (~2 nm). On the other hand, when the interparticle separation is large (~14 nm), the particle size decreases.

Elongation of Au nanoparticles under SHI irradiation
The Au-silica NC thin film, with metal fraction of 15 at.% annealed to get large size particles when irradiated with 120 MeV Au ions at a fluence of 3×10^{13} ions cm^{-2}, showed the elongation of Au particles along the beam direction [70]. The particle size in this case is rather large as compared to track size. There are smaller particles as well in the pristine film. Figure 5.8 shows a cross sectional TEM picture of the (a) pristine and (b) irradiated Au-silica nanocomposite, showing the elongation of Au nanoparticles along the beam direction. It is observed that all the large size particles elongate whereas small size particles remain spherical. There are several studies in this area [71-78].

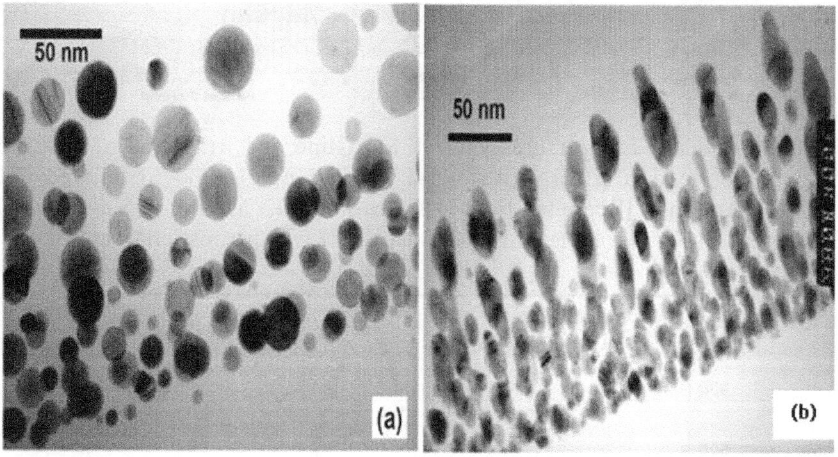

Figure 5.8: The cross sectional TEM pictures of the (a) pristine and (b) irradiated Au-silica nanocomposite, showing the elongation of Au nanoparticles along the beam direction. [Reprinted with permission from Y.K. Mishra, F. Singh, D.K. Avasthi, J.C. Pivin, D. Malinovska and E. Pippel, *Appl. Phys. Lett.*, **91** (2007) 063103 copyright (2007), American Institute of Physics.]

Awazu et al. [71] observed aspect ratio of ~2 for 110 MeV Cu and ~3 for 110 MeV Br ion irradiation at fluence of 10^{14} ions cm^{-2}. Au aspect ratio of ~2.3 is achieved by 5 GeV Pb ion irradiation [73], and ~5 for 54 MeV Ag ion irradiation at 8×10^{14} ions cm^{-2} [75].

The transformation of Au nanoparticles (NPs) embedded in SiO_2 from spherical to rod-like shapes (with an aspect ratio of ~5) induced by swift heavy ion irradiation has been studied with ^{197}Au ions at energies between 54 and 185 MeV. The saturation width of the NP rods as well as a minimum size required for the NPs to elongate depends on the energy [78]. The saturation width of NP is correlated with the ion track diameter in the SiO_2.

The irradiation of Au particle surrounded by silica shell results in the flattening of the silica perpendicular to the incident ion direction, whereas the Au particle gets elongated along the ion beam direction [76]. The flattening of silica shell is due to the hammering effect caused by the viscoelastic flow of material [79]. The silica material during its expansion causes a pressure to the Au particle squeezing in plane perpendicular to the ion beam resulting in elongation of particle in plane along the beam direction. The summary of results so far on the irradiation of Au-silica system is given in Table 5.1.

Table 5.1: Summary of the effect of SHI irradiation on Au nanoparticles embedded in silica matrix system by different groups

Metal-matrix	*Growth/ reduction/ elongation*	*NP size (aspect ratio, AR)*	*Energy and ion*	*Energy loss (S_e, S_n) (keV/nm)*	*Fluence (ϕ) ions /cm²*	*$S_e \times \phi$ (keV/ nm³)*	*Refer-ences*
Au NPs in silica 3D NC	Reduction	11 to 5 nm	100 MeV Ag	11.7, 0.06	1×10^{13}	1.17	[68]
Au NPs in silica 3D NC	Growth	4 to 9 nm	90 MeVNi	7.2, 0.01	1×10^{14}	7.2	[69]
Au NPs in silica, 3D NC	Elongation	Pri- 14 nm AR= 3.5	120 MeV Au	14.1,0.02	3×10^{13}	4.2	[70]
Au NPs in silica 2D NC	Elongation	12 nm AR=5.0	185 MeV Au	17.6, 0.15	2×10^{14}	35.2	[78]
Au-silica core shell	Elongation	14 nm AR= 9.0	30 MeV Se	6.7, 0.07	2×10^{14}	13.4	[76]
Au NPs in silica 2D NC	Elongation	AR=1.99	100 MeV Cu	3.6, 0.02	1×10^{14}	3.7	[71]
Au NPs in silica 2D NC	Elongation	20 nm AR = 3.0	110 MeV Br	9.7, 0.02	1×10^{14}	9.7	[71]
Au NPs n silica 2D NC	Elongation	8 nm AR=2.3	5 GeV Pb	17, 0.01	5×10^{13}	8.5	[73]
Au NPs in silica 2D NC	Elongation	15 nm AR=5	54 MeV Ag	9.6, 0.09	8×10^{14}	76.8	[75]

Three types of Au silica nanocomposite are possible namely (i) Au core silica shell, (ii) two dimensional nanocomposite (2D NC) where a single layer of nanoparticles is embedded in silica matrix and (iii) three dimensional (3D NC) nanocomposite where the nanoparticles are dispersed throughout the thickness of the film in three dimensions. On the basis of several experiments, reported by different groups, it appears that the aspect ratio introduced by SHI for the same electronic energy deposition (electronic energy loss × fluence) is highest for core shell followed by three dimensional and two dimensional nanocomposite as shown in Fig. 5.9 [77].

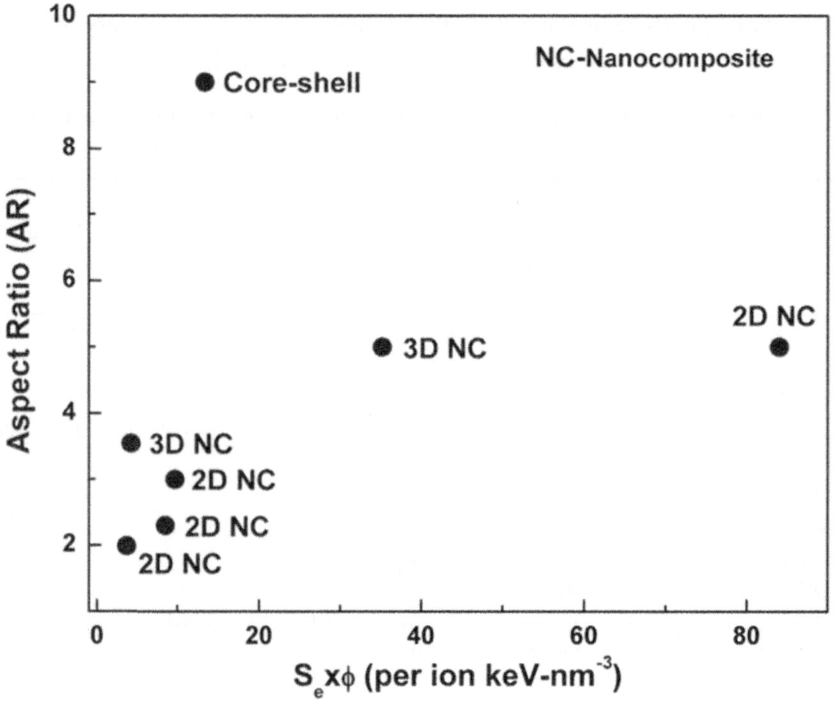

Figure 5.9: The figure shows the aspect ratios achieved for different type of Au silica nanocomposite thin films by different groups. [Reprinted from D.K. Avasthi, Y.K. Mishra, F. Singh and J.P. Stoquert, *Nucl. Instr. and Meth.*, **B 268** (2010) 3027 with permission from Elsevier.]

The ion induced elongation of Au nanoparticles is explained by reasoning that the silica gets compressed in the beam direction and expands in the plane perpendicular to the beam, which is a consequence of hammering effect, as seen in Au-silica core shell experiment [76]. The pressure exerted by silica results in compressing the Au nanoparticle causing shape transformation from spherical to oblate. However, if we consider this mechanism, then irrespective of nanoparticle size, the elongation should take place, except for very big size particles, where the pressure is insufficient to cause effect on shape. It was observed that the small size particles do not elongate and therefore this particular mechanism is not appropriate for explaining the ion beam induced elongation [70]. If the particle size is smaller or comparable to the track diameter in silica, there is isotropic growth or reduction in particle size because the particle is surrounded by transiently melt silica, which explains the fact that the small size nanoparticles do not elongate. The growth of the particle occurs at the interface of the nanoparticle and silica within the ion track region nanoparticle

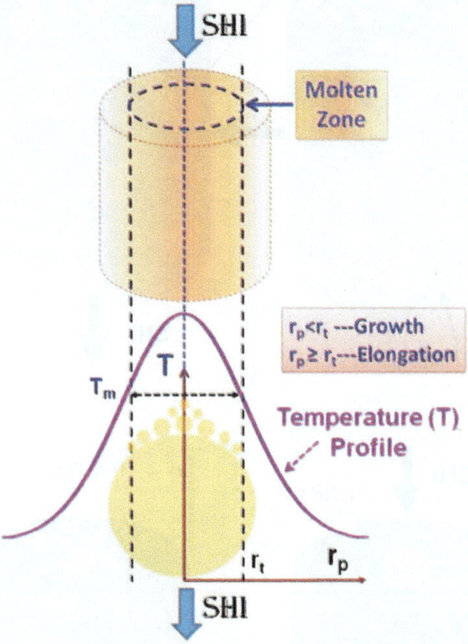

Figure 5.10: The schematic shows growth of Au nanoparticle, occuring in ion track, resulting in elongation of the particles. r_t is track radius, r_p is radius of nanoparticles, r is distance from the ion path, T is temperature of track and T_m is melting point of material in track which is silica in the present case. [Reprinted from D.K. Avasthi, Y.K. Mishra, F. Singh and J.P. Stoquert, *Nucl. Instr. and Meth.*, **B 268** (2010) 3027 with permission from Elsevier.]

during transient melt state of silica, as shown in Fig. 5.10. There are large number of overlaps of ion tracks in the nanocomposite sample and what is observed is the cumulative effect of these. If the particle size is smaller than the track size, there is isotropic growth or reduction in particle size (depending on the interparticle separation), because the nanoparticle is surrounded by transiently molten silica, as shown in Fig. 5.11. This explains why small size nanoparticles do not elongate.

Another possible explanation is based on the volume expansion coefficient of Au and silica in molten state. The volume expansion of Au in transformation from solid to molten state, is more than that for silica under similar transformation. Therefore the elongation of Au nanoparticle along the ion track is expected. Possible mechanisms of reduction, growth and elongation of nanoparticles embedded in dielectric matrix under swift heavy ion irradiation in terms of ion beam and nanocomposites parameters are discussed in framework of thermal spike model [71, 77, 80, 81] using the two coupled

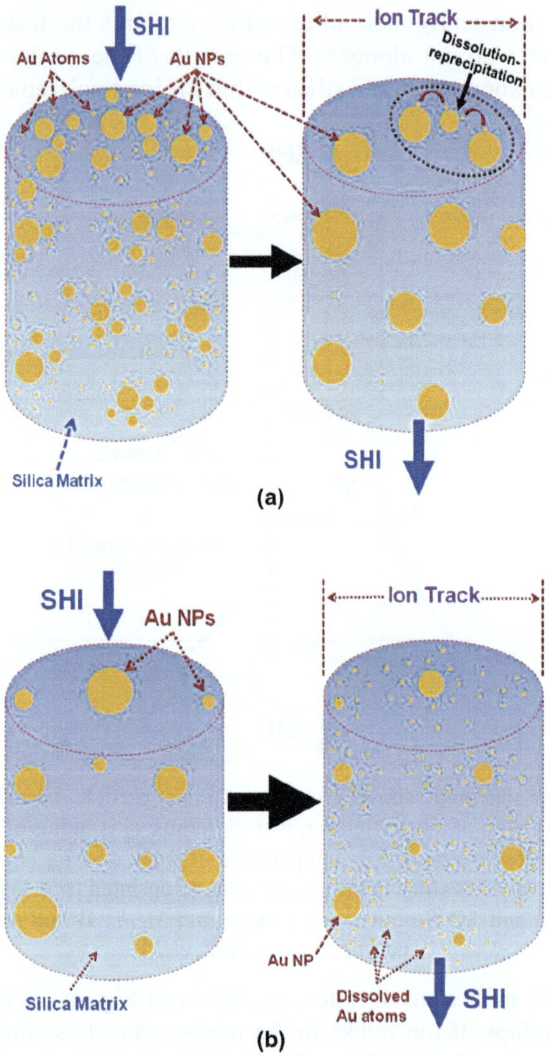

Figure 5.11: The schematic shows (a) increase and (b) decrease of nanoparticle size in the ion track, depending on the inter particle separation. In (a) the inter particle separation is small and the big particles grow at the expense of smaller ones while in (b) the inter particle separation is large and dissolved atoms remain in the matrix. [Reprinted from D.K. Avasthi, Y.K. Mishra, F. Singh and J.P. Stoquert, *Nucl. Instr. and Meth.*, **B 268** (2010) 3027 with permission from Elsevier.]

differential equation for the electronic and lattice subsystem, considering a cylinder of SiO_2, with one Au nanoparticle at the centre of the cylinder. The ion track is assumed to be along the cylinder axis. To simplify the calculations, the parameters for Au are taken same as that of bulk without considering the size effect. The calculations qualitatively predict size and shape change of Au nanoparticle as observed in experiments.

Shape Change of Ag, Pt, Sn, Co and Ge Nanoparticles

A summary of the shape change of the nanoparticles of elements other than Au is given in Table 5.2 and each case is discussed.

Table 5.2: Summary of the effect of SHI irradiation on the shape change of the Ag, Pt, Sn, Co and Ge nanoparticles embedded in silica matrix

NPs	*Growth/ reduction/ elongation*	*NP size (nm) (aspect, ratio AR)*	*Ion and energy*	*Energy loss (S_e, S_n) (keV/nm)*	*Fluence (ϕ) ions /cm^2*	*$S_e \times \phi$ (keV/ nm^3)*	*Refer- ences*
Ag	Elongation	9.6 AR=1.47	Au 120 MeV	13.9/0.02	3×10^{13}	4.17	[80]
Pt	Elongation	14.5 AR=1.87	185 MeV Au	54.9/0.81	1×10^{14}	54.9	[84]
Sn	Elongation	>11 AR=2.3	185 MeV Au	17.6/0.15	1×10^{14}	17.6	[85]
*Co	Elongation	10 AR=3.9	200 MeV I	14.8/0.048	1×10^{14}	14.8	[91]
#Co	Elongation	4.8 AR=6	200 MeV I	14.8/0.048	1×10^{14}	14.8	[90]
Ge	Elongation	26±5 AR=1.8	38 MeV I	7.7/0.18	1×10^{15}	77	[96]

*Note: Sample heated at 873K during Co implantation.
#Note: Sample at 295K during Co implantation.

The SiO_2:Ag films deposited by magnetron co-sputtering, with volume fractions of particles (10-20%), annealed at high temperature for promoting a growth of the particles were irradiated with 100 MeV Au. In this case, an elongation of the particles parallel to the beam direction was evidenced by the analysis of the optical response and TEM [80, 81]. The initial size of particles and their volume fraction play a deciding role in the effect of swift heavy ions on the nanoparticles. The Ag nanoparticles (of about 10 nm) in silica matrix get aligned along the beam direction as a consequence of SHI irradiation [82]. The aligned metal nanoparticles have optical property quite similar to that expected from metal nanorod or nanowire.

Calculations based on a thermal spike model for a single particle of metal embedded in an insulating matrix with a lower thermal conductivity and electron-phonon coupling constant indicate that the temperature in the particle is more when the particle is small, which causes a transient melting [81]. Atoms from particles close to the surface are volatilized more easily. Ag nanoparticles (larger than the size of ion track) in silica matrix elongate under 120 MeV Ag ion irradiation which is explainable by thermal spike model [80].

The Pt nanoparticles embedded in silica under 185 MeV Au ion irradiation elongates along the beam direction with an aspect ratio of about 5, at a fluence of 2×10^{14} ions cm^{-2}. The width (minor axis) of elongated nano particles embedded in silica matrix saturates beyond certain fluence of SHI irradiation, as was evident by experiments on Co, Pt and Au nanoparticles [83, 84]. Pt nanoparticle in silica matrix undergo shape transformation from spherical to prolate under SHI irradiation (27-185 MeV Ar ion). The minor axis of prolate particles saturates with fluence [84]. However, the elongated nanoparticles fragment via Raleigh instability.

The shape change from spherical to rod-like along the beam direction occurs in Sn nanoparticles embedded in silica matrix by 185 MeV Au ion irradiation [85]. The Sn nanoparticles more than 11 nm undergo this shape transformation whereas the smaller particles remain spherical as reported in the case of Au nanoparticle in silica matrix [70]. The experiment also proved that elongation needs a molten ion track. The elongation of nanoparticle does not take place in Si matrix [85] where it is known that the track formation does not occur by mono energetic ion beam of any species and any energy. This is a very strong proof that the basic condition of the shape engineering of nanoparticles is that the nanoparticles must be embedded in a matrix where ions can create ion tracks.

It has been shown that SHI irradiation causes the elongation of Zn nanoparticles embedded in silica and the elongation is induced by each incident ion [86].

Shape change from spherical to elongated along the beam direction is seen in several metal nanoparticles (Au, Ag, Pt, Zn and Sn) embedded in silica matrix, where the aspect ratio of the particles can be controlled by the ion beam parameters. The common findings are that the nanoparticles below certain size (smaller than the track diameter in silica) and above certain size, do not undergo shape transformation. The metal nanoparticles above certain diameter behave like bulk material for the incident ion beam.

Effect of SHI Irradiation on Magnetic Nanoparticles in Silica

Swift heavy ion irradiation on silica films containing Fe in the form of solute ions indicate that the electronic excitations induce reduction of solute ions and precipitation of metal particles. The hammering effect of ions on the silica matrix causes a tilt of easy magnetization axis when the volume fraction of particles is of a few percent, but this axis remains in the film plane for the Fe silica nanocomposite with metal fraction over 10% [87]. TEM investigations did not reveal elongation or alignment of particles. It is believed that the stress on the particles due to the hammering effect can cause the tilt of magnetization plane. Normally the magnetization of a thin film is parallel to the surface rather than perpendicular to it because of the shape anisotropy: the hysteresis M(H) curves are more squared and residual magnetization under zero applied field is larger parallel to the surface. For application in magnetic recording with high density, an easy axis of magnetization perpendicular to the surface is desirable. It has been shown that this can be achieved for films made of silica embedded with small Fe particles with a low volume fraction by irradiation with swift heavy ions [87]. SHI irradiation results in tilting the easy magnetic plane from in plane to perpendicular to plane. This is explained on the basis of the stress exerted on the particles by the matrix, due to the hammering effect of SHI on amorphous targets. This change in the orientation of easy magnetization from parallel to perpendicular to film plane was evident from electron spin resonance measurements. The resonance field is minimum when magnetic field is parallel to the surface in pristine films and becomes perpendicular to it after irradiation with 10^{13} Au ions of 100 MeV as shown in Fig. 5.12 for a silica film containing 3 vol.% of super paramagnetic Fe particles with a size of ~5 nm. The minimum resonance field occurs for the easy axis. The films of Fe-silica containing 14% Fe under 100 MeV Ag ion irradiation shows Fe nanoparticles aligned like a string along the beam direction as shown in Fig. 5.13. However, in this case, the easy magnetic axis does not become perpendicular to film plane and remains in the film plane, due to strong dipolar interaction between the arrays of nanoparticles [81].

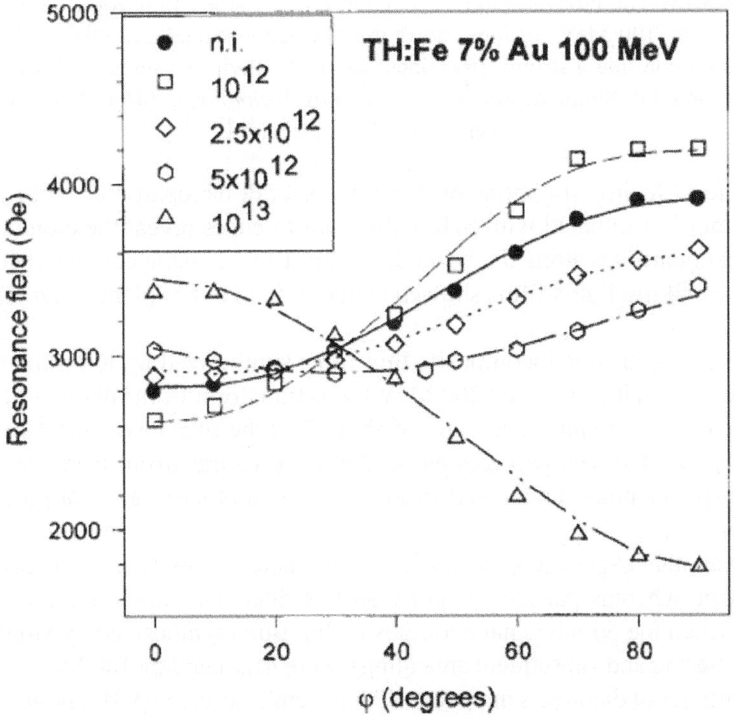

Figure 5.12: The resonance field at different angles with respect to the magnetic field axis for the Fe-silica nanocomposite irradiated by 100 MeV Au ions at different fluences. [Reprinted with permission from J.C. Pivin, S. Esnouf, F. Singh and D.K. Avasthi, *J. Appl. Phys.*, **98** (2005) 023908, copyright (2005) American Institute of Physics.]

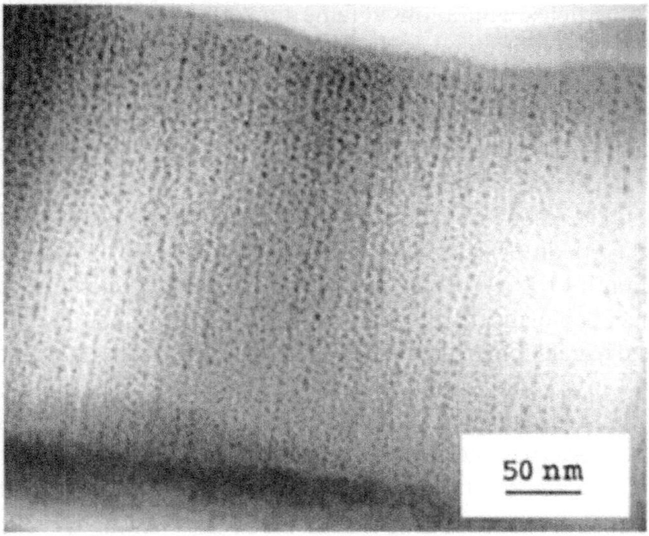

Figure 5.13: The cross section TEM picture of Fe-silica film containing 14% Fe irradiated by 100 MeV Ag ions shows that the nanoparticles get aligned along the beam direction like a string. [Reprinted from J.C. Pivin, F. Singh, Y. Mishra, D.K. Avasthi and J.P. Stoquert, *Surface & Coatings Technology*, **203** (2009) 2432 with permission from Elsevier.]

The TEM investigations of the films of FePt nanoparticles embedded in silica matrix irradiated with SHI at different fluences reveal the elongation of FePt nanoparticles along the beam direction. The magnetic characterization of pristine and irradiated films showed an increase in out of plane magnetization [89].

The Co-silica nanocomposite thin films synthesized by ion implantation of silica, when irradiated by 200 MeV I ions transform from spherical shape to prolate shape with an aspect ratio of about 7, at the fluence of $\sim 10^{14}$ ions cm^{-2} [90-91]. The Co nanoparticles embedded in silica transform from crystalline to amorphous phase under SHI irradiation which otherwise is not possible in bulk phase [92].

In another experiment, Ni particles of diameter below 10 nm did not change the shape whereas particles larger than this diameter transformed to prolate shape when the Ni-silica nanocomposite thin film (synthesized by atom beam co-sputtering and subsequent annealing), were irradiated by 100 MeV Ag ions. The particles of diameters more than 40 nm were unaffected [93]. These findings of ion beam induced shape transformation in case of FePt, Co and Ni particles embedded in silica are similar to the findings of ion beam induced shape changes in Au, Ag, Sn, Pt and Zn nanoparticles embedded in silica.

Ferromagnetism in Non-magnetic Nanomaterials

Ferromagnetism in non-magnetic materials is a subject of interest, which is also observed in nano dimensional systems. It is shown that embedded Ge nanostructures formed by ion irradiation of GeO_x show magnetic behaviour [94]. The Pd-C nanocomposite (Pd nanoparticles embedded in C matrix), synthesized by atom beam co-sputtering exhibits ferromagnetic behaviour and the saturation magnetization increases about twenty times, by 100 MeV Ag ion irradiation [95]. Since the size of Pd nanoparticles increase as a result of SHI irradiation, it is expected that due to decrease in the surface to volume ratio (because of increase in particle size) the ferromagnetic behaviour should decrease as it is considered to be occurring only in the nano dimensional system and not in the bulk. An experiment has shown that the ferromagnetic nature in the irradiated films increases due to the defects created by ion beam especially at the interface of Pd nanoparticle and the matrix [95].

Growth/Shape Change of Ge Nanoparticles in Silica

Spherical Ge nanoparticles (embedded in silica) can be transformed into disks and rods by SHI irradiation. Irradiation with low fluences (10^{14} ions cm^{-2}) of 38 MeV I^{7+} ions shapes medium-sized Ge nanospheres into disks, whereas smaller ones become rod-like and larger ones do not change. At higher fluences, the larger Ge nanospheres shrink due to Ge loss and also shape into disks [96].

Spheroidal silica SiO_2 and ZnS colloidal particles undergo transformation to oblate ellipsoidal shapes under 4 MeV Xe ion irradiation [97].

The formation of Ge crystallites as a result of swift heavy ion (SHI) irradiation was studied using Ge nanoparticles embedded in SiO_2 matrix prepared by atom beam co-sputtering of Ge and SiO_2 on Si substrate. The samples were annealed and irradiated with 100 MeV Au ions with fixed fluence. The pristine and irradiated samples were characterized by Raman, X-ray diffraction and atomic force microscopy (AFM). The nucleation and growth of Ge nanoparticles occur during the transient melt state of ion track of silica created by incident ion [98, 99].

Engineering the Optical Properties of Embedded Nanoparticles

ZnO quantum dots embedded in polyvinyl alcohol (PVOH) irradiated with 100 MeV Cl ions at different fluence ranging from 1×10^{11} to 1×10^{13} ions cm^{-2} show a red shift of the energy-gap parameter with respect to unirradiated quantum dots [100]. The narrowing of the energy gap of nanoparticles indicates particle growth under ion irradiation which is confirmed from transmission electron microscope images.

SHI (80 to 100 MeV O, Cl, and Au ion) irradiation of CdS embedded in PVOH matrix modifies the optical properties [101]. PL of O ion irradiated sample shows band edge emission, whereas band edge emission bleached in Cl ion irradiation. Au ion irradiation causes broad PL. Thus PL in CdS-PVOH nanocomposite can be tuned by electronic energy deposition.

SHI (80 MeV O ion) irradiation of Mn doped CdS nanoparticles embedded in polymer matrix results in alignment and elongation of nanotracks. PL of the irradiated film show tunablity with ion fluence [102]. SHI irradiation affects the photoluminescence properties of semiconductor-conducting polymer nanocomposites (SnO_2 nanoparticles embedded in polypyrrole). The PL intensity is significantly increased in the case of irradiated samples [103]. Grain boundaries act as colour centres and fragmentation of grains caused by SHI irradiation increases the density of these grain boundaries which enhances the PL emission intensity. The excitonic recombination is possible due to thermal detrapping of charge carriers owing to the localized temperature spike generated during SHI irradiation.

Making Poly(HFP) Layered Silicate Nanocomposites Radiation Resistant

Poly (vinylidene fluoride-co-hexafluoropropylene) (HFP) nanocomposites with layered silicate investigated with swift heavy ions in a wide range of fluence, and the nanostructure, crystalline structure, morphology etc. of the nanocomposites have been studied and compared critically with pristine polymer [104]. The nanoclay induces the piezoelectric β-phase in bulk HFP, and the structure remains intact upon SHI irradiation. SHI irradiation degrades pure polymer, but the degradation is suppressed radically in nanocomposites. The heat of fusion of pristine HFP gets reduced drastically upon SHI irradiation, whereas there are relatively minute changes in nanocomposites. The degradation has been considerably suppressed in nanocomposites through cross-linking of polymer chains, providing a suitable high-energy, radiation-resistant polymeric material.

Enhancement in Antioxidant Activity of PANI Nanofibres

Swift heavy ion irradiation increases the antioxidant activity and biocompatibility of polyaniline (PANI) nanofibres [105]. Irradiation leads to resizing of nanofibres. They undergo fragmentation and morphology changes. PANI nanofibres are known to have antioxidant activity. To improve their antioxidant activity, Kumar et al. doped PANI nanofibres with hydrochloric acid and camphor sulphonic acid (CSA), and irradiated with 90 MeV O ions. SHI irradiation decreased the diameter of HCl doped nanofibres from 40 nm to about 10 nm and that of CSA doped nanofibres from 50 nm to 15 nm. The enhancement in antioxidant activity of PANI nanofibres is due to the reduction in their size which increases their surface to volume ratio and conformational changes in the nanofibres.

5.5 TAILORING THE MECHANICAL PROPERTIES OF Si NANOROD STRUCTURES

Swift heavy ions provide a way to engineer the mechanical properties of nanostructures needed for MEMS/NEMS. Hardness of arrays of slanted and straight Cr nanorods on Si(100) substrates irradiated with 100 MeV Ag^{+8} ions at liquid nitrogen temperature increase with ion fluence. For slanted nanorods, the samples show an almost 300% increase in hardness as compared to their pristine counterparts. The corresponding increase in the case of straight nanorods is found to be 77% [106].

Ion irradiation induces anisotropic deformation of Si nanosprings [107]. Si nanosprings grown on Si (111) substrates and irradiated with 1.2 MeV Ar^{+8} ions at different ion fluences varying from 10^{15} ions cm^{-2} to 10^{17} ions cm^{-2} decrease in length and the width of their arms increases after the irradiation. The improved mechanical strength of irradiated Si nanosprings indicate that these coatings are capable of withstanding higher loads without getting delaminated. The role of inelastic electronic excitations and elastic nuclear stopping on the stiffness of Si nanosprings was investigated by using different ion beams at various energies such that the energy deposited per unit volume ranged from less than 1 eV nm^{-3} to ~keV nm^{-3}. For moderate values of energy densities, a logarithmic dependence of the stiffness change was observed [108]. The stiffness of the nanosprings increased up to 32% without any visible deformation due to densification of structures caused via electronic excitations induced by energetic ions. However, for very high energy densities (~MeV nm^{-3}), predominantly an anisotropic deformation of nanosprings initiated by nuclear energy losses occurred and the stiffness of the nanosprings was observed to increase up to 170%.

100 MeV Ag^{+8} ions have been used to study the modification in the mechanical properties of slanted silicon nanorods grown by glancing angle deposition technique [GLAD] on a patterned Si(1 0 0) substrate [109]. Atomic force microscope (AFM) based force–distance (F–z) spectroscopy was employed to measure the k-values of individual nanorods before and after irradiation. The average spring constant (k) of the nanorods was found to decrease to $65.6 \pm 20.8\ Nm^{-1}$ as compared to $174.2 \pm 26.5\ Nm^{-1}$ for pristine nanorods. Scanning electron microscope (SEM) micrographs show bending of the Si nanorods after irradiation. Raman and high-resolution transmission electron microscope (HRTEM) studies on pristine and irradiated Si nanorods showed the transformation of nanocrystalline regions present in pristine nanorods to amorphous phase on irradiation. This structural transformation along with the bending of the nanorods leads to a decrease in the spring constant of the nanorods. This technique offers a simpler possibility of tailoring the mechanical properties of a variety of structures such as nanorods, nanosprings and nanowires structures by ion beam irradiation.

5.6 SUMMARY

Nanostructuring of materials holds the key for engineering the properties of materials for specific applications in futuristic nanosystems. Present status of multifarious studies in this area points to the need to focus the attention to understand ion matter interactions in the nano region and find answers to some of the basic questions such as: How does the electronic energy loss (S_e) affect the surface morphology, shape, size and size distribution, mechanical, optical, magnetic properties etc. for nanodimensional system? There are some recent reviews and books on the aspects of synthesis and engineering of nano structures by energetic ions [110-115].

A summary of the possible ways of nanostructuring by SHI is given in Fig. 5.14. The electronic energy loss brings in nucleation and growth of the nanoparticles in semiconductor suboxides, metal suboxides, Si based gels and metal dissolved glasses. Specific feature of swift heavy ions of creating an ion track plays role in synthesizing nanowire-like structures. The ion tracks can be utilized in different applications. The nanoripples at surface can be created by swift heavy ion irradiation. The swift heavy ion irradiation of metal-silica nanocomposite can result in increase or decrease in particle size depending on the inter particle separation. If the embedded particles are larger than the track size in silica, the elongation of particle occurs whereas the particles larger than certain size (typically >40 nm) are unaffected. Contrary to the finding in metal nanoparticles, the Ge nanoparticles embedded in silica matrix gets flattened instead of getting elongated as a result of SHI irradiation. There are indications of getting aligned arrays of nanoparticles in specific cases of Ag and Fe in

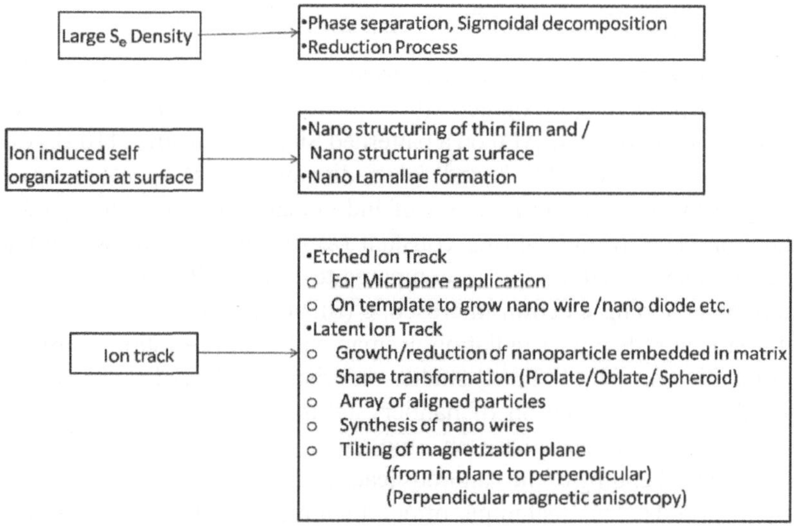

Figure 5.14: A summary of the possible ways (discussed in chapter) of nanostructuring by SHI is given.

silica. The shape transformation of metal and semiconducting nanoparticles embedded in silica (or any other matrix) is one of the most important aspects of role of ion beams in nanostructuring. Such a shape engineering of embedded nanoparticles in a matrix is not feasible by any other technique.

REFERENCES

1. M. Zaiser, Y. Lyutovich and F. Banhart, *Phys. Rev.*, **B 62** (2000) 3058.
2. S. Dhar, R.P. Davis and L.C. Feldman, *Nanotechnology*, **17** (2006) 4514.
3. J. Ghatak, B. Satpati, M. Umananda, P.V. Satyam, K. Akimoto, K. Ito and T. Emoto, *Nucl. Instr. and Meth.*, **B 244** (2006) 64.
4. Y. Batra, D. Kabiraj, S. Kumar and D. Kanjilal, *J. Phys. D: Appl. Phys.*, **40** (2007) 4568.
5. W.M. Arnoldbik, D. Knoesen, N. Tomozeiu and F.H.P.M. Habraken, *Nucl. Instr. and Meth.*, **B 258** (2007) 199.
6. W.M. Arnoldbik, N. Tomozeiu, E.D. van Hattum, R.W. Lof, A.M. Vredenberg and F.H.P.M. Habraken, *Phys. Rev.*, **B 71** (2005) 125329.
7. Shyama Rath, D. Kabiraj, D.K. Avasthi, A. Tripathi, K.P. Jain, Manoj Kumar, H.S. Mavi and A.K. Shukla, *Nucl. Instr. and Meth.*, **B 263** (2007) 419.
8. P.S. Chaudhari, T.M. Bhave, D. Kanjilal and S.V. Bhoraskar, *J. Appl. Phys.*, **93** (2003) 3486.
9. Y. Batra, T. Mohanty and D. Kanjilal, *Nucl. Instr. and Meth.*, **B 266** (2008) 3107.
10. Y. Batra, D. Kabiraj, S. Kumar and D. Kanjilal, *Surface and Coatings Technology*, **203** (2009) 2415.
11. Y. Batra, S.A. Khan, D. Kabiraj, S. Kumar and D. Kanjilal, *Nucl. Instr. and Meth.*, **B 266** (2008) 1697.
12. Neeti Tripathi, Shyama Rath, P.K. Kulriya, S.A. Khan, D. Kabiraj and D.K. Avasthi, *Nucl. Instr. and Meth.*, **B 268** (2010) 3335.
13. J.E. Hillard, *Trans. Met. Soc. ASME*, **242** (1968) 168.
14. J.C. Pivin, D.K. Avasthi, F. Singh, A. Kumar, E. Pippel and G. Sagon, *Nucl. Instr. and Meth.*, **B 236** (2005) 73.
15. J.C. Pivin, G. Roger, M.A. Garcia, Fouran Singh and D.K. Avasthi, *Nucl. Instr. and Meth.*, **B 215** (2004) 373.
16. B. Balamurugan, B.R. Mehta, D.K. Avasthi, Fouran Singh, Akhilesh K. Arora, M. Rajalakshmi, G. Raghavan, A.K. Tyagi and S.M. Shivaprasad, *J. Appl. Phys.*, **92** (2002) 3304.
17. E. Valentin, H. Bernas, C. Ricolleau and F. Creuzet, *Phy Rev. Lett.*, **86** (2001) 99.
18. Mukesh Kumar, P.G. Ganesan, V.N. Singh, B.R. Mehta and J.P. Singh, *Nanotechnology*, **19** (2008) 175606.
19. T. Mohanty, A. Pradhan, S. Gupta and D. Kanjilal, *Nanotechnology*, **15** (2004) 1620.
20. T. Mohanty, P.V. Satyam and D. Kanjilal, *J. Nano Sci & Nano Tech*, **6** (2006) 2554.
21. Madhavi Thakurdesai, A. Mahadkar, P.K. Kulriya and D. Kanjilal and Varsha Bhattacharyya, *Nucl. Instr. and Meth.*, **B 266** (2008) 1343.

22. Wolfgang Bolse, Thunu Bolse, Christian Dais, Derejee Etissa-Debissa, Ammar Elsanousi, Ando Feyh, Melih Kalafat and Hartmut Paulus, *Surface & Coatings Technology*, **200** (2005) 1430.
23. Thunu Bolse, Ammar Elsanousi, Hartmut Paulus, Wolfgang Bolse et al., *Nucl. Instr. and Meth.*, **B 244** (2006) 115.
24. W. Bolse, *Nucl. Instr. and Meth.*, **B 244** (2006) 8.
25. T. Bolse, H. Paulus and W. Bolse, *Nucl. Instr. and Meth.*, **B 245** (2006) 264.
26. R.S. Chauhan et al., *Vaccum* (Accepted).
27. F. Seitz and J.S. Kochler, *Solid State Phys.*, **2** (1956) 305.
28. Z.G. Wang, Ch. Dufour, E. Paumier and M. Toulemoude, *J. Phy. Cond. Matter*, **6** (1994) 6733.
29. M. Toulemonde, Ch. Dufour, Z. Wang, E. Paumier, *Nucl. Instr. and Meth.*, **B 112** (1996) 26.
30. H. Triankaus and A.I. Ryazanov, *Phy Rev. Lett.*, **74** (1995) 5072.
31. H. Triankaus, *Nucl. Instr. and Meth.*, **B 146** (1998) 204.
32. M.A. Grinfield and D.A. Nank, *SSSR*, **290** (1986) 1358.
33. D.J. Srolovitz, *Acta Mater*, **37** (1989) 621.
34. S. Klaumünzer and G. Schumacher, *Phy. Rev. Lett.*, **51** (1983) 1987.
35. S. Klaumunzer, Ming dong Hou and G. Schumacher, *Phy. Rev. Lett.*, **57** (1986) 850.
36. S. Klaumünzer and A. Benyagoub, *Phys. Rev.*, **B 43** (1991) 7502.
37. D.K. Avasthi, W. Assmann, H. Huber, H.D. Mieskes and H. Nolte, *Nucl. Instr. and Meth.*, **B 142** (1998) 117.
38. M. Kumar, S.A. Khan, F. Singh, A. Tripathi, D.K. Avasthi and A.C. Pandey, *Nucl. Instr. and Meth.*, **B 256** (2007) 328.
39. D.C. Agarwal, R.S. Chauhan, D.K. Avasthi, S.A. Khan, D. Kabiraj and Sulania, *J. Appl. Phys.*, **104** (2008) 024304.
40. J. Krauser, A.-K. Nix, H.-G. Gehrke, H. Hofsäss, C. Trautmann, A. Weidinger, F. Wünsch and J. Bruns, *Jl. Vacuum Science Techn.*, **B 26** (2008) 2468.
41. Jens-Hendrik Zollondz and A. Weidinger, *Nucl. Instr. and Meth.*, **B 225** (2004) 178.
42. J. Krauser, J.-H. Zollondz, A. Weidinger and C. Trautmann, *J. Appl. Phys.*, **94** (2003) 1959.
43. A. Tripathi, Amit Kumar, D. Kabiraj, S.A. Khan, V. Baranwal and D.K. Avasthi, *Nucl. Instr. and Meth.*, **B 244** (2006) 15.
44. Amit Kumar, D.K. Avasthi, A. Tripathi, D. Kabiraj, F. Singh and J.C. Pivin, *J. Appl. Phys.*, **101** (2007) 014308.
45. Amit Kumar, D.K. Avasthi, A. Tripathi, L.D. Filip, J.D. Carey and J.C. Pivin, *J. Appl. Phys.*, **102** (2007) 044305.
46. Amit Kumar, F. Singh, J.C. Pivin and D.K. Avasthi, *J. Phys. D: Appl. Phys.*, **40** (2007) 2083.
47. Amit Kumar, F. Singh, A. Tripathi, J. Pernot, J.C. Pivin and D.K. Avasthi, *J. Phys. D: Appl. Phys.*, **41** (2008) 095304.
48. S.K. Srivastava, D.K. Avasthi and E. Pippel, *Nanotechnology*, **17** (2006) 2518.
49. Amit Kumar, Fouran Singh, D.K. Avasthi and J.C. Pivin, *Nucl. Instr. and Meth.*, **B 244** (2006) 221.
50. S. Ghosh, V.V. Sivakumar, A. Tripathi, S. Khan, V. Ganesan, A. Gupta, A. Nath and D.K. Avasthi, *Nucl. Instr. and Meth.*, **B 248** (2006) 71.

51. Shubhra Kala, B.R. Mehta, S.A. Khan and D.K. Avasthi, *Applied Phys Lett.*, **90** (2007) 153121.
52. Lasse Vines, Edouard Monakhov, Bengt G. Svensson, Jens Jensen, Anders Hallén and Andrej Yu. Kuznetsov, *Phys. Rev.*, **B 73** (2006) 085312.
53. C.R. Martin and P. Kohli, *Nature Reviews Drug Discovery*, **2** (2003) 29.
54. J. Vetter and R. Spohr, *Nucl. Instr. and Meth.*, **B 79** (1993) 691.
55. D. Fink, A.V. Petrov, K. Hoppe, W.R. Fahrner, R.M. Papaleo, A.S. Berdinsky, A. Chandra, A. Chemseddine, A. Zrineh, A. Biswas, F. Faupel and L.T. Chadderton, *Nucl. Instr. and Meth.*, **B 218** (2004) 355.
56. A. Biswas, D.K. Avasthi, Benoy K. Singh, S. Lotha, J.P. Singh, D. Fink, B.K. Yadav, B. Bhattacharya and S.K. Bose, *Nucl. Instr. and Meth.*, **B 151** (1999) 84.
57. S. Amirthapandian, P. Magudapathy, B.K. Panigrahi, K. Saravanan, K.G.M. Nair and C.S. Sundar, *Surface and Coating Technology*, **203** (2009) 2687.
58. Vijayalakshmi Rao, J.V. Amar, D.K. Avasthi and R. Narayana Charyulu, *Rad. Measurements*, **36** (2003) 585.
59. Y.K. Vijay, Vaibhav Kulshrestha, K. Awasthi, N.K. Acharya, A. Jain, M. Singh, S.N. Dolia, S.A. Khan and D.K. Awasthi, *Jl. of Polymer Research*, **13** (2006) 357.
60. M. Grasselli and N. Betz, *Nucl. Instr. and Meth.*, **B 236** (2005) 501.
61. E. Ferain and R. Legras, *Nucl. Instr. and Meth.*, **B 208** (2003) 115.
62. C.R. Martin, M. Nishizawa, K. Jirage, M. Kang and S.B. Lee, *Adv. Mater.*, **13** (2001) 1351.
63. Sunita Rattan, Inderjeet Kaur, Nitika Gupta, Devesh Kumar Avasthi, Ambuj Tripathi and Saif Ahmad Khan, *Chemistry & Chemical Technology*, **3** (2009) 123.
64. S. Chawla, A.K. Ghosh, D.K. Avasthi, P. Kulriya and S. Ahmad, *J. Appl. Polym. Sci.*, **105** (2007) 3578.
65. A. Biswas, D.K. Avasthi, D. Fink, J. Kanzow, U. Schürmann, S.J. Ding, O.C. Aktas, U. Saeed, V. Zaporojtchenko, F. Faupel, R. Gupta and N. Kumar, *Nucl. Instr. & Meth.*, **B 217** (2004) 39.
66. Ajay Gupta and D.K. Avasthi, *Phys. Rev.*, **B 64** (2001) 155407.
67. Y.K. Mishra, V.S.K. Chakravadhanula, U. Schürmann, Hardeep Kumar, D. Kabiraj, S. Ghosh, V. Zaporojtchenko, D.K. Avasthi and F. Faupel, *Nucl. Instr. and Meth.*, **B 266** (2008) 1804.
68. Y.K. Mishra, D. Kabiraj, D.K. Avasthi and J.C. Pivin, *Rad. Eff. Def. Solids*, **162** (2007) 207.
69. Y.K. Mishra, D.K. Avasthi, P.K. Kulriya, F. Singh, D. Kabiraj, A. Tripathi, J.C. Pivin, I.S. Bayer and A. Biswas, *Appl. Phys. Lett.*, **90** (2007) 073110.
70. Y.K. Mishra, F. Singh, D.K. Avasthi, J.C. Pivin, D. Malinovska and E. Pippel, *Appl. Phys. Lett.*, **91** (2007) 063103.
71. Koichi Awazu, Xiaomin Wang, Makoto Fujimaki, Junji Tominaga, Hirohiko Aiba, Yoshimichi Ohki and Tetsuro Komatsubara, *Phys. Rev.*, **B 78** (2008) 054102.
72. Koichi Awazu, Xiaomin Wang, Tetsuro Komatsubara, JunWatanabe, Yuki Matsumoto, Shin'ichiWarisawa and Sunao Ishihara, *Nanotechnology*, **20** (2009) 325303.
73. G. Rizza, E.A. Dawi, A.M. Vredenberg and I. Monnet, *Appl. Phys. Lett.*, **95** (2009) 043105.

74. V. Rodrguez-Iglesias, O. Pea-Rodrguez, H.G. Silva-Pereyra, L. Rodrguez-Fernndez, G. Kellermann, J.C. Cheang-Wong, A. Crespo-Sosa and A. Oliver, *J. Phys. Chem.*, **C 114** (2010) 746.
75. E.A. Dawi, G. Rizza, M.P. Mink, A.M. Vredenberg, F.H.P.M. Habraken, *J. Appl. Phys.*, **105** (2009) 074305.
76. S. Roorda, T. van Dillen, A. Polman, C. Graf, A. van Blaaderen and B.J. Kooi, *Adv. Mater.*, **16** (2004) 235.
77. D.K. Avasthi, Y.K. Mishra, F. Singh and J.P. Stoquert, *Nucl. Instr. and Meth.*, **B 268** (2010) 3027.
78. P. Kluth, R. Giulian, D.J. Sprouster, C.S. Schnohr, A.P. Byrne, D.J. Cookson and M.C. Ridgway, *Appl. Phys. Lett.*, **94** (2009) 113107.
79. S. Klaumünzer, C.L. Li, S. Löffler, M. Rammensee, G. Schumacher and H.C. Neitzer, *Rad. Eff. and Def. in Solids*, **108** (1989) 131.
80. Fouran Singh, S. Mohapatra, J.P. Stoquert, D.K. Avasthi and J.C. Pivin, *Nucl. Instr. and Meth.*, **B 267** (2009) 936.
81. J.C. Pivin, F. Singh, Y. Mishra, D.K. Avasthi and J.P. Stoquert, *Surface and Coatings Technology*, **203** (2009) 2432.
82. L.A. Sweatlock, S.A. Maier, H.A. Atwater, J.J. Penninkhof and A. Polman, *Phys. Rev.*, **B 71** (2005) 235408.
83. M.C. Ridgway, P. Kluth, R. Giulian, D.J. Sprouster, L.L. Araujo, C.S. Schnohr, D.J. Llewellyn, A.P. Byrne, G.J. Foran and D.J. Cookson, *Nucl. Instr. and Meth.*, **B 267** (2009) 931 and M.C. Ridgway et al., *Phys. Rev. Lett.*, **106** (2011) 095505.
84. R. Giulian, P. Kluth, L.L. Araujo, D.J. Sprouster, A.P. Byrne, D.J. Cookson and M.C. Ridgway, *Phys. Rev.*, **B 78** (2008) 125413.
85. R. Giulian, F. Kremer, L.L. Araujo, D.J. Sprouster, P. Kluth, P.F.P. Fichtner, A.P. Byrne, M.C. Ridgway, *Phys. Rev.*, **B 82** (2010) 113410.
86. H. Amekura et al., International Conference on Ion Beam Modifications of Materials (2010) Canada.
87. F. Singh, D.K. Avasthi, O. Angelov, P. Berthet and J.C. Pivin, *Nucl. Instr. and Meth.*, **B 245** (2006) 214.
88. J.C. Pivin, S. Esnouf, F. Singh and D.K. Avasthi, *J. Appl. Phys.*, **98** (2005) 023908.
89. J.C. Pivin, F. Singh, O. Angelov and L. Vincent, *J. Phys. D: Appl. Phys.*, **42** (2009) 025005.
90. C.D. Orleans, J.P. Stoquert, C. Estournes, J.J. Grob, D. Muller, J.L. Guille, M. Richard-Plouet, C. Cerruti and F. Haas, *Nucl. Instr. and Meth.*, **B 216** (2004) 372.
91. C. D'Orleans, J.P. Stoquert, C. Estournes, C. Cerruti, J.J. Grob, J.L. Guille, F. Haas, D. Muller and M. Richard Plouet, *Phys. Rev.*, **B 67** (2003) 220101.
92. D.J. Sprouster, R. Giulian, L.L. Araujo, P. Kluth, B. Johannessen, N. Kirby, K. Nordlund and M.C. Ridgway, *Phys. Rev.*, **B 81** (2010) 155414.
93. Hardeep Kumar et al., *Nanoscale Research Letters* (Accepted).
94. Y. Batra and D. Kanjilal, *International Journal of Nanotechnology*, **6** (2009) 456.
95. P.K. Kulriya, B.R. Mehta, D.K. Avasthi, D.C. Agarwal, P. Thakur, N.B. Brookes, A.K. Chawla and R. Chandra, *Appl. Phys. Lett.*, **96** (2010) 53103.
96. B. Schmidt, K.-H. Heinig, A. Mücklich and C. Akhmadaliev, *Nucl. Instr. and Meth.*, **B 267** (2009) 1345.

97. E. Snoeks, A. van Blaaderen, T. van Dillen, C.M. van Kats, K. Velikov, M.L. Brongersma and A. Plman, *Nucl. Instr. and Meth.*, **B 178** (2001) 62.
98. N. Srinivasa Rao, S. Dhamodaran, A.P. Pathak, D. Kabiraj, S.A. Khan, B.K. Panigrahi, K.G.M. Nair, B. Sundaravel, J.C. Pivin and D.K. Avasthi, *Radiation Effects & Defects in Solids*, **164** (2009) 452.
99. N. Srinivasa Rao, A.P. Pathak, N. Sathish, G. Devaraju, S.A. Khan, K. Saravanan, B.K. Panigrahi, K.G.M. Nair and D.K. Avasthi, *Nucl. Instr. and Meth.*, **B 268** (2010) 1741.
100. D. Mohanta, S.S. Nath, N.C. Mishra and A. Choudhry, *Bull Mater Sci.*, **26** (2003) 289.
101. Dambarudhar Mohanta, Fouran Singh, D.K. Avasthi and Amarjyoti Choudhury, *Central Eur Jl of Phys*, **4** (2006) 187.
102. D. Mohanta, G.A. Ahmed, A. Choudhury, F. Singh, D.K. Avasthi, G. Boyer and G.A. Stanciu, *Eur Jl of Appl Phys*, **35** (2006) 29.
103. S. Samarah and A. Kumar, *Phy. Stat. Sol.*, **A207** (2010) 2279.
104. Vimal K. Tiwari, Pawan K. Kulriya, Devesh K. Avasthi and Pralay Maiti, *J. Phys. Chem.*, **B 113** (2009) 11632.
105. A. Kumar, S. Banerjee, J.P. Saikia and B. Konwar, *Nanotechnology*, **21** (2010) 175102.
106. Rupali Nagar, B.R. Mehta, J.P. Singh, D. Jain, V. Ganesan, S.V. Kesapragada and D. Gall, *Journal of Vacuum Science and Technology*, **A 26** (2008) 887.
107. Rupali Nagar, C. Patzig, B. Rauschenbach, V. Sathe, D. Kanjilal, B.R. Mehta and J.P. Singh, *J. Phys. D: Appl. Phys.*, **42** (2009) 145404.
108. Rupali Nagar, B.R. Mehta, J.P. Singh, C. Patzig, B. Rauschenbach and D. Kanjilal, *J. Appl. Phys.*, **107** (2010) 094315.
109. Rupali Nagar, D. Kanjilal, B.R. Mehta and J.P. Singh, *Nucl. Instr. and Meth.*, **B 267** (2009) 3617.
110. M. Toulemonde, C. Trautmann, E. Balanzat, K. Hjort and A. Weidinger, *Nucl. Instr. and Meth.*, **B 216** (2004) 1.
111. D.K. Avasthi and J.C. Pivin, *Current Science*, **98** (2010) 780.
112. D.K. Avasthi, "Ion beam in nanostructuring: An overview". *In:* Synthesis and engineering of nanostructures by energetic ions. Edited by D.K. Avasthi and J.C. Pivin (2011) Nova Science Publishers, USA.
113. Engineering thin film and nano structures ion beams. Edited by Emile Knystautas (2005) Taylor & Francis group.
114. J.C. Pivin, Use of ion beam to produce or modify nanostructure in materials. *In:* Nanomaterials: New Research (ed. Caruta, B.M.), Nova Science Publishers Inc (2005) pp. 81-113.
115. P. Mazzoldi et al., Metal Alloy nanostructures by ion implantation in silica. Edited by Emile Knystautas (2003) Taylor & Francis group.

6

Materials Engineering with Swift Heavy Ions

High energy heavy ions are proving to be important tools in the efforts to modify the properties of materials in a controlled fashion to provide possibility of making them functional for specific applications in planned developments towards futuristic technologies. Presently, research in this field is mostly of exploratory kind. Various types of materials are being irradiated with swift heavy ions and the effects of irradiation are being studied using the sophisticated characterization tools, such as XRD, RBS, ERDA, High resolution transmission microscopy, Photoluminescence spectroscopy, Fourier transform infrared and Raman spectroscopy, SQUID, AFM, SEM, etc.

The changes in the properties of the materials are being correlated with the ion beam parameters to provide information about the irradiation required to modify the desired property of the material. With the emergence of nano-technology, the importance of the ion beams in the engineering of the nanomaterials and producing nanostructures has been developing in several directions. On-line characterization of materials under irradiation have been used for quite some time. Now the efforts are to see whether we can use the characterization facility in dynamic mode to enable us to see the development of the required property of the material on-line and stop the irradiation when the required condition is achieved.

Efforts made towards the engineering of the properties of various types of materials using swift heavy ions are described here. There are several other types of materials where effect of high energy heavy ions in changing the properties has been studied. It has not been possible to cover all the materials.

6.1 CARBON

Introduction

Carbon based systems are fascinating materials of nature. They occur in several different forms such as graphite, diamond, fullerenes, carbon nano tubes (CNT), grapheme etc. Diamond-like Carbon (DLC) is amorphous in structure and contains both sp^2 and sp^3 bonded carbon. It has both diamond-like and graphitic properties depending upon the ratio of sp^2 to sp^3. A new molecule Buckminster fullerene (C_{60}), discovered in 1985 [1], is a molecular cage made of 60 identical carbon atoms which appears like a ~7 Å soccer ball. Multi-walled carbon nanotubes (MWNTs) were discovered in 1991 [2], followed by the observation of single-walled carbon nanotubes (SWNTs), i.e., a very long cylindrical cage of carbon with only a single wall of C-atoms [3]. Carbon also exists in several other forms [4-7]. A new member of the family, graphene is a sheet of crystalline carbon just one atom thick. In diamond each carbon atom is bonded to four others, and all four electrons in its outer shell are involved in joining the atoms together, making it a good insulator. In contrast, graphene has one electron left over when each carbon atom bonds with three others, thus making it an electrical conductor.

Controlled Formation of Conducting Nanowires

The controlled growth of a single nanowire or an ensemble of nanowires, growth direction, and their suitable alignment on a substrate is of paramount importance for nanotechnology applications. Conducting carbon nanowires in fullerene, diamond-like carbon films and Si based polymer films have been demonstrated using ion beams [8-16]. Swift heavy ions provide unique way to control the growth because they deposit large electronic energy in a confined cylindrical zone in a controlled fashion, thus inducing modifications in the materials in nanometre channels of controllable diameter. These narrow cylindrical zones (ion tracks) have properties that are different from the surrounding virgin material.

Thin film of fullerene (C_{60} and C_{70}) transforms to amorphous carbon (a:C) and graphitic carbon under ion irradiation. At lower fluence (10^{10} ions cm^{-2}), SHI irradiation of fullerene thin films results in the formation of conducting C nanowires embedded in fullerene matrix [11, 12]. It is because each ion creates a track within which the fullerene molecules are transformed into conducting form of carbon. At low fluence these nanocylinders are well isolated and each of them is conducting. Thus, one can have the conducting nanowires embedded in fullerene film, simply by the irradiation of the films with SHI. The ions providing higher value of electronic energy (S_e) create conducting wires with larger diameter.The conductivity of carbon nanowire such formed is about 7 orders of magnitude more than the conductivity of virgin C_{60} film. The nanowires are parallel to each other and their length can be tuned by the thickness of fullerene film. Further, the orientation of nanowires can be changed by simply changing the incidence angle of ion beam. The formation of conducting C nanowires also occurs in SHI irradiation of DLC films [8].

The field emission from these conducting carbon nanowires of diameter 40 to 120 nm in C_{60} matrix has been investigated by Amit Kumar et al. [13]. Figures 6.1 and 6.2 show the schematic of the nanowire formation and the results of the field emission measurements. A threshold field of 9 V/μm is observed for field emission in these nanowires. The pristine fullerene film shows breakdown at 51 V/μm. Since the ion beam hits at the film are random, the nanowires are randomly located in the film. However, if the SHI irradiation is performed using a mask, the nanowires with regular and desired spacing can be synthesized.

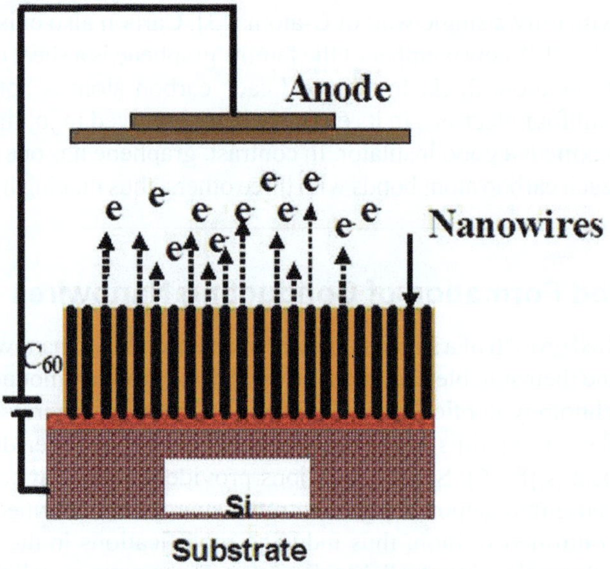

Figure 6.1: The schematic sketch of field emission measurements is shown. Dark black lines are C nanowires formed by SHI irradiation of fullerene. [Reprinted with permission from Amit Kumar, D.K. Avasthi, A. Tripathi, L.D. Filip, J.D. Carey and J.C. Pivin, *J. Appl. Phys.*, **102** (2007) 044305, copyright (2007), American Institute of Physics.]

The carbon nano-dot and nano-wires synthesized using energetic ion beams in inorganic polymers/gels have been studied using high resolution Transmission microscopy [14]. The diameter of the ion track (which is conducting nanowire in present case) is also determined using online ERDA data on H release, which is in agreement with the TEM results. The photoluminescence in the C nanodots samples is maximum at a particular fluence. The nature of the formed carbon nanostructures, which have amorphous sp^3 nature (-C:H, hydro generated carbon) was investigated with the Fourier transform infrared and Raman spectroscopy [15-17].

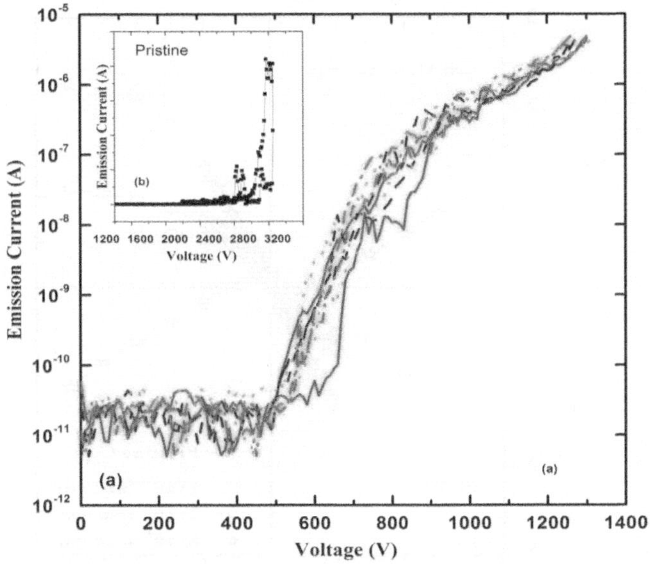

Figure 6.2: The field emission measurements for the synthesized C nanowires are shown. Inset shows the breakdown of the pristine film at high field. [Reprinted with permission from Amit Kumar, D.K. Avasthi, A. Tripathi, L.D. Filip, J.D. Carey and J.C. Pivin, *J. Appl. Phys.*, **102** (2007) 044305, copyright (2007), American Institute of Physics.]

Defect Mediated Magnetism in Carbon Based Materials

Magnetic moment formation in the solids is associated with the elements containing partially filled 3*d* or 4*f* sub shells. However, there has been increasing evidence that localized defect states (or surface or edge states) in *sp* materials may form local moments and exhibit magnetism. The study of magnetization of fullerene films irradiated with 90 MeV Au ion at different fluences indicated [18] ferromagnetic ordering induced by ion irradiation. The pristine film shows a diamagnetic behaviour, and the ferromagnetic contribution increases with increasing ion fluences shown in Fig. 6.3 [18]. The experimental findings of Amit et al. [18-20] clearly show that the magnetism in carbon based materials is mainly defect mediated, which provides information on the controversial topic "origin of magnetism in carbon phases".

It is known that the fullerene molecules at high temperature and pressure form dimmers. This process is also referred to as polymerization. The ferromagnetic behaviour was shown in dimerized fullerene [21]. Since the dimmer formation was observed in SHI irradiated fullerenes (C_{60}) at low fluence, it was expected that fullerenes irradiated at low fluence will be ferromagnetic in nature. Indeed the SHI irradiated fullerene showed the ferromagnetism. It was found that the ferromagnetism increased more at higher fluence where dimmer formation vanishes which indicated that the ferromagnetism is not due to fullerene dimmers. It was proposed to be due to

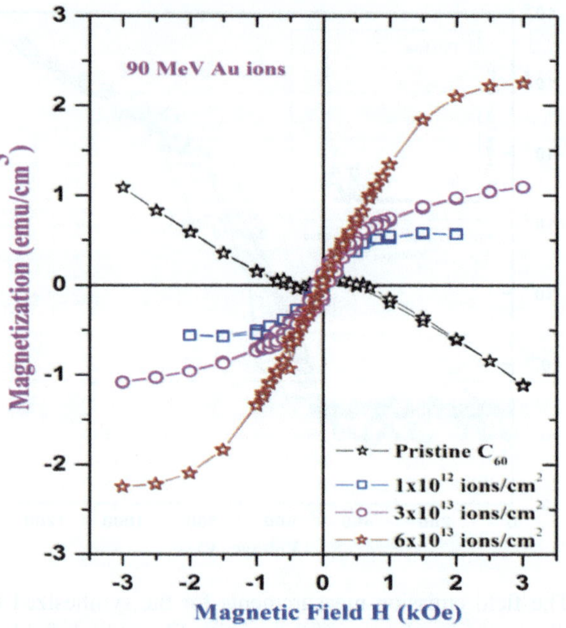

Figure 6.3: The magnetic behaviour of the pristine and irradiated C_{60} films are shown. [Reproduction from Amit Kumar, D.K. Avasthi and J.C. Pivin, *Appl. Phys. Express*, **1** (2008) 125002 with permission from Japan Society of Applied Physics.]

some ordering of sp^2 and sp^3 bonded atoms having the correlation between unpaired spins [20]. Increase in O was observed in irradiated sample, indicating that the trapping of O atoms could also be a possible cause of ferromagnetic behaviour.

The ferromagnetism is also observed in low energy ion irradiated graphite and nanocrystalline diamond [22, 23]. There have been calculations [24, 25] predicting the magnetic ordering in graphene but there is no experimental evidence so far. It is felt that the controlled defect creation in graphene layer can be achieved by ion irradiation, but the challenge is to detect the small magnetic signal.

Ordering of Carbon Nanostructures

SHI irradiation of carbon nanotubes at low fluence ($<5 \times 10^{11}$ ions cm^{-2}) results in the ordering of carbon nanostructures but destruction at higher fluence [26]. The degree of modification/damage in irradiated CNTs is usually quantified by a disorder parameter $a = I_D/I_G$, where I_D and I_G are the integrated intensity of the D (Distorted Graphite) and G (Graphite like) modes respectively. Figure 6.4 shows the first order Raman scattering modes of pristine and 60 MeV Ni ion irradiated MWCNTs. The inset of the figure shows the variation of disorder parameter (I_D/I_G) as a function of ion fluence. It is evident from the figure that the disorder parameter decreases at low fluence and increases at high fluence.

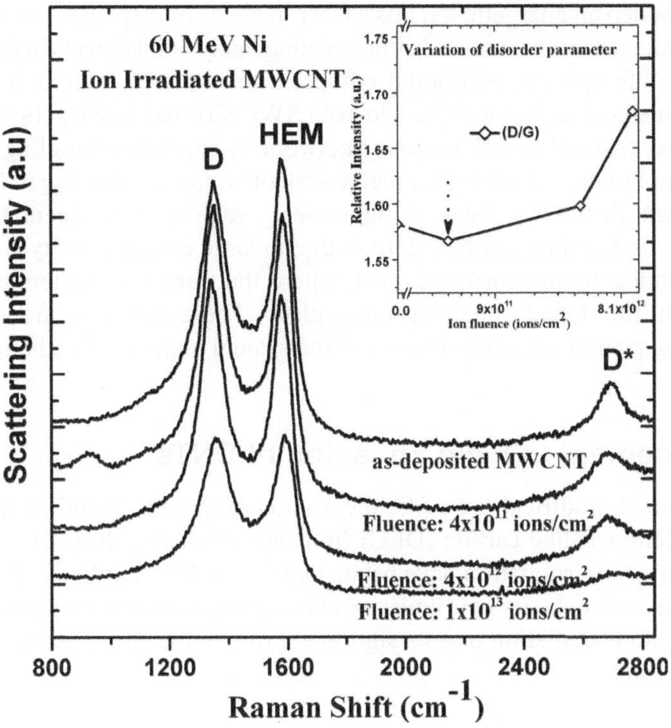

Figure 6.4: First order Raman scattering modes of pristine and 60 MeV Ni ion irradiated MWCNTs are shown. The inset shows the variation of disorder parameter (I_D/I_G) with fluence. [Reprinted with permission from Amit Kumar, D.K. Avasthi, J.C. Pivin and P.M. Koinkar, *Appl. Phys. Lett.*, **92** (2008) 221904, copyright (2007), American Institute of Physics.]

Low disorder parameter, for low fluence ion irradiated CNTs indicates the ordering of CNTs. The ordering or purification of CNTs at low fluence irradiation can be explained by energy dissipation process in ion-solid interaction.

Stability of CNTs both single-wall CNTs (SWCNTs) and multiwalled CNTs (MWCNTs) under irradiation of a 55 MeV carbon ions analyzed using Raman spectroscopy shows interesting phenomenon of healing or annealing of CNTs under ion beam irradiation. The annealing process appears to begin at the lowest value of fluences and persists for quite a good range of fluence values. As the irradiation fluence increases to ~1 × 10^{14} ions cm^{-2} the MWCNTs begin to amorphize whereas the SWCNTs continue to heal [27]. At low values of S_e the defect annealing continues till high fluences whereas the defect annealing at high S_e values starts vanishing at lower fluence, nevertheless, it does take place. Similar irradiation effects are observed on multiwalled CNTs (MWCNTs) with swift heavy Au ions of energy 120 MeV [28].

Raman results indicate that the net effect on the system is a result of two competing processes occurring simultaneously—defect production and defect healing. When an energetic ion hits a CNT, it ejects or displaces one or several C atom from the atomic network, thus creating many point defects in the system. In MWCNTs system, additional complex defects like intershell covalent bonding also occur. Point defects in both SWCNTs and MWCNTs are healed through reconstruction of the atomic network by saturating dangling bonds of the system but the complex intershell covalent bonding occurring in MWCNT systems are difficult to heal. In high energy regime, it is speculated that a narrow cylindrical region gets defects due to large energy density whereas in larger annular region surrounding it, where the transient temperature is not very high, the defect annealing takes place. These results demonstrate the interesting phenomenon of ordering of the system under ion irradiation at low fluences.

Enhancement in Field Emission of CNTs

Improvement/modifications of field emission properties in carbon nanotubes (CNT), diamond-like carbon (DLC) films and nanocrystalline diamond film under heavy ion irradiation have been studied [29-32]. It is found that the ion irradiation leads to better field emission characteristics and the structural damage caused by ion irradiation plays a significant role in the emission behaviour of diamond films.

Reorientation of Ni Planes in Ni Encapsulated by CNT

Reorientation of the crystalline planes occurs in confined single crystal nickel nanorods in carbon nanotubes under SHI irradiation [31, 33, 34]. The effects of irradiation are less when the numbers of tube walls are large. It reveals that the multi walled carbon nanotubes can be effectively used as radiation resistant coatings.

Figure 6.5 shows the high resolution TEM images of pristine and 100 MeV Au ion irradiated MWCNT filled with Ni. Figure 6.5 (a and b) shows the pure/crystalline walls of the unirradiated CNTs and the nickel planes with lattice spacing of 1.96 ± 0.04 Å. Figure 6.5 (c) is after irradiation at fluence 5×10^{12} ions cm^{-2}. The irradiation leads to the change in inclination of nickel planes with tube axis. The damage on the tube walls and reorientation of Ni planes making angles of 15° and 30° at two consecutive locations for fluence of 5×10^{12} ions cm^{-2} are shown in Fig. 6.5 (d). At a fluence of 3×10^{13} ions cm^{-2} [Fig. 6.5 (e)] the Ni inside CNT is completely amorphous. The 100 MeV Au ions irradiation induces an electronic energy loss of 32.5 keV/nm, which is not sufficient to damage the nickel lattice in bulk whereas it is shown here that Ni is completely amorphized when it is in nanodimensional form. This could also be due to formation of NiC which can be amorphized.

The simulation of response of CNT's under extreme conditions has been useful to provide an insight to the effects of ion beams on CNT discussed in detail along with other interesting aspects in recent review [35].

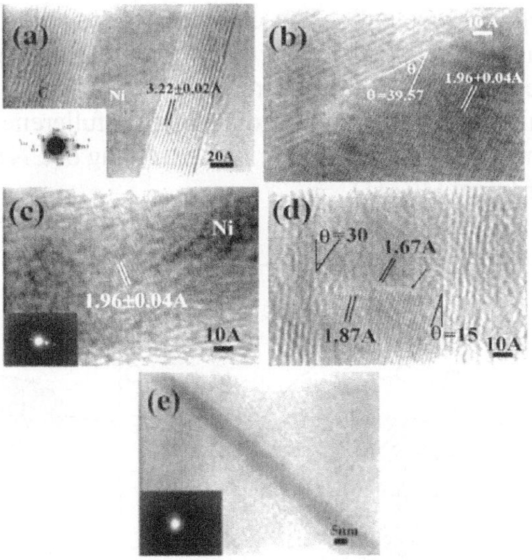

Figure 6.5: (a) shows the pure/crystalline walls of the unirradiated CNTs. (b and c) show the pristine and film irradiated at fluence 5×10^{12} ions cm^{-2} and the lattice spacing between nickel planes (1.96±0.04 Å). The Ni planes make an angle of 39.6° with respect to the CNT planes. (d) shows the irradiation (at fluence of 1×10^{12} ions cm^{-2}) induced changes in inclination of nickel planes with tube axis, making angles of 15° and 30° at two consecutive locations. (e) shows completely amorphous Ni nanorod at a fluence of 3×10^{13} ions cm^{-2}. [Reprinted with permission from Abha Misra, Pawan K. Tyagi, Padmnabh Rai, D.S. Misra, Jay Ghatak, P.V. Satyam and D.K. Avasthi, *Appl. Phys. Lett.*, **89** (2006) 091907, copyright (2006), American Institute of Physics.]

Modifications of the Fullerene Molecules (C_{60} and C_{70})

Study of polymerization of irradiated C_{60} films shows the formation of dimmer of fullerene (polymerization) [36-39] at low fluences and that fullerene molecules are destroyed at high fluences. Raman spectroscopy is very effective tool in characterizing various allotropes of C. Perfect crystalline graphite has characteristic sharp Raman vibrational mode at 1588 cm^{-1}. The Raman modes for a-C are distinctly different from that of graphite and appear as two broad features as D (distorted graphite) and G (graphite-like) bands at 1350 cm^{-1} and 1500 cm^{-1}, whereas Raman mode signatures for C_{60} and C_{70} appear dominantly at 1467 cm^{-1} and 1567 cm^{-1}.

Each incident ion produces a cylindrical zone with high temperature beyond a S_e threshold of the material. The molecules within this cylindrical region get vapourized, due to low sublimation temperature of fullerene (~700 K). The quenching of the vapourized state results in transformation of fullerene to a-C. While the fullerene in cylindrical region is in vapour state, the surrounding annular region experiences a high temperature and pressure. The high pressure in annular region is due to pressure exerted by the vapours in cylindrical region (due to volume increase resulting from change of state from solid to vapour).

The annular region surrounding the cylindrical ion track undergoes transformation to dimmer, as shown in Fig. 6.6. It is well established that fullerene molecules under high pressure and temperature form dimmer. The formation of dimmer can be considered as welding of two fullerene balls under the influence of ion beam, which is similar to the welding of CNTs under electron and ion bombardment [40, 41].

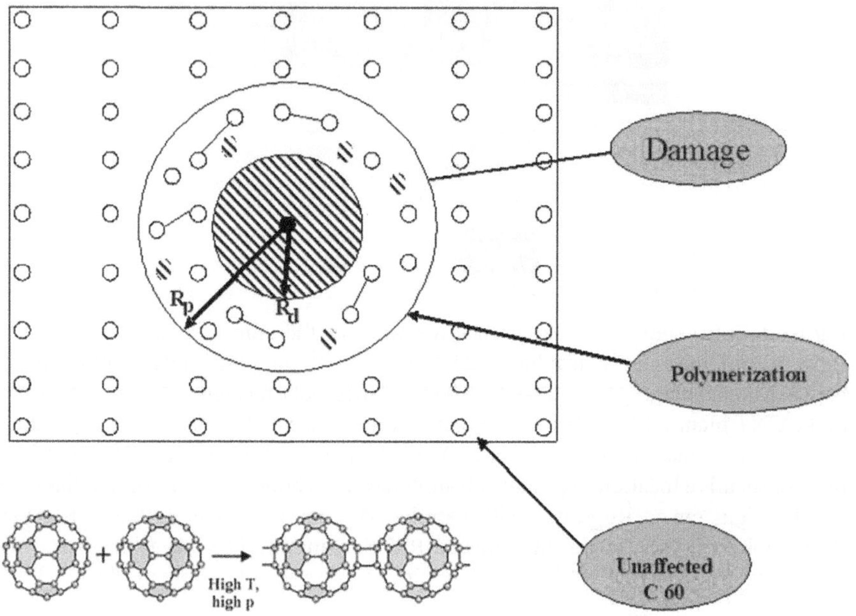

Figure 6.6: The annular region with radius (R_p-R_d) surrounding the cylindrical ion track radius (R_d) is shown which undergoes transformation to dimmer. The schematic below the figure shows the formation of dimmer when two fullerene molecules are together at high temperature and pressure. [Concept from Ref. 38.]

Structural phase transitions in low temperature resistivity measurements and optical molecular transitions using photoluminescence spectroscopy and modifications of these under heavy ion irradiation have been investigated [42, 43]. The 90 MeV Si ion irradiation at low fluence (~10^{12} ions cm^{-2}) results in enhancement of PL emission, whereas the PL intensity decreases at high fluence (~10^{13} ions cm^{-2}). Most of the optical molecular transitions could be observed in PL of the sample irradiated at ~10^{12} ions cm^{-2} due to significant enhancement in PL [43]. The UV-Vis spectrum of this particular sample also reveals the molecular transition, which are not so clear in the pristine sample.

The amorphization and polymerization of C_{60} molecules by low energy ion irradiation has been investigated in detail [44]. Modifications of properties of C_{70} thin films following ion beam irradiation with 120 MeV Au ions have been investigated. The energetic ion impacts lead to the destruction of the C_{70}

molecules. The radius of damaged cylindrical zone is about 2.9 nm. The resistivity for pristine C_{70} is of the order of 105 ohm-cm, which decreases with increasing ion fluence. Measurements show a steep increase in the conductivity of irradiated fullerenes beyond a fluence of ~10^{12} ions cm^{-2} which is ascribed to the transformation of C_{70} into a-C within each ion track, with higher conductivity than the surrounding fullerene matrix. At a fluence of 3×10^{13} ions cm^{-2}, the molecule is transformed into a-C [45]. A study on the comparison of behaviour of C_{60} and C_{70} under the SHI impingement revealed that C_{60} fullerenes form dimmer whereas C_{70} fullerenes do not form dimmer under SHI bombardment [46].

Modification in DLC and Graphite

The DLC film can be prepared by microwave discharge of acetylene-hydrogen and methane-hydrogen gas mixture. The effects produced by SHI irradiation in these two type of films are different [47], due to different microstructure of the film. The formation of diamond and C onions occurs in the irradiation of graphitic carbon soot by 3 MeV Ne^+ ions at different fluences [48]. Discontinuous ion tracks are formed in SHI irradiation of HOPG [49]. Hillocks and craters are formed in SHI irradiation of graphite [50-52]. Strong signature of formation of CNT also exists as a result of SHI irradiation of graphite [50, 51].

The behaviour of thin films of different form of C under extreme conditions created by ion beam, have been discussed in review articles [35, 38, 41, 53].

It is expected that the techniques connected with ion beams will help in the following directions:

- Controlled formations of conducing channels for field emission devices in fullerene matrix.
- Defect mediated magnetism in carbon based nanostructures.
- Electronic properties of graphene are not fully understood in terms of scattering mechanism. Controlled defect creation using ion beam impacts and simultaneously electrical transport measurement will be interesting to understand the defect induced scattering mechanism in low dimensional systems.
- It will be interesting to investigate the ion track formation in single atom thick carbon sheet, in insulating and metallic state. In metallic case (when doping is higher or Fermi level is far from Dirac point), graphene sheet should work as ion resistant (as energy dissipation will be much faster as carrier dynamics is relativistic and no track formation should occur) therefore graphene can be used as radiation resist material.
- It is proposed [54] that fullerene or DLC film can be made with a few percent of noble metals dissolved in it. The SHI irradiation of such a system may form small metal clusters in the ion track and enhance the conductivity of the conducting ion track.

- CNT has been in discussion for bending the energetic ion beam in channeling condition, with a possibility of replacing huge and expensive magnets used in bending the ion beam in accelerator laboratories [55, 56].

6.2 POLYMERS

Introduction

Polymers are highly radiation sensitive materials. Ion beams can modify molecular structure in polymers leading to changes in their chemical, electronic, electrical, tribological and optical properties in a tunable way. Ionization trail produced by swift heavy ions (SHI) causes bond cleavages and produce free radicals. These are responsible for most of the chemical transformations in polymers: chain scission, cross linking and bond formation. The structural modifications in polymers depend on the electronic energy loss (S_e) and ion fluence. Bond cleavage and chain scission leads to formation of alkyne and allene groups and this occurs only under SHI irradiation above a S_e threshold [1]. Beyond a threshold S_e value SHI produces zone of reduced density along its trajectory (ion track). What happens to the material inside the track depends critically on the primary parameter, S_e.

The electronic energy loss along the trajectory of the particle occurs discretely, not continuously because of the quantized energy levels and certain potential barrier for excitation and ionization processes. Such discrete energy loss [2] is called a '*spur*'. The spur energy for polymers is of the order of 30-40 eV, which is the energy required to produce one ion or radical pair. As a consequence, a significant volume around impinging ion is affected and results in the formation of several active chemical species like cations, anions, radicals, secondary electrons, etc. along the polymeric chains. The Coulomb forces, among these vibrant species, introduce segmental motion and vibrant bond stretching resulting in the bond-cleavage as well as cross-linking in the polymer chains. In the case of lower values of linear energy transfer (LET) by the ion, the spurs develop far apart and remain independent and the deposited energy tends to be confined in one chain leading to chain-scission. With increasing LET, spurs may connect and overlap, resulting in a high radical concentration gradient. Thus, the probability for two radical pairs to be in the neighbouring chains is enhanced and the cross-linking is facilitated. The nuclear energy loss, which displaces atoms from their original location, can lead to the chain-scission. In this case, the possibility of cross-linking is small because of the fact that nuclear energy loss is an independent damage and the probability of simultaneous creation of two radicals in adjoining chains is small.

Both nuclear and electronic energy losses lead to chain-scissioning and cross-linking during ion irradiation. However, the relative magnitude of chain-scissioning and cross-linking depends on the structure of polymer and the nature of the incident ion parameters. Such changes in the structure of the polymers are responsible for the changes in their optical, electrical, mechanical properties, etc.

The properties of polymers can be engineered as per need with ion–polymer interaction by correlating the induced changes in the properties of the polymers with the energy loss process. This requires the knowledge of energy loss in polymers and associated fluctuations. These have been measured for various thicknesses of polypropylene to establish the validity of the energy loss and straggling formulations [3].

A comparative study of various stopping power tables and codes for heavy ions in polymers has been made by comparing the computed stopping power values with the corresponding experimental values [4]. The energy loss of different heavy ions in the energy domain of ~0.19–3.14 MeV/n has been studied in different polymers used as particle track detectors viz., Mylar (Polyethylene terephthalate), PEN (Polyethylene napthalate), Polycarbonate, CR-39 (Polyallyl diglycol carbonate), Kapton (Polypyromellitimide) and LR-115 (Cellulose nitrate) to compare with the simulated calculations based on available computer codes. There have been extensive measurements for various types of ions in different types of polymers such as: Polypropylene, Mylar, Polycarbonate, Kapton, etc. covering a wide energy range of swift heavy ions [5-7].

The Latent Track and Etched Track

Swift heavy ion passing through an electrical insulator produces a narrow region of radiation effected material, known as a 'latent track'. The process of track formation can be viewed as follows. A positively charged particle knocks out the orbital electrons of the atoms lying along its trajectory producing cylindrical region full of positive ions. The positive ions so formed produce a more or less cylindrical modified-region which can be seen with a transmission electron microscope [8]. The use of an 'etching process' is made to develop them as enlarged version of the original damaged trail which can then be seen easily under an ordinary optical microscope. Etched track from a few tens of nm to a few μm can be made by controlling the etching time.

Etched tracks can have conical, cylindrical or funnel-like shapes. These can be used for various applications by depositing materials within the etched tracks, such as metals, semiconductors, liquids or solid electrolytes. The material deposited within the etched tracks can have various structures such as single crystalline, polycrystalline, or amorphous, and one can have dispersed nanoparticles or nanoparticle/polymer composites. Etched track membranes can be filled with materials having sensitivity to temperature, pH value, electric field etc. This provides immense possibilities of making sensitive sensors and efficient control devices.

With high energy heavy ion beams it is possible to adjust the energy deposited per unit length (S_e) by choosing an ion and its energy, and then with controlled etching one can produce micro-channels which provide the nuclear track filters [9] capable of filtering bacterias. A moderate level of S_e is preferable for achieving tracks in polymers [10]. Efforts are now being made to use these channels to provide devices or sensors. Electro-deposition of pores of 800 nm with copper and selenium was successfully done to produce diodes [11]. It is now giving rise to a development of a new type of nanoelectronics based on swift-heavy ions. These novel structures, denoted by the acronyms "TEAMS" (tunable electrically anisotropic material on semiconductor) and "TEMPOS" (tunable electronic material with pores in oxide on semiconductor), may exhibit properties of tunable electronic components sensors and transistors [12-15]. Behaviour of the devices fabricated using the TEAMS/TEMPOS structures depends upon the track dimension, their density, type of material embedded in the tracks and the contact geometry. Devices like gateable resistors, capacitors, diodes, transistors, sensors (light, temperature, humidity, ammonia and alcohol), amplifiers, oscillators and logic gates have been made using these structures. The TEMPOS structures have been used to obtain highly sensitive sensors of ammonia and humidity by filling in polymer electrolytes and their composites in the ion tracks [16, 17]. The nano-porous polymers are also used for gas separation, ion separation, controlled drug delivery and low cost purification process [18].

The micro filter can be used for slow drug delivery at suitable places for medical treatment. Normally, a drug is given at regular interval, say once a day. Controlled release of the drug or medicine through the micro filter of optimum size offers the possibility of providing the drug or medicine continuously through its pores at the infected region. Controlled drug release through micropores has been demonstrated by Rao et al. [19].

Micro filters also find applications in the biomedical field in filtering the bacteria of specific dimensions. Normally, in filter application, the micro holes get clogged. The clogging can be removed by expansion of these micro holes to use them again. Such a possibility has been demonstrated by coating the inner walls of the micro holes with gel [20]. The gel expands at elevated temperature of 80°C, which allows control on the size of micro holes. A similar possibility of variable micron size pores exists in piezoelectric polymer materials, where control on the pore size can be achieved by applying an electric field. Determination of the dimensions of the ion damaged zone is of interest from the application point of view and for the understanding of the ion-polymer interaction.

Measurement of the Ion Track Radius

The track diameter is a quantity of interest for the understanding of basic ion insulator interaction. It can be measured by scanning force microscopy and other state-of-the-art surface morphology probing equipments. A novel approach was used to determine the track diameters in polymers by online measurement of hydrogen loss during ion irradiation by on-line elastic recoil detection analysis (ERDA) [21]. Hydrogen is liberated due to electronic excitation of constituent atoms causing the breaking of bonds associated with H. Free H atoms combine to form hydrogen molecules. These molecules, being the lightest gaseous molecules having high diffusivity, escape from the polymer causing reduction in H content due to ion irradiation. Thus the incident ion along its path releases H and it diffuses out of the polymer as hydrogen molecules. Each ion is effective over a larger area than its own size, releasing H from a cylindrical zone of damaged polymer, whose radius can be determined. The H concentration in the film decreases due to the loss of H from the cylindrical damage zone. Thus the concentration of H in the polymer decreases with the fluence, which is measured on-line by ERDA.

On Line Analysis of the Track with Quadrupole Mass Analyzer

Apart from the measurement of ion track dimensions it has been possible to look into varying damage zones within an ion track by on-line measurements using quadrupole mass analyzer [QMA]. A thin (20 mm) polymer film when irradiated with 200 MeV Ag ions ($S_e \sim 10$ keV/nm), produces damage along its path. Due to high electronic excitation, various gases are formed within the ion damage core. These gases diffuse out, resulting in an increase in the partial pressure of the gases, which reaches a saturation value and decreases, when the overlap of the track diameter begins. This decrease in the partial pressure is because of the fact that the amount of gas release is decreased when the ion goes into an already ion-damaged track zone. The decay of the partial pressure of the gases with the ion fluence thus gives an idea of the cylindrical zone [22] through which the gases are released. The partial pressure measurements were performed using QMA. The changes in the partial pressure of gases H_2, CH_4 and CO were measured as a function of ion fluence. The estimation of damage cross-section by partial pressure measurement is similar to that done by on-line ERDA. The only difference is that ERDA measures the H left behind in the sample whereas in the partial pressure measurement technique, the hydrogen coming out of the sample is measured. It was observed [22] that the release of different gases followed different curves indicating that different gases get released from the cylindrical zones of different diameters as shown in Fig. 6.7.

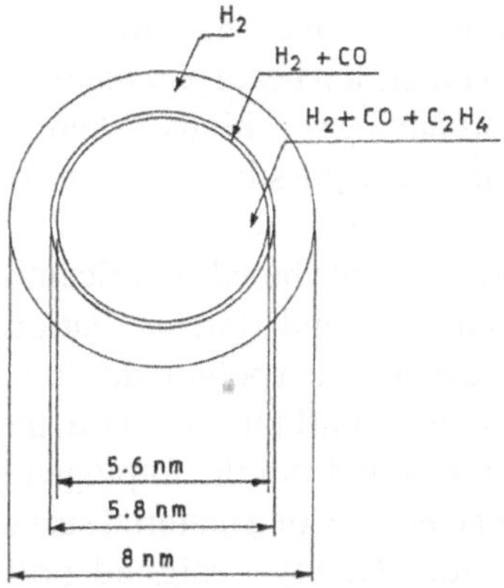

Figure 6.7: Damage zones in polymer, from where the evolution of various gases takes place under the impact of swift heavy ions. [Reprinted from D.K. Avasthi, J.P. Singh, A. Biswas and S.K. Bose, *Nucl. Instr. and Meth.*, **B 146** (1998) 504 with permission from Elsevier.]

The inner damage core produced three different gases whereas the outer zone gave only hydrogen. This can be understood from the fact that the influence of the ion reduces with the radial distance and therefore more gases are evolved from the inner zone (of diameter 5.6 nm) having a larger damage. In this particular case, the damage by the incident ion beyond a radial distance of 5.8 nm is so small that only H is evolved in the outer zone of 5.8 nm to 8 nm. The effect of the impinging ion beyond a radial distance of 8 nm is negligible and therefore no gas is evolved beyond this distance.

Modification of the Polymers Properties with Ion Beams

Ion beams modify electronic, electrical, tribological and optical properties. In certain polymers new carbonaceous material is formed with enhanced electrical conductivity and at the same time the optical absorption of material also increases. Enhanced electrical properties of as much as eight order of magnitude have been observed [23].

Irradiation can enhance the adhesion property [24], which improves subsequent metallization of polymer with elements such as copper or aluminum by an order of magnitude. Ion beam surface modification has shown great potential for improving the tribological behaviour as well as surface mechanical properties of polymeric materials. The wear resistance of ultra high molecular weight polyethylene used for hip joints replacements shows dramatic improvement following ion beam bombardment [25].

Heavy ions increase reflective index to permit optical guidance in the irradiated layer. It is observed that the variation in reflective index depends on the parameters S_e and fluence and not on doping effect. Change of index of refraction with S_e has been measured in poly-methyl-meth-acrylate (PMMA) when irradiated with 5.6 Mev He ions [26]. There are possibilities of engineering waveguides with ion beams and optimizing and controlling the loss of channelling but there is need to investigate the nature of damage created by ions and its correlations with the ion beam parameters. Etched track membranes can be filled with materials having sensitivity to temperature, pH value, electric field etc. This provides immense possibilities of making sensitive sensors and efficient control devices.

Polymeric materials have contributed significantly to the development and improvement of medical devices and systems. Various interactions take place between biomaterials and a body when polymeric biomaterials come into contact with a living organism. Ion implantation is a physicochemical surface modification process that results from the impingement of a high-energy ion beam. The chemical characteristics of a material surface are changed when ion implantation is performed. Newly designed polymers using ion-beam modification methods have been studied to improve blood and tissue compatibility.

Further studies on the effects of heavy ions on different types of polymers and their composites are opening up with the developments in ion beam technologies providing new application possibilities. Nano-focussed ion beams are opening a new horizon by providing a technique to create one- and two-dimensional nanoscale features on the surface of polymers. Controlled variation of the ion intensity and angle at each surface point of the polymer can provide structural surface features with desired shape and morphology.

Investigations [27] of the dielectric properties and average surface roughness of polymethyl methacrylate (PMMA) Polymer Matrix filled with organometallic complex irradiated with 120 MeV Ni^{10+} ions at the fluences of 1×10^{11} and 1×10^{12} ions cm^{-2} showed that the ion irradiation promotes the metal to polymer bonding, and converts the polymeric structure into hydrogen depleted carbon network due to the emission of hydrogen gas and/or other volatile gases.

Effects of SHI at the interface of two media such as metal-polymers, polymers-semiconductors are of interest. The polymer surface on semiconductor shows drastic surface modifications resulting in formation of ~100 nm diameter holes due to dewetting process when polycarbonate Makrofol KG coated on GaAs substrate is irradiated with 50 MeV Li ions [28].

Polymer Gas Sensors

Particle track membrane (PTM) which provide cylindrical pores of uniform size can be modified to possess characteristics which depend on physical and/or chemical properties, according to the environmental factors such as temperature, pH value, electric field etc. This provides [29] possibilities of making sensitive sensors with modified polymer materials.

Swift heavy ion modified insulating polymer-conducting polymer (IPCP) [polyvinyl chloride-polyaniline composites] was shown as very sensitive and fast response sensors for ammonia gas [30]. Room Temperature Ammonia Gas Sensing has been demonstrated by using mixed conductor based TEMPOS structures [17].

Swift heavy ion irradiated polyvinylchloridepolyethylenterephthalate (PVC-PET) composites have been tested for their hydrogen gas sensitivity [31].

Radiation Induced Grafting

Polymers are considered to be good electrical insulators and their applications have exploited this property along with its chemical inertness, durability etc. They are now classified as commodity polymers, engineering polymers and specialty polymers. Radiation tailored polymers is an area which opens up large number of possibilities.

In some situations, specific polymer properties such as solvent resistivity, hydrophilicity or biocompatibility often limit their applications. Radiation grafting provides a way to modify some of these properties. In grafting process a monomer is polymerized in presence of a polymer which can be initiated by ionizing radiation:

- Pre-irradiation method: The material is first irradiated. The created radicals react with oxygen during storage forming hydroperoxide. In a second step monomer is added and temperature increased which starts the graft polymerization.
- Simultaneous irradiation and grafting: The polymer (kapton) is first saturated with styrene for many hours and then irradiated [32]. Thus the grafting and irradiation occur simultaneously. For grafted particle track membranes, reduced pore size are observed which increase the separation ratio of H_2/D_2 [33].

Functionalization of Industrial Polypropylene Films via SHI Induced Grafting of Glycidyl Methacrylate

Swift silver ion irradiation was explored as a means of forming chemically active sites on the surface of biaxially oriented polypropylene films [34]. The active species, formed in air, were used to induce the graft copolymerization of glycidyl methacrylate in an aqueous solution. The contact angle of the modified films decreased with the grafting percentage of glycidyl methacrylate on the polypropylene. The swift silver ions induce significant grafting only in small regions (i.e., the latent tracks) of the polymer. When the fluence of swift heavy ions increases beyond an optimum value, the overlapping of the latent tracks reduce the grafting yield. This can provide a way to develop an initiator-free grafting system.

Swift Heavy Ion Induced Graft Polymerization in Track-etched Membrane's Submicroscopic Pores

Grafting of styrene on the pore walls of nuclear track membranes was shown by Mazzei et al. [35]. SHI irradiated foils of poly vinylidene difluoride (PVDF) were chemically etched to produce nuclear track membranes (NTM) with submicroscopic pores, and were grafted with styrene using the remaining active sites produced by the heavy ion beams. Experimental curves of grafting yield were measured by weight difference as a function of ion fluence and etching time. Grafted foils were also analyzed by Fourier transform infrared spectroscopy (FTIR). Both measurements suggest that the styrene was grafted on the pore wall of the NTM using the active sites left by the ion beam.

Development of "Heparin-like" Polymers

The development of ideal antithrombogenic polymers, a major problem in biomaterials sciences, is a primary objective in the fields of cardiovascular prostheses, artificial hearts, and other devices. To decrease their thrombogenicity, Porte-Durrieu et al. [36] developed polymeric materials endowed with a specific affinity for antithrombin III (ATIII) to catalyze the inhibition of thrombin by ATIII, like heparin. Sulfonate and sulfonamide groups are introduced onto phenyl rings belonging to styrene residues, which are radiation grafted (using swift heavy ion and gamma radiation) onto poly (vinylidene difluoride) (PVDF) and also onto poly(vinylidene fluoride/

hexafluoropropylene) [P(VDF-HFP)]. In contrast to gamma radiation, which leads to a homogeneous modification, the advantage of swift heavy ion grafting is that only small regions are modified; thus, the surface may present hydrophilic (corresponding to the modified areas) and hydrophobic micro-domains (corresponding to the unmodified areas) of different sizes, depending on the absorbed dose and grafting yield. The amount of fluorine decreases as polystyrene (PS) is grafted, whatever the kind of radiation and polymer. When the polymers are functionalized, the amount of fluorine also decreases. Functionalization seems to increase the roughness of the surface.

Engineering the Properties of Electroactive Polymers

Electroactive Polymers (EAPs) are polymers whose shapes are modified when a voltage is applied to them. The fact that certain types of polymers can change shape in response to electrical stimulation has been known for decades but the induced strain was found to be small. Since early nineties, electroactive polymers (EAPs) have emerged which can be stimulated to produce a significant shape or size change, which can be used as actuators or sensors. Due to the similarities with biological tissues in terms of achievable stress and force, they are often called artificial muscles, and have the potential for application in the field of robotics. Ion irradiation of polymers can be used to induce modifications of their molecular structure, in a controlled way, leading to the desired changes in the polymer properties [37, 38].

Conducting Polymers

Polymers are being developed which can be conductors or semiconductors. It is generally accepted that metals conduct electricity well and that organic compounds are insulating, but this class of materials combines the properties of both. Conductive polymers are also plastics and therefore can combine the mechanical properties (flexibility, toughness, malleability, elasticity, etc.) of plastics with high electrical conductivities. The mobility of conductive polymers is very low compared to inorganic semiconductors. This difference is diminishing with the invention of new polymers and the development of new processing techniques.

The conjugated polymers in their undoped, pristine state have >2 eV energy gap which is too large for thermally activated conduction. Therefore, undoped conjugated polymers, such as polythiophenes, polyacetylenes have a low electrical conductivity of around 10^{-10} to 10^{-8} S/cm. Even at a very low level of doping (<1 %), electrical conductivity increases several orders of magnitude up to values of around 10^{-1} S/cm. Subsequent doping of the conducting polymers results in a saturation of the conductivity at values around 100-10000 S/cm for different polymers.

Engineering the Conductivity

Conducting polymers show enhancement in electronic conductivity upon exposure to SHI irradiation, due to the electronic energy loss which produces large number of charged and active chemical species, cations, anions, radicals and electrons along the ion track. Coulombic interactions among these active charged species may give rise to either cross-linking or bond breaking depending upon the energy and fluence of the irradiating beam. Inter-chain electron hopping required for conduction between two chains, which increases the resistance of the polymer, is facilitated due to the cross-linking of the polymer chains after irradiation. The increase of crystallinity of the polymer films upon SHI irradiation also contributes to the increase in conductivity. At high energy (~100 MeV) irradiation of polymer, cross linking/recrystallization processes dominate over chain scission and bond breaking processes, which increase the conductivity by enhancing the carrier mobility through the polymer chains which otherwise occur by the inter chain hopping mechanism required for conduction between the chains. Defect sites in the molecular structure of the polymer chain created by SHI irradiation also contributes to higher dc conductivity as charge accumulation takes place which produces charge carriers, the electrons.

The dc conductivity of polyaniline films increases very significantly on heavy ion irradiation [39]. Measurements were done using Ni^{12+} and Si^{9+} ions. The increase in conductivity is less in case of Si^{9+} as compared to Ni^{12+} due to the lower value of electronic energy loss of Si^{9+} ion than that of Ni^{12+} ion. In case of poly(3-methylthiophene) the dc conductivity increases with increase in ion fluence [40, 41].

Study of 50 MeV Li^{3+} and 90 MeV C^{6+} ions on structural, electrical and morphological properties of the free-standing PANI films indicated that the crystalline nature of the Polyaniline (PANI) film increases with SHI fluence, followed by a decrease beyond the critical ion fluence. The conductivity increases with the formation of clusters and craters at higher fluences [42]. Swift heavy ion irradiation affects the interchain conductivity of polypyrrole. The conductivity of irradiated samples is stable under atmospheric conditions for more than nine months. The results open up the scope for the applicability of irradiated conducting polymers as microstructures with defined conductivity for sensor applications. The mechanism of conduction in pristine as well as irradiated polypyrrole is by 3D variable range hopping [43]. Poly(3-hexyltiophene) films irradiated with 100 MeV Ag ions show an increase in room temperature conductivity from 2.39×10^{-8} to 1.65×10^{-6} Ohm^{-1} cm^{-1} at the fluence of 10^{11} ions cm^{-2} [44]. Blended conducting polymer films irradiated with swift heavy ions also show increase in conductivity [45].

Crystallinity

Conducting polymers are semi-crystalline materials. Their crystallinity is due to chain folding or formation of multiple helices. SHI irradiation can change the crystallinity of the conducting polymer films manifold. Collective excitations (plasmons) produced upon SHI irradiation, which produces a large excited volume, can rotate the backbone bonds and adopt a variety of conformations, which come to the lower-energy positions as they cool. Stereo-regular chains will then favour regular helical shapes. At the same time, these regular sections of chain may pack together to form regions of crystallinity. Upon SHI irradiation the density of the polymer increases making the polymer more compact, which may produce closely packed regions by chain folding, cross-linking of polymer chain or by the formation of single or multiple helices, which also contributes to the production of more crystalline regions in the polymer films resulting in increase in degree of crystallinity. The unirradiated polypyrrole films are semicrystalline. The increased degree of crystallinity also leads to higher conductivity due to decreased scattering of charge carriers. The degree of crystallinity increases with the increase in fluence of SHI irradiation [46].

Optical Properties

SHI irradiation affects the UV-visible optical properties of conducting polymers. The UV-Vis spectra of conducting polypyrrole films doped with $LiCF_3SO_3$ before and after SHI irradiation at different fluences show that the intensity of the carrier absorption peak goes on increasing with the increase in fluence which indicates an increase in carrier concentration upon ion irradiation [47]. The electronic structure and carrier density can be visualized from the study of UV–Vis spectra. The increase in absorption towards higher wavelength is also observed in irradiated polypyrrole films, which indicates high dc conductivity of the conducting polymer due to lowering of band gap [47].

The UV-visible absorption spectra of irradiated and pristine poly(m-toluidine)-poly vinyl chloride (PmT-PVC) polymer blend reveal that the optical density varies with fluence [48]. It is observed that there is no absorption peak in the irradiated samples after a fluence of 10^{12} ions cm^{-2}. The band gap increases up to a fluence of 10^{12} ions cm^{-2} and then decreases with further increase in fluence. The optical band gap increases when the sample gets amorphized and decreases as the new crystallite formation starts. This means the structural changes have more impact on the optical properties of the polymer, which in-turn show the effect of irradiation on the optical properties. With irradiation, a phase change takes place resulting in changes in the optical band gap.

Microstructural Changes

Conducting polymers exhibit changes in their grain size upon irradiation with swift heavy ions. The surface morphological studies of the poly(3-methylthiophene) conducting polymer films doped with $LiCF_3SO_3$ were carried out by taking SEM images before and after irradiation with 120 MeV Si^{9+} ions at different fluences of 5×10^{10}, 5×10^{11} and 3×10^{12} ions cm^{-2} [40]. A decrease in porosity of the polymer films is observed due to densification owing to the displacement of polymer molecules. The surface morphological studies of the polypyrrole conducting polymer films doped with $LiCF_3SO_3$ under similar irradiation show cauliflower like flaky surface morphology. As the fluence of

SHI irradiation increases the grain size decreases and at the fluence of 3×10^{12} ions cm^{-2}, the surface of the polymer films becomes smooth and dense. The changes observed in the morphology of the polymer films may be ascribed to the displacement of the material from hilly to valley regions under the impact of the incident swift heavy ions making the film surface dense and smooth [40].

Electrochemical Stability

SHI irradiation affects the electrochemical stability of supercapacitors fabricated from the conducting polymer films [46]. The supercapacitor with unirradiated polypyrrole films on both sides as electrode initially show higher capacitance (205.5 F/gm) but with continuous charging and discharging the capacitance goes down and after 10,000 cycles it becomes 136.8 F/gm. The initial capacitance of the supercapacitors with irradiated polymer films on both sides show less capacitance but the capacitance after 10,000 cycles is higher than that of the unirradiated films depicting increase in cycle life of the irradiated samples as shown in Fig. 6.8 [46]. From the comparison of stability and capacitance it can be seen that 5×10^{11} ions cm^{-2} is the most suitable fluence for the use of polypyrrole films in supercapacitors. The decrease in capacitance occurs due to some irreversible reactions in the electrode materials which decreases the charge capacity.

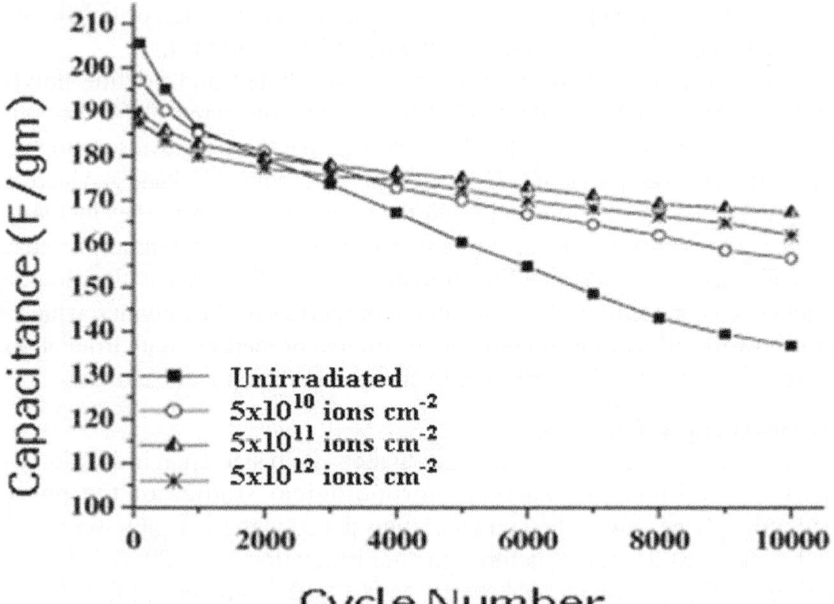

Figure 6.8: Stability plot of supercapacitors with $LiClO_4$ doped polypyrrole electrode for pristine and irradiated cases. [Reprinted from A.M.P. Hussain, D. Saikia, F. Singh, D.K. Avasthi and A. Kumar, *Nucl. Instr. and Meth.*, **B 240 (4)** (2005) 834 with permission from Elsevier.]

Polymer Electrolytes

SHI irradiation of polymer electrolyte membranes leads to decrease in crystallinity of the membranes at lower fluences (5×10^{10} and 10^{10} ions cm^{-2}) and increase at higher fluences (5×10^{12} and 10^{12} ions cm^{-2}) [49, 50]. Degree of crystallinity of C^{5+} and Li^{3+} ions irradiated P(VDF – HFP) – (PC + DEC) – $LiClO_4$ gel polymer electrolytes decreases to 11.5% and 10.48% respectively after low fluence (5×10^{10} and 5×10^{9} ions cm^{-2}) ion irradiation and increases to 18.75% and 17.1% respectively after high fluence (5×10^{12} and 10^{12} ions cm^{-2}) ion irradiation as compared to 13.64% of unirradiated P(VDF – HFP) – (PC + DEC) – $LiClO_4$ gel polymer electrolytes. In polymer electrolyte upon SHI irradiation chain scission and bond breaking occurs at low fluence, which amorphizes the polymer. However, at higher fluence, reordering and cross-linking of polymer chains take place forming the new crystalline regions, which leads to increased degree of crystallinity. The increased degree of crystallinity leads to decrease in ionic conductivity as ion transport in polymer electrolyte takes place through the amorphous rich phase.

Surface Morphology

SHI on polymer electrolytes changes the surface morphology completely. Upon ion irradiation the porosity of the electrolytes membranes increases as shown in Fig. 6.9, which helps to trap more liquid electrolyte in the same volume of polymer resulting in higher ionic conductivity [49].

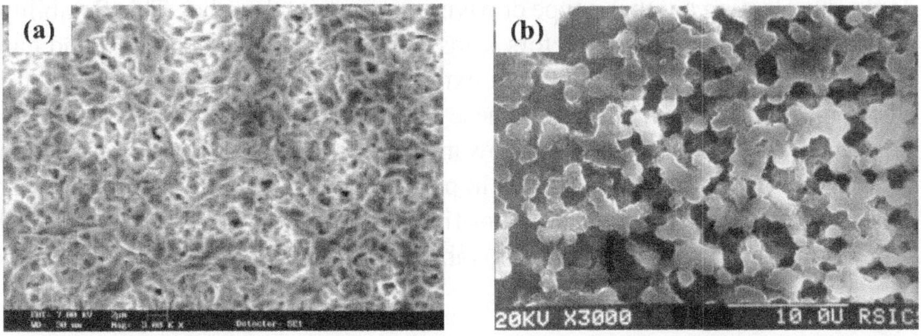

Figure 6.9: SEM image of (a) unirradiated and (b) Li^{3+} ion irradiated P(VDF–HFP)–(PC+DEC)–$LiClO_4$ gel polymer electrolyte. [Reprinted from D. Saikia, A. Kumar, F. Singh and D.K. Avasthi, *J. Phys. D: Appl. Phys.*, **39** (2006) 4208 with permission from Institute of Physics Publishing Ltd.]

The SHI can cause creation of ion track, amorphization, chain scissioning and cross linking. The measurement of released gases gave a clue that ion track has different zones of cylindrical damage. SHI-induced grafting and etched ion track have been shown to have potential applications. The use of SHI irradiated doped polypyrrole as the electrodes of supercapacitor improve its stability in terms of number of cycles (of use).

6.3 SEMICONDUCTORS

Introduction

The semiconductors are group of elements and compounds that are neither good electrical conductors nor insulators at normal room temperatures. Ideally at absolute zero temperature, these are insulators but at any finite temperature, their conductivity is in between conductors and insulators and hence the name Semiconductors. The pure material semiconductors, such as germanium and silicon, are referred to as intrinsic semiconductors. Semiconductors can also be made by adding a small number of impurity atoms to certain substances. Such a semiconductor is referred to as an extrinsic semiconductor.

The two most extensively used intrinsic semiconductors, namely Si and Ge are Indirect Band Gap semiconductors. In these materials the bottom of conduction band and top of the valence band do not fall at same value of wave vector (i.e. momentum). Electron cannot shift from the top of the valence band to the bottom of the conduction band without a change in momentum. By contrast, there are several compound semiconductors (III-V as well as II-VI) like GaAs, AlAs, InAs, GaN, AlN, InN, CdTe, CdSe, etc. which are Direct Band Gap semiconductors with bottom of conduction band falling at the same value of wave vector as top of valence band. Most of these compound semiconductors have been extensively studied because of their direct band gap nature on the one hand and possibility to engineer the band gaps by suitably doping/implanting another n type or p type element, on the other. This flexibility of III-V and II-VI semiconductors leads to possibility of electronic band gap engineering and has been exploited extensively during last several years for electronic device applications of these compounds. More recently, some of these have also been investigated towards their role in energy applications in general and photo-voltaic materials in particular, thus breaking the monopoly of silicon-based solar cells. Another novel outcome of compound semiconductors has been found in fabrication of super-lattices and multi-quantum wells, as discussed below.

Effect of High Electronic Excitation in Semiconductors

Experimental evidence for the formation of continuous and discontinuous amorphous tracks during SHI irradiation with monoatomic ion beams exists in the crystalline semiconductors GeS [1], InP [2], InAs, InSb and GaSb [3], Ge [4] as well as in crystalline SiGe alloys [5], but not in Si and GaAs. Swift-heavy-ion-induced modifications in III-V binary semiconductors InP, GaP, InAs, GaAs and the related ternary alloys $Ga_{0.50}In_{0.50}P$ and $Ga_{0.47}In_{0.53}As$ were studied with 185 MeV Au ions using Rutherford backscattering spectroscopy in channeling configuration, transmission electron microscopy, and small-angle X-ray scattering [6]. Despite nearly identical ion-energy loss in these materials, their behaviour under swift-heavy-ion irradiation is found to be strikingly different. InP and $Ga_{0.50}In_{0.50}P$ are readily amorphized, GaP and GaAs remain almost undamaged and InAs and $Ga_{0.47}In_{0.53}As$ exhibit intermediate behaviour. Irradiation-induced changes in the materials depend not only on the electronic energy deposition but varies with properties of the material.

Irradiation of single-crystalline InP with swift heavy ions (SHIs) causes the formation of ion tracks for certain irradiation temperatures if the electronic energy deposition exceeds a threshold value. Kamarou et al. [7] showed that the experimental data obtained can be qualitatively and quantitatively described on the basis of the inelastic thermal spike (TS) model. It offers a self-consistent way to explain the influence of various irradiation conditions (ion mass, ion energy, irradiation temperature, etc.) on the ion track formation and damage accumulation [7]. Further, they found that there is no universal RT threshold for track formation in InP, but it is noticeably higher for lighter ions (12.0 and 14.8 keV/nm for irradiations with Au and Xe, respectively).

Structural Modification of SHI Irradiated Amorphous Si and Ge Layers

Swift heavy ion (SHI) irradiation of amorphous Si (a-Si) at non-perpendicular incidence leads to non-saturable plastic flow [8]. The positive direction of flow suggests that a liquid phase of similar density to that of the amorphous solid must exist and accordingly a-Si behaves like a conventional glass under SHI irradiation. For room-temperature irradiation of a-Si, plastic flow is accompanied by swelling due to the formation of voids and a porous structure.

Similar effect of SHI irradiation at room temperature is observed on amorphous Ge (a-Ge). Like a-Si, positive plastic flow is apparent, demonstrating that liquid polymorphism is common to these two semiconductors. Porosity is also observed, again confined to the amorphous phase as a result of electronic energy deposition. Enhanced plastic flow coupled with a volume expansion is responsible for the structural modification of both a-Si and a-Ge irradiated at room temperature with swift heavy ions.

High Energy Heavy Ions in Semiconductor Devices

High energy ion implantation plays important role in the fabrication of special device structures. The defects are produced due to nuclear collisions by high energy ions at the end of their projected ranges in semiconductors, where they are implanted. There are also some defects produced in surface and subsurface regions in crystalline Si exposed to high energy Si ions [9, 10]. The devices produced by high energy ion implantation are sensitive to electrically effective defects. The effect of annealing of 70 MeV ^{120}Sn implanted GaAs on the electrical properties of the implanted layer have been investigated [11]. An effective passivation of the surface states in n-GaAs has been achieved by 50 MeV Si ion irradiation, which produced an enhancement in the intensity of the band edge photoluminescence (PL) [12].

The switching time of silicon diodes is reduced from 1000 ns to less than 200 ns by introducing ion induced defects into the n-region of silicon junction diodes. It has been observed that the defects produced near the junction are more effective in reducing the turn-off time as compared to those produced near the surface [13].

Semiconductor devices irradiated with 100 MeV heavy ions show a change of conductivity type i.e. from n to p, at a depth of approximately the stopping range of the ions in silicon [14]. Si implantation using 100 MeV ^{28}Si ion in GaAs shows an increase in mobility and a decrease in sheet resistivity [15]. Electrical behaviour of n-GaAs substrates implanted with 70 MeV ^{120}Sn ions after annealing at 850°C shows I-V characteristics typical of a p-n junction, indicating a p-type conductivity in the implanted layer [16].

Fourier Transform Infrared reflectivity (FT/IR) and low temperature (298-133 K) ESR measurements on 70 MeV $^{56}Fe^{5+}$ ion implanted samples of p-type Si show the presence of interference fringes in high fluence ($>1 \times 10^{13}$ ions cm^{-2}) implanted samples [17], providing information on the ion induced modified layers.

SHI irradiation of silicon Schottky diodes results in saturated values of the barrier height and other parameters after a critical ion fluence [18, 19]. It is established that ion irradiation into semiconductors causes structural damage, which in turn results in electrically active defects. These electrically active defects produced in Si can result in pinning of the Fermi level to a dominant deep level [20, 21].

The Schottky barrier height in Ni/n-Si (100) changes from a value of 0.59 eV for unirradiated diode to 0.68 eV after irradiation with 100 MeV oxygen ions at a fluence of 1×10^{13} ions cm^{-2}, and the leakage current decreases by about two orders of magnitude [22]. The Schottky barrier in Au/n-Si structure decreases after irradiation by 100MeV Si^{7+} ion at a fluence of 1×10^{11} ions cm^{-2} and remains immune to further irradiation upto a fluence of 1×10^{12} ions cm^{-2}. A combination of in situ deep level transient spectroscopy and current-voltage measurements of Au/n-Si diodes demonstrates that 100 MeV Si ion irradiation introduces hydrogen related defect complexes, which has a major influence on the Schottky barrier height and the leakage current in the irradiated structure [23].

SHI Induced Changes in Porous Silicon

Anodically etched porous silicon samples irradiated with 85 MeV Ni ions results in a complete suppression of the major photoluminescence (PL) peak at 697 nm and the shoulder peak at 6.7 nm, and a weak PL peak appears at 588 nm. The changes in the photoluminescence spectra are correlated with the resulting changes in the concentration of hydrogen and chemical complexes containing Si, H and O in the porous silicon [24].

Irradiation of Porous Silicon (PS) samples with 10 MeV Si ions shows improvement in the efficiency of photoluminescence (PL) and its stability with time [25]. Grazing angle X-ray diffraction (XRD) analysis reveals that preferential recrystallisation occurs in the irradiated region. The virgin PS sample exhibited only the (111) peak in the XRD pattern, whereas the irradiated PS sample showed a (311) peak along with the (111) peak. FTIR studies of boron doped p-type (111) porous silicon irradiated with 10 MeV Si ions showed considerable increase in the PL yield [26].

Role of Microstructure in the Energy Relaxation Processes in Thin Film Semiconductors

In thin film semiconductors, the energy relaxation of swift heavy ions is greatly influenced by the material properties, especially the microstructure [27-29]. The energy relaxation channels like defect annealing, defect generation, phase transformation, etc., differ in their S_e thresholds for their occurrence [30].

Study of CdS and CdTe thin films of different microstructures synthesized by using thermal evaporation and spray pyrolysis after irradiation with 100 MeV Ag^{7+} at the fluence of 3×10^{10} to 1×10^{13} ions cm^{-2} show that they have different energy relaxation channels [27]. In evaporated CdS, for lower energy depositions, defect-annealing processes dominate, but at higher fluences, the energy deposition results in many atomic displacements together with their migration and development of a significant amount of compressive strain. The atomic migration results in a transformation of the metastable arrangement in these films to stable hexagonal phase with a change of bandgap from 2.34 eV for the pristine film to 2.43 eV for the sample irradiated at the highest fluence. However, for the same energy deposition range, in spray deposited CdS films, ion irradiation results in an improvement of crystallinity and a relaxation of tensile strain [27]. The studies performed on CdTe confirm the role of microstructure [28, 29].

As in the case of evaporated CdS, evaporated CdTe also undergo defect annealing at lower energy depositions, but at higher energy dissipations, defect generation dominates. An increase of lattice constant with fluence due to an increase of tensile strain is observed in these films. CdTe films grown by spray deposition without electric field have stable cubic phase while the films deposited in presence of the electric field possess metastable hexagonal regions [31]. For the samples deposited by spray pyrolysis without electric field, defect generation always dominates and no change of lattice parameter is observed as a result of SHI irradiation. In the films deposited with electric field, the dense electronic excitation results in a transformation of the metastable hexagonal regions in the film to stable cubic arrangements together with a change of bandgap from 1.47 eV for the pristine sample to 1.44 eV for the sample irradiated at the fluence 1×10^{13} ions cm^{-2} [29].

Superlattices and Multi Quantum Wells (MQW)

Introduction of another column III or column V element in a III-V compound leads to small change in lattice parameter accompanied by corresponding variation in energy band gap of original semiconductor. For example introduction of small amounts of Indium or Aluminum in GaAs or GaN results in a change of lattice parameter and band gap. If such layers are deposited on the parent substrate, the small lattice mismatch induces tensile or compressive strain in the deposited epilayer. This strain increases with thickness of the deposited layer or with composition of the doped/implanted amount of In or Al. Beyond a certain critical value, the strain relaxes into misfit defects (mostly dislocations). The optoelectronic device performance deteriorates drastically due to generation of these misfit defects. It is possible to fabricate Strained Layer Superlattice (SLS) with equal thickness of alternate layers of small lattice mismatch with thickness below critical value for strain relaxation.

Semiconductor superlattices have potential device applications for high performance detectors, high speed and high frequency digital and analogue circuits. The usefulness of these structures is that, they offer precise control over the states and motions of charge carriers in semiconductors, which is possible because of the ability to tailor the band structure of these materials. As a result, the electronic and optoelectronic properties are enhanced many-fold. With the advent of epitaxial growth techniques, it is now possible to grow multilayers with a small lattice mismatch leading to a tensile or compressive strain in the alternating layers called strained layer superlattice (SLS). The strain produced in SLS improves the device performance and is one of the important parameters for tailoring the band structure. Spatial band gap tuning of heterostructures is important because the integration of photonic circuits demand different band gaps for different devices. Meeting such band gap requirements is quite difficult during the growth. The alternative way is to alter the band gap after growing the structures. Compositional disordering and mixing at the interface by ion implantation and subsequent thermal annealing is normally employed. Swift heavy ion (SHI) beam induced mixing is more suitable for the integration of optoelectronic devices because of the advantage that the mixing can be confined to a narrow region at the interface as against the lateral straggling effects in low energy ion irradiation. Monolithic integration of optoelectronic circuits is crucial for fabricating stable photonic devices with submicron alignment of various optical components. These periodic structures with alternating small and large energy band gap lead to realization of Multi Quantum Wells (MQW) offering ideal situations for quantum confinement of conduction/valence electrons at nanoscale. Such structures have in fact been extensively investigated using ion beams, both for their modifications (leading to band gap engineering) as well as characterizations.

Band Gap Engineering

Band gap engineering, the process of controlling or altering the band gap of materials, is a powerful technique for the design of new semiconductor materials and devices. For certain semiconductor alloys, such as GaAlAs, InGaAs and

InAlAs, it can be done by controlling the composition. Advances in the growth technology of semiconductors have made it possible to fabricate complicated semiconductor microstructures, which can be used to regulate the motion of electrons. In particular, one dimensional semiconductor superlattices consisting of alternating ultrathin layers of different composition or different levels of doping have been obtained using molecular beam epitaxy (MBE) and metal-organic chemical vapour deposition (MOCVD). The thickness of the layers is on the scale of the de Broglie wavelength of an electron.

Multi Quantum Wells (III-V Systems)

SHI irradiation provides a way to induce a tensile strain in an initially lattice matched system and to decrease the compressive strain in an initially compressive strained system, without much loss of crystalline/interface quality [32-38]. Rapid Thermal Annealing (RTA) is used to remove irradiation induced defects in the samples. The study on InGaAs/GaAs based materials indicates a gradual diffusion of In from surface and the migration of Ga or As like atoms to the surface regions due to the SHI irradiation and/or annealing processes. The study also shows an increase in the superlattice period after the irradiation. The SHI can be used to tune layer composition, thickness and the strain of an SLS which in turn will enable the spatial band-gap tuning for the integration of optoelectronic devices. The multi-quantum wells are of total thickness in sub micron range; within which the energy loss of swift heavy ions are more or less uniform. This results in uniform modification throughout the entire thickness of the multi-quantum well.

Band Gap Engineering of Multi Quantum Wells (MQW)

The band gap engineering of quantum well with SHI irradiations was studied for $In_{0.55}Ga_{0.45}As$ (20 nm)/InP and $In_{0.23}Ga_{77}As$ (5 nm)/InP [39, 40]. In the first case InGaAs is nearly lattice matched to InP. It was irradiated with 150 MeV Ag^{12+} ions at fluence of 1×10^{13} ions cm^{-2}. For the second case the InGaAs layer is highly tensile strained to InP which was irradiated using 100 MeV Au^{13+} ions of fluence 5×10^{12} and 1×10^{13} ions cm^{-2}. All the samples were annealed (rapid thermal annealing at 700°C for 60 sec in N_2 atmosphere) before and after irradiation for comparison and subsequently characterized using low-temperature photoluminescence. For nearly lattice matched sample and with the irradiation conditions a band gap change of about 23 nm was observed [39]. In case of the highly strained samples, band gap changes observed were 48 nm and 82 nm for 100 MeV Au^{13+} ions of 5×10^{12} and 1×10^{13} ions cm^{-2} fluence respectively [40]. This increase of band gap has been attributed to increase of tensile strain upon irradiation [41]. The dependence of the band gap change with ion beam parameters can provide a way to engineer the MQW properties in a controlled fashion.

Band-gap Engineering in III-Nitride Compound Semiconductors

III-Nitride compound semiconductors have applications in optoelectronics, high power and high frequency devices. Band gaps of such compound semiconductors can be tailored from 6.2 eV to 0.7 eV by changing alloy composition and can be used in UV, IR and solar applications. These materials show excellent luminescence properties in spite of huge defect densities that arise from lattice mismatch and thermal mismatch with underling substrates such as sapphire, silicon and silicon carbide. MOCVD and MBE are advanced growth techniques in realizing hetero structures and MQW semiconductors. Swift heavy ion irradiation of such grown MQWs for structural modifications is a post-growth technique to alter band gap spatially. Interaction of such ion with matter occurs on time scale of picoseconds and length scales as low as nanometres.

Irradiation with 150 MeV Ag ions at fluence of 5×10^{12} on GaN bulk epitaxial crystal to study the effects of irradiation on III-Nitrides, showed the epitaxial reconstruction of GaN layers [42-44]. Studies on AlGaN/GaN heterostructures showed that AlGaN is more radiation resistant than GaN and strain inducted in the AlGaN layer due to reconstruction of AlGaN layer, is more pronounced than GaN layers. The defect density in GaN bulk epitaxial layer increases on irradiation, which may be due to the reconstruction of the low temperature GaN buffer layer at GaN/Sapphire interface [45]. Irradiation of tensile strained AlGaN/GaN MQWs with higher Se ions (200 MeV Au ions) and low fluence (5×10^{11} ions cm^{-2}) showed that there is shift of band gap of AlGaN from 4.4 eV to 3.78 eV [46, 47].

Band Gap Engineering of Chalcogenides

A chalcogenide is a chemical compound consisting of chalcogenide elements (Group 16 in the periodic table e.g. sulphur, selenium, tellurium). Research on these compounds is leading towards applications in various fields. Swift heavy ion based research is providing ways to enhance the band gap, photoconductivity etc. of these materials. Effects of SHI on γ indium selenide [γ-In_2Se_3] films were studied using 90 MeV Si ions with a fluence of 2×10^{13} ions cm^{-2} [48]. It was observed that the photosensitivity increases and the negative photoconductivity of the materials turns into positive photoconductivity on irradiation. The increase in the optical band gap was attributed to the annihilation of the localized defect band, near the conductance and valence band edge on irradiation.

Optical properties of CdTe films show significant variations with Swift Heavy Ion (80 MeV O^{6+}) irradiation. The band gap decreases and the refractive index increases [49]. The band gap of the films get red shifted and the grain size increases with SHI (100 MeV Au ions) leading to the decrease in resistivity and the increase in thermo-emf of the film [50].

The copper chalcogenide (CuX, X = S, Se) thin films irradiated with SHI (100 MeV Au ions) show increase in the particle size, electrical conductivity, and PL intensity and the optical band edges are red shifted [51].

From the technological viewpoint, technique to engineer the optical properties using ion beams is not fully established in comparison to engineering of the electrical properties. There is a need to investigate dependence of the band gap change with ion beam parameters, which will provide a way to engineer the MQW properties in a controlled fashion. The post irradiation processes need to be optimized to achieve the device quality of the engineered structures.

6.4 TRANSPARENT CONDUCTING OXIDES (TCO)

Introduction

Transparent conducting oxides (TCOs) have high electrical conductivity and optical transparency. These simultaneous properties enable numerous applications, such as flat-panel displays, photovoltaic devices, and light emitting diodes (LED). Coexistence of transparency and conductivity is normally not possible. For a material to be transparent, it must not absorb light in the wavelength range 380 nm (3.26 eV) to 780 nm (1.59 eV). It should have an optical band gap greater than 3.26 eV. It was found that binary oxides of tin, indium and zinc could exhibit high conductivity, while possessing very large optical band gaps of greater than 3 eV. Most common TCOs are ZnO, SnO_2, In_2O_3, CdO, Ga_2O_3 etc. They are insulators in stoichiometric form but can become semiconductor in their off-stoichiometric form, with high optical transparency and low electrical resistivity [1-3]. Functional oxides have two structural characteristics: cations with mixed valence states and anions with deficiencies (vacancies) [4]. Charge carriers in these oxides arise from three fundamental sources: interstitial metal ion impurities, oxygen vacancies, and doping ions. The first two sources are inherent and always act as electron donors. Thus, these systems are typically n-type with a carrier concentration on the order of 10^{20} cm^{-3}, provided by interstitial metal ions and oxygen vacancies. When an oxygen vacancy is present in the lattice it acts as a doubly-charged electron donor. These oxides exhibit n-type free electron conductivity and doping with higher valence cations or lower valence anions can help further reduce the resistivity. Shallow donors near the conduction band (n-type) allow electrons to be thermally excited into the conduction band (partially filled by degenerate electron gas), while acceptors near the valence band (p-type) allow electrons to jump from the valence band to the acceptor level, populating the valence band with holes. There is a limit on resistivity that can be achieved by doping of TCOs.

TCOs doped with transition metals form novel diluted magnetic semiconductors (DMS). Engineering the DMS materials with ion beams is discussed in Section 6.6.

Zinc Oxide as TCO

Zinc oxide (ZnO), a II-VI semiconductor, has several favourable properties such as: good transparency, high electron mobility, wide bandgap, strong room-temperature luminescence etc. It has a relatively large direct band gap (~3.3eV) at room temperature making pure ZnO colourless and transparent.

The radiation resistance of ZnO makes it suitable for space applications. Mostly ZnO has n-type character, even in the absence of intentional doping. Reliable p-type doping of ZnO is very difficult. ZnO nanostructures can be synthesized into a variety of morphologies including nanowires, nanorods, etc. For spintronic applications, if doped with 1-10% of magnetic ions (Mn, Fe, Co, V, etc.), ZnO can become ferromagnetic even at room temperature. It has high biocompatibility and fast electron transfer kinetics. Among the functional oxide materials, ZnO is one of the most widely studied materials for electronic, optoelectronics and optical application [5, 6]. It is unique in having semiconducting, piezoelectric and pyroelectric properties and is an ideal candidate for fabricating electromechanical coupled devices. It is biodegradable and possibly biocompatible material suitable for medical and biological applications [7].

Swift Heavy Ion Modification of ZnO Thin Film

Swift heavy ion beams can be used very effectively to modify the properties of ZnO thin film [8-11]. Figure 6.10 shows the XRD spectra of pristine and 100 MeV Ag ion irradiated ZnO films (synthesized by atom beam sputtering) at different fluences [8]. Reflections (100), (002) and (101) of wurtzite phase are observed at 31.7°, 34.5° and 36.4° respectively and the ratio of their intensities accounts for a preferential growth of the ZnO crystals, with the C axis of the hexagonal cell perpendicular to the surface. This type of texture is usual for ZnO films, grown by physical vapour deposition techniques [12]. After low fluence irradiation of evaporated films, the intensity of the (100) and (002) peaks increases and full width at half maximum (FWHM) decreases, while the intensity of (101) peak decreases. This shows that the irradiation at low fluence increases the texturing along lower energy states and enables the atoms to move to their stable orientation [13]. At higher fluences, a significant decrease of peaks intensity but little broadening of all peaks is observed. The XRD pattern of as-deposited sputtered film shows large value of FWHM and shifting of peak towards the smaller 2θ value with respect to the actual value, which reveals the formation of compressive strained nanocrystalline ZnO thin film. The crystallinity and the quality of the film is improved by ion irradiation for all ion fluences. The irradiated films show the preferred orientation along (0 0 2) at the expense of other reflections together with the release of compressive strain.

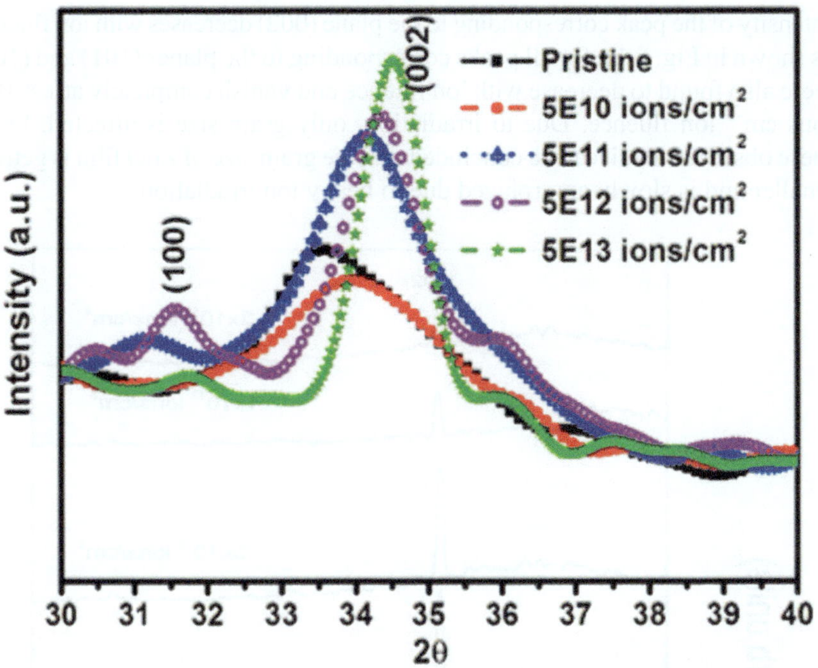

Figure 6.10: X-ray diffraction spectra of pristine and irradiated ZnO thin films (deposited by atom beam sputtering) with 100 MeV Ag ions at different fluences. [Reprinted from D.C. Agarwal, F. Singh, D. Kabiraj, S. Sen, P.K. Kulriya, I. Sulania, S. Nozaki, R.S. Chauhan and D.K. Avasthi, *J. Phys. D: Appl. Phys.*, **41** (2008) 045305 with permission from Institute of Physics Publishing Ltd.]

Raman and FTIR results of both films substantiated XRD results. From the optical absorption study of evaporated film it is clear that there is no significant change in band edge after the irradiation which implies that the basic crystal structure is not changed. The lack of irradiation effect on the vibration mode and electronic gap indicates that the atomic order at long distance is not affected significantly. The lattice is not amorphized since this structural transformation generally leads to a reduction in the gap of semiconductors. The band-gap of the film is not modified by irradiation. However, significant decrease in band-gap of the sputtered films, from 3.42 eV to 3.2 eV, occurs after the irradiation, which could be due to increase in crystallite size.

The SHI (100 MeV Au ions) irradiation of ZnO thin film (synthesized by e-gun evaporation) leads to improvement in crystallinity at lower fluence (upto 5×10^{12} ions cm^{-2}) and loss of crystallinity and disorder at higher fluence [10].

The amorphization of ZnO thin film (prepared using chemical spray pyrolysis CSP technique) occurs under the swift heavy ion (120 MeV Ag ions) irradiation [14]. Optical absorption edge remains unaffected after irradiation.

Intensity of the peak corresponding to the plane (002) decreases with ion fluence as shown in Fig. 6.11. Small peaks corresponding to the planes (101) and (103) were also found to decrease with ion fluence and vanish completely at 1×10^{13} ions cm^{-2} ion fluence. Due to irradiation, only grain size is affected. From these observations, it can be concluded that the grain size of ZnO film is getting smaller and is slowly amorphized due to heavy ion irradiation.

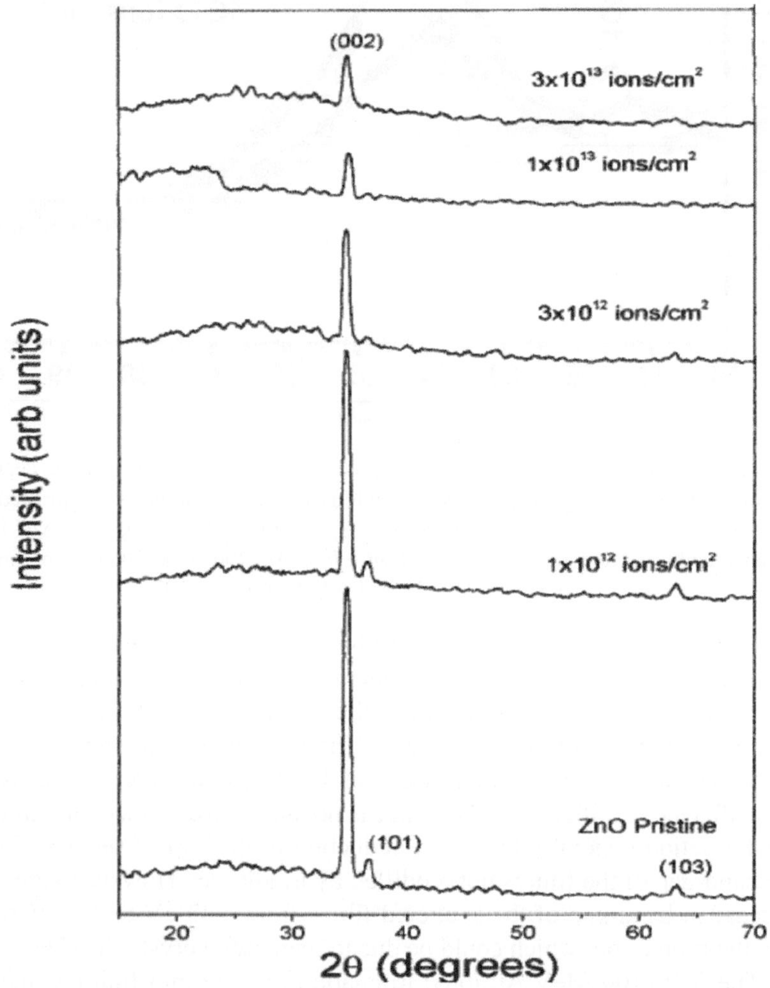

Figure 6.11: X-ray diffraction spectra of pristine and 120 MeV Au ion irradiated ZnO thin films (deposited by CSP method) at different fluences. [Reprinted from P.M. Ratheesh Kumar, C. Sudha Kartha, K.P. Vijayakumar, F. Singh, D.K. Avasthi, T. Abe, Y. Kashiwaba, G.S. Okram, M. Kumar and Sarvesh Kumar, *J. Appl. Phys.*, **97** (2005) 013509, copyright (2005), American Institute of Physics.]

All samples showed characteristic photoluminescence (PL) peak at 517 nm, which corresponds to the blue-green emission due to the defects related to oxygen and zinc (vacancies, interstitial etc.) as shown in Fig. 6.12. These emissions were found to be extremely intense and broad, and decrease with ion fluence. Conductivity of film increases with ion fluence.

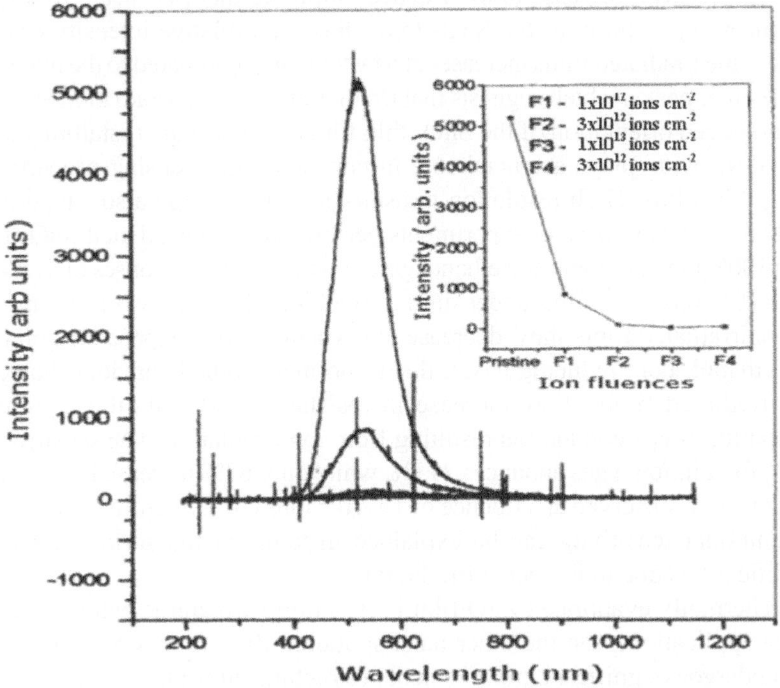

Figure 6.12: PL spectra of pristine and 120 MeV Au ion irradiated ZnO thin films (deposited by CSP method) at different fluences. [Reprinted from P.M. Ratheesh Kumar, C. Sudha Kartha, K.P. Vijayakumar, F. Singh, D.K. Avasthi, T. Abe, Y. Kashiwaba, G.S. Okram, M. Kumar and Sarvesh Kumar, *J Appl. Phys.*, **97** (2005) 013509, copyright (2005), American Institute of Physics.]

Tin Oxide as TCO

Tin oxide SnO_2, is of great scientific interest because of its wide range of applications. It has been studied as overcoat for thin film magnetic recording media, as transparent electrode in panels and other electrochromic devices, in photoelectrochemical studies as well as those related to the development of solar cells, and as material for Li-ion batteries. The study of SnO_2 as gas sensing material is due to its suitable physicochemical properties as, for example, it has a high reactivity to reducing gases at relatively low operating temperatures due to the easy adsorption of oxygen at its surface because of the natural non-stoichiometry of SnO_2. The research on these materials is focussed to increase their sensitivity, selectivity and stability.

Swift Heavy Ion Induced Enhancement in Gas Sensing Properties of SnO_2

The gas sensing property of SnO_2 thin film improves with the swift heavy ion irradiation [15, 16]. SHI (75 MeV Ni ions) irradiation of SnO_2 films prepared by solution cast method on corning glass results in increase in intensities of XRD peaks with increase in ion fluence, implying that SHI beam irradiation enhances crystallinity of the SnO_2 thin films. The relative intensity of (110) peak in the irradiated films increases at a faster rate as compared to the intensities of the other peaks, which suggests that irradiation of Ni ion irradiation enhances the preferred orientation of the SnO_2 thin films. Increase of crystallinity along the (110) plane plays a crucial role in enhancing gas sensing properties of SnO_2 thin films. High resolution transmission microscopy also supports the XRD results. Gas sensing experiments performed on unirradiated SnO_2 films with 1000 ppm ammonia and ethanol gases compares the responses of irradiated (with 10^{12} ions cm^{-2}) films under similar conditions. Here it is worth mentioning that unirradiated films show decrease of resistance when exposed to ammonia and ethanol, both reducing gases, due to normal *n*-type behaviour; however, the irradiated films show increase in resistance under similar conditions indicating *p*-type conduction resulting from ion irradiation. The sensitivity of SnO_2 for ammonia gas enhances 213%, while only 63% increase in sensitivity for ethanol is observed at a fluence of 1×10^{12} ions cm^{-2}. Such an enhancement of ammonia sensitivity can be explained in terms of the surface chemistry modifications due to ion beam irradiation.

Thermally evaporated ZnO film is radiation hard and is better suited for space applications. On the other hand sputtered ZnO film is nanocrystalline and undergoes significant improvement in structural and optical properties with ion irradiation [8-11], suitable for other applications. This illustrates how ion beams can enhance the functionality of surfaces without affecting other properties of ZnO. Films deposited by CSP method [14], having large amount of defects as characterized from PL spectra, can be amorphized under swift heavy ion irradiation. Sol-gel derived film show enhancement in gas sensing properties of SnO_2 film after the ion irradiation.

6.5 TRANSITION METAL OXIDES (TMO)

Introduction

The Transition metal oxides is a class of materials that contain transition elements and oxygen, exhibiting a variety of structures and electrical, magnetic and optical properties. The nature of metal-oxygen bonding can vary between nearly ionic to highly covalent or metallic. The unusual properties of transition metal oxides are due to unique nature of the outer d-electrons. The 4d oxides are expected to be more ionic than the 3d oxides because the 4d electrons are less tightly bonded to metal atoms than 3d electrons. With increased ionicity, the oxides tend to deviate from stoichiometry. In the transition-metal ions, for example, d electrons experience competing forces: Coulombic repulsion tends to localize individual electrons at atomic lattice sites, while hybridization with the oxygen p electron states tends to delocalize the electrons. The subtle balance makes many of the transition-metal oxides excellent resources for studying and taking advantage of the metal-insulator transition that can accompany dramatic changes in a system's electronic properties.

Engineering the properties of a few transition metal oxides—nickel oxide, titanium oxide and magnetite—with swift heavy ions are discussed here.

Nickel Oxide (NiO)

Nickel oxide is an attractive material for use as a stable antiferromagnetic layer [1] in various magnetic structures and is a p-type transparent conducting material due to holes generated by Ni vacancies. Stoichiometric NiO is an insulator at room temperature. Its resistivity can be lowered by an increase of Ni^{3+} ions resulting from addition of monovalent ions such as lithium, by nickel vacancies, or by the presence of interstitial oxygen in the NiO crystallites. It is used in electrochromic devices (information display, smart window, variable reflectance mirror etc.) due to its ability to reversibly and persistently change its optical properties under the applied electric field [2]. It can also be used in functional sensor layers for chemical sensors and resistive switching application [3]. Nonvolatile resistive memories, which employ resistance switching initiated by applied voltage is extensively used in devices for memory applications.

Texturing and Surface Evolution in NiO Thin Films

The swift heavy ion induces grain re-orientation in NiO thin film [4-6]. NiO thin films grown on Si (100) substrate by electron beam evaporation and sintered at 500 and 700 °C, irradiated with 120 MeV Au ions at room temperature, show two prominent peaks ((111) and (200)) corresponding to the FCC structure of NiO in the XRD patterns of the pristine and irradiated films. The FCC structure is retained at all fluences of irradiation in both films [4]. This is in contrast to the case of zirconia and hafnia where SHI induces transition from monoclinic to tetragonal phase [7]. Since NiO retains its bunsenite structure when heated till its melting point, no structural phase transition is expected in this system due to SHI irradiation. The evolution of texture and crystallinity with irradiation was studied. 120 MeV Au ions can create columnar defects [6], since value of S_e for these ions is 31.8 keV/nm, which is higher than S_{eth} (30 keV/nm) [8]. These ions thus should suppress the crystalline volume fraction with increasing irradiation fluence. Interestingly, improvement in crystallization is observed at some intermediate fluences in both the films. Subsequent irradiation at higher fluences indicates SHI induced suppression of crystallinity. Improvement in crystallization at low fluence and loss of crystallinity at high fluence give an indication that crystallization and damage of crystallinity are competing processes. The former occurs in the annular zone surrounding the ion track and latter in the ion track.

SHI Induced Modification of the Microstructure of NiO Thin Films

Microstructure of NiO nanoparticle films (200 nm thick) grown on Si substrates was studied on irradiation with 200 MeV Ag^{15+} ions. Though electronic energy loss of 200 MeV Ag ions in NiO matrix was higher than the threshold electronic energy loss for creation of columnar defects, films remained crystalline with the initial fcc structure even up to a fluence of 5×10^{13} ions cm^{-2}, where ion tracks are expected to overlap. Irradiation however modified the microstructure of the NiO films considerably. The grain size decreased with increasing ion fluence, which led to reduced surface roughness and increased optical band gap due to quantum confinement [6, 9]. These results correlate well with variation of the power spectral density exponent with ion fluence, which indicate that at high ion fluences, the evolution of surface morphology is governed by surface diffusion.

Titanium Dioxide (TiO_2)

TiO_2 has a number of attractive properties, which includes high refractive index, high dielectric constant, semiconducting properties and chemical stability. It is known to exhibit three major crystalline structures: rutile tetragonal, anatase tetragonal, and brookite rhombohedral with decreasing order of stability.

Titania thin films are of immense importance and there is need to optimize their performance for wide range of applications as optical waveguides [10], gas sensors [11], solar cells [12], thin film capacitors [13], electrochromic materials [14, 15] etc. Titania acts as a photosensitiser for photovoltaic cells, and when used as an electrode coating in photoelectrolysis cells can enhance the efficiency of electrolytic splitting of water into hydrogen and oxygen. Even in mildly reducing atmospheres titania tends to lose oxygen and become sub-stoichiometric. In this form the material becomes a semiconductor and the electrical resistivity of the material can be correlated to the oxygen content of the atmosphere to which it is exposed. Hence titania can be used to sense the amount of oxygen (or reducing species) present in an atmosphere.

Swift Heavy Ion Modification of TiO_2 Thin Film

Thermal annealed TiO_2 films contain both the rutile and anatase phase whereas the SHI irradiated TiO_2 has dominantly rutile phase [16]. A study on evolution of the XRD pattern with 200 MeV Ag ion irradiation for the films containing both anatase and rutile phases, shows suppression of anatase phase at 3×10^{12} ions cm^{-2}. The rutile phase on the other hand shows a complex variation with irradiation fluence. Its peak decreases at the initial fluence of 5×10^{11} ions cm^{-2} where additional peaks corresponding to the tetragonal and hexagonal phases of TiO appear. At higher fluence (1×10^{12} ions cm^{-2}), the intensity of the XRD peaks due to the unstable phases decreases considerably. At the highest fluence of 3×10^{12} ions cm^{-2}, the peak due to rutile phase is most intense and peaks due to all other phases are suppressed as shown in Fig. 6.13. The inset shows the variation in intensities of different phases of TiO_2 with ion fluence.

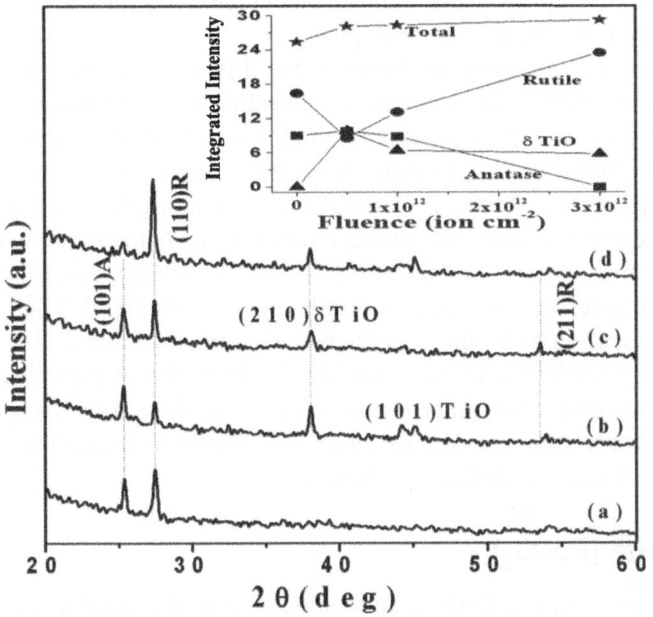

Figure 6.13: GAXRD of TiO_2 thin films annealed at 1000 °C and irradiated with 200 MeV Ag ion (a) pristine, (b) 5×10^{11}, (c) 1×10^{12}, and (d) 3×10^{12} ions cm^{-2}. The inset shows the evolution of total integrated intensity of XRD peaks and the intensity due to individual phases with irradiation fluence. [Reprinted from H. Rath, P. Dash, T. Som, P.V. Satyam, U.P. Singh, P.K. Kulriya, D. Kanjilal, D.K. Avasthi and N.C. Mishra, *J. Appl. Phys.*, **105** (2009) 074311, copyright (2009), American Institute of Physics.]

Raman study shows that complete conversion of anatase to rutile phase of TiO_2 does not occur even at 1000 °C annealing temperature. However, complete conversion from anatase to rutile phase occurs after the irradiation.

The SHI induced thermal spike in nanoparticle thin film of TiO_2, which leads to grain growth and improvement in crystallinity with increasing irradiation fluence, also seems to be responsible for transformation of anatase to rutile phase. Ordinarily, the transient temperature rise in the wake of SHI leads to the formation of amorphized latent tracks in a crystalline medium if the melted zone along the ion path is quenched very fast. However, if the quenching is hindered, the resulting slower cooling can lead to recrystallization starting from the track periphery [17]. This process can even lead to crystalline-to-crystalline phase transition if the thermal spike temperature T_s exceeds the phase transition temperature T_p [7]. At intermediate fluences, XRD shows formation of other TiO_2 phases out of the mixed anatase/rutile phase in 1000 °C annealed film. Evolution of these phases may be a consequence of the instability of TiO_2 permitting the formation of a series of intermediate Ti_nO_{2n-1} crystalline phases [18] on SHI irradiation. At the highest fluence of 3×10^{12} ions cm^{-2}, where the thermal spike induced cylindrically excited regions of 5 nm radii [19] start to overlap, the anatase and all other phases are completely suppressed and the film contains mostly rutile phase.

Though high temperature provides the driving force for transition from anatase to rutile phase in both thermal annealing and in SHI irradiation, there is distinct difference between the two processes. In thermal annealing rutile evolves through the growth from the surface of anatase grains. In case of energetic ions, they can pierce through the grains along their path and convert the anatase to rutile within the grain. Thermal spike temperature in oxide film has been estimated to be ~1000 °C [20], which is within the grains and facilitates complete conversion of anatase to rutile phase as observed. Thus due to the intense interaction of the incident ion with the target atoms the SHI irradiation is a unique post-deposition treatment for the formation of pure phase of rutile TiO_2. There have been other works on SHI irradiation of TiO_2 thin films [21-26], where transformations from amorphous to crystalline and poly-crystalline to nanocrystalline are shown.

Magnetite (Fe_3O_4)

Magnetite [Fe_3O_4] is a ferrimagnetic mineral, one of several iron oxides, also known as ferrous-ferric oxide [$FeO.Fe_2O_3$]. It is a member of the spinel group with the formula $A(B)_2O_4$. A and B are usually different metal ions that occupy specific sites in the crystal structure. In magnetite, A metal is Fe^{+2} and the B metal is Fe^{+3}. It is a metallic ferrimagnet at ambient conditions but there is a discontinuous drop in the conductance on cooling the sample below 122 K, known as Verwey transition [27]. One third of the Fe ions with valence state +3 occupy the A site, which is tetrahedrally coordinated by four oxygen ions. The remaining two-thirds with equiproportion of +2 and +3 valence state are octahedrally surrounded by six oxygen ions, known as B site. Fe_3O_4 is an important ferromagnetic material having 100% spin polarization at Fermi level [28]. Some of its interesting properties are its high Curie temperature (~850 K), low electrical resistivity at room temperature and charge ordering at 120 K (metal to insulator Verwey transition [27]. At 120 K structural transition from room temperature cubic phase to monoclinic phase [27, 28] and across this transition charge ordering also take place in the lattice.

Swift Heavy Ion Modification of Fe_3O_4 Thin Film

A detailed analysis of the pristine and 190 MeV Ag ions irradiated Fe_3O_4 thin films at different fluences (from 5×10^{10} to 1×10^{13} ions cm^{-2}) shows that at low fluence the films are partially relaxed, whereas at higher fluence the mixed phase and finally the disordered *maghemite* phase appears [29]. The resistivity measurements show that the Verwey transition temperature T_V of these films increase with the ion fluence values from 109 K for pristine to 117 K for the film irradiated with 5×10^{11} ions cm^{-2}. This may be due to changes in the structure of the anti phase boundaries by SHI irradiation. However, at higher fluences the films do not show Verwey transition down to 77 K and the resistivity value is higher than that of the pristine film again representing that the SHI irradiation induced structural disorder in the film in agreement to XRD results.

The study of the isothermal dc magnetization hysteresis at 300 K for pristine Fe_3O_4 and 190 MeV Ag ion irradiated films shows that the magnetization for the irradiated film is higher than that of the pristine film. As the fluence is increased up to 5×10^{12} ions cm^{-2} the magnetization decreases rapidly, which is consistent with the formation of iron oxide phases other than magnetite [29].

These experimental results show that at low fluence the swift heavy ion irradiation modifies the strain of the films and cation distribution in the vicinity of APBs. Whereas, at higher fluence, it introduces structural disorder and transforms the phase from magnetite to maghemite. It indicates that by selecting appropriate irradiation parameters, the material's properties can be tuned to get an optimum Verwey transition temperature and the maximum magnetization in this system. The SHI irradiation is a sensitive tool to modify the various parameters of these materials [29, 30].

SHI irradiation results in texturing, surface evolution, phase transformation and modification in Verwey transition temperature in TMO's (NiO, TiO_2 and Fe_3O_4). It is shown that complete rutile TiO_2 film can be achieved and Verwey transition temperature in magnetite can be tuned by SHI irradiation.

6.6 DILUTED MAGNETIC SEMICONDUCTORS (DMS) AND OTHER MAGNETIC MATERIALS

Introduction

The semiconductor based electronics uses the charge property of the electron. But apart from the charge, electrons also possess spin. In the conventional electronics, the spin of the electron is not utilized. Mass storage of information carried out by magnetic recording use spin of electrons in ferromagnetic materials. In order to use 'spin' degree of freedom in electronic devices one needs to create, sustain, control and detect the spin polarization of electrons. 'Diluted magnetic semiconductor (DMS)' has strong potential for use in the spin based devices. The semiconductors used for devices, such as Si and GaAs, do not contain magnetic ions and are nonmagnetic. The crystal structures of magnetic materials are usually quite different from that of the semiconductors used in electronics, which makes these materials incompatible with each other. Magnetic semiconductors (MS) are materials which exhibit both ferromagnetism and semiconductor properties. These materials are of interest as they can provide a new type of control of conduction. The traditional electronics is based on the control of charge carriers (n- or p-type). The magnetic semiconductors can allow control of quantum spin state (up or down).

The usefulness of semiconductors resides in the ability to dope them with impurities to change their properties, usually to p- or n-type. Similar approach can be followed, i.e. to dope magnetic elements into nonmagnetic semiconductors to make them magnetic. This category of semiconductors are called Diluted Magnetic Semiconductors (DMS) [1] or Semi Magnetic Semiconductors (SMSC). These are formed by replacing a fraction of the cations in a range of compound semiconductors by the transition metal ions. Some examples of DMS are $Cd_{1-x}Mn_xTe$, $Zn_{1-x}Fe_xSe$ etc. A wide range of devices can be produced using these DMS materials by exploiting their semiconducting and magnetic properties. For example, the energy gap, lattice parameters, defects etc. can be varied in a controlled fashion by varying the component x.

The possibility of using spins of the electrons in addition to their charge in information technology has created a new field of electronics known as "spintronics." Manipulating an electron's magnetic state in a semiconductor device is the key to successful spintronics, and the simplest way to do that is to incorporate magnetic elements in a semiconductor material. An intensely studied approach to obtain spin-polarized carriers for data-storage devices is the use of diluted magnetic semiconductors created by doping ions like Mn, Fe, or Co having a net spin into a semiconducting host such as GaAs, ZnO, or GaN. The interaction among these spins leads to ferromagnetic order at low temperatures, which is necessary to create spin-polarized carriers. A major challenge to develop such devices is creating magnetic semiconductor materials that work at room temperature.

To realize diluted magnetic semiconductor (DMS) and transparent magnetic semiconductor (TMS) materials based magneto-electronic or magneto-optical device is a daunting task. Significant research efforts have been made so far by different scientific groups throughout the globe. A brief survey of this is given here.

Oxide semiconductors being wide band gap materials are transparent in visible region and if optimally doped with n-type carriers become potential candidate for transparent conductors [2]. Research interest is focussed to realize optical as well as magnetic properties in a single host matrix [3, 4] for future opto-spintronic devices, such as magnetic random access memories, optical isolators, quantum computers etc. Zinc oxide (ZnO) doped with transition metals has been identified as one of the potential candidate in this regard [5, 6].

The basic requirements of these materials are ferromagnetism with high Curie temperature (T_c) value (room temperature and above) and good optical transmittance. Ferromagnetism in transition metal doped ZnO was theoretically predicted by Sato et al. [5] using abinitio calculations based on local density approximation. Both physical and chemical synthesis routes have been adopted to grow these materials. Molecular Beam Epitaxy (MBE) [7], Pulsed Laser Deposition (PLD) [8] and RF magnetron sputtering [9] are some of the important physical processes to develop these materials. Spin coating of high quality colloidal DMS quantum dots as solution precursors [6], sol-gel technique etc. are among the chemical synthesis routes of these materials. Implantation of transition metal ions in wide band gap semiconductor host has also emerged as another potential technique [10, 11]. Good control over depth, concentration and radiation induced defects by implantation helps tuning the magnetic and transport properties of these materials.

Swift Heavy Ions for DMS

Ion implantation gives clustering of transition metals in the matrix and hence needs high temperature post-annealing processes. Swift heavy ions (SHI) provide a different way to synthesize DMS materials. Diluted magnetic semiconductors (DMSs) with reduced dimensions, which have ferromagnetic phase transition temperatures (T_C) above 300 K, are seen as the prime candidate materials for spintronic technologies. Integration of DMS materials in future electronics will require very low dimension, in order to make real use of the advantage offered by the spin. Efforts are being made to identify and prepare transition-metal-doped semiconductor nanostructures that exhibit ferromagnetic ordering above room temperature, high charge carrier concentration and mobility, and are transparent to visible light for integrated optoelectronic applications [12]. Swift heavy ions can provide a way to produce appropriate nanostructures to achieve the objective.

Dissolution of Co Nanoclusters in Co Doped ZnO Thin Film

The effect of swift heavy ion irradiation on Co doped ZnO thin film for spintronics application has been investigated [13]. Co implanted ZnO films were irradiated with 200 MeV Ag ions to fluence of 1×10^{12} ions cm^{-2}. X-ray diffraction (XRD) spectra indicated that Co clusters dissolve in the ZnO matrix on irradiation. The presence of Co clusters of nanosize indicate that in the Co implanted ZnO films, Co does not go to the Zn site. Further, it was found that the Co cluster size increases as the doping concentration is increased. Co clusters and other phases of Co oxide are dissolved with the ZnO matrix after the ion irradiation. SHI irradiation has been very effective in taking the sample to very high temperature through electron-phonon coupling in a short span of time, which essentially dissolves the Co clusters and does not allow them to grow. The electrical resistivity of the irradiated samples is lowered to half. The magnetization hysteresis measurements show ferromagnetic behaviour at 300 K.

Room temperature ferromagnetism and a metal–semiconductor transition was observed at 227 K in 200 MeV Ag^{15+} ion irradiated thin films of Fe-implanted ZnO. The single phase nature of Fe-doped ZnO after ion irradiation was confirmed by X-ray diffraction. Magneto-resistance measurements show spin polarization below 150 K. X-ray absorption spectroscopy and X-ray magnetic circular dichroism studies at room temperature reveal that Fe is oxidized in a mixed valence (Fe^{2+} and Fe^{3+}) state and the magnetic signal is due to the Fe^{2+} state [14].

SHI in Magnetic Tunnel Junctions

Apart from the synthesis of DMS, SHI has been used in other spintronics materials. A trilayer system consisting of two ferromagnetic layers separated by an insulating layer acts like a magnetic tunnel junction (MTJ) [15]. Tunneling magnetoresistance (TMR) is a specific feature of MTJ, dealing with change of resistance in presence of magnetic field. The 70 MeV O ion irradiation of magnetic tunnel junction causes reduction of TMR whereas the magnetic tunnel junction is completely damaged by 200 MeV Ag ion irradiation [16]. Effects of SHI bombardment on functional properties of magnetic tunnel junction (MTJ) have been investigated with various ions and fluences in 10 MeV/amu energy range [17]. It is observed that the tunnel magnetoresistance of magnetic tunnel junctions (MTJ) decreases irreversibly with increasing ion fluence, with little or no impact on overall resistance for all ions with different fluence thresholds.

SHI in Giant Magnetoresistance (GMR)

Srivastava et al. [18] investigated Giant magnetoresistance (GMR) in swift heavy ion irradiated Fe films on c-silicon. Fe/c-Si devices irradiated with 100 MeV ions of Fe^{7+} at a dose of 10^{14} ions cm^{-2} show large increase in magnetoresistance (MR), up to 2400%. Unirradiated devices do not show any effect of the magnetic field. M–H study of the irradiated devices shows a behaviour of coupled magnetic nanograins. The results are explained by considering the formation of a nanogranular magnetic silicide phase (of Fe_5Si_3) due to intermixing at the interface (as evidenced from XRD and SEM features). The observed strong GMR could be due to the spin dependent interface scattering in the presence of the strong AF coupling across the tunnelling barrier.

Defect Mediated Ferromagnetism in MgO

The samples of crystalline MgO doped with Ni and Co impurities have small concentration of equilibrium vacancies and show perfect paramagnetic behaviour. The SHI irradiation of these sample results in ferromagnetic ordering [19], indicating the defect mediated ferromagnetic behaviour.

Materials with Perpendicular Magnetic Anisotropy (PMA)

SHI irradiation induces PMA in ferromagnetic amorphous $F_{18}B_{13.5}Si_{3.5}C_{12}$ ribbon due to the stress generated between the unirradiated regions and anisotropic deformed irradiated regions [20]. The Fe/Tb multilayers exhibits a large PMA which decreases with 80 MeV Si ion irradiation [21]. The decrease in PMA is due to stress relation in bulk film and increase in interfacial roughness [21]. SHI irradiation of $Fe_{0.85}N_{0.15}$ results in gradual removal of PMA which is due to the relaxation of stress in the film [22]. Role of SHI in PMA in magnetic nanoparticles embedded in insulating matrix is discussed in Chapter 5.

SHI irradiation of granite ceramic $Y_3Fe_5O_{12}$ leads to the formation of composite of ferromagnetic crystal and paramagnetic amorphous ion tracks [23].

SHI is emerging as a tool which will provide ways to develop spintronics materials. It is demonstrated that they play crucial role in dissolving Co clusters in ZnO where it is important not to have the magnetic elements as clusters. It has been used in synthesizing magnetic tunnel junction and material for GMR. SHI created defects induce ferromagnetic ordering in doped MgO. SHI irradiation can induce PMA.

6.7 HIGH TEMPERATURE SUPERCONDUCTORS (HTSC)

Introduction

Electric resistance in simple free electron metals is caused by the scattering of electrons by phonons (quantized lattice vibrations) and the impurities or imperfections. As the temperature is reduced, the first component (intrinsic resistivity) starts decreasing. At sufficiently low temperature, around 20 K, this part of resistivity nearly disappears. The second component called residual resistivity, however, is temperature independent and stays constant down to zero degree. Kammerlingh Onnes, soon after liquefying helium in 1908, started probing into the temperature behaviour of the resistivity of pure metals at very low temperatures available with the liquefaction of helium. He observed complete disappearance of electrical resistivity in mercury just below 4.2 K. This phenomenon named superconductivity was discovered by him in 1911. Superconductors have the ability to conduct electrical current with no loss of energy. Later the phenomenon was observed in a large number of metals when cooled below a certain characteristic temperature, called the transition temperature, T_c limited to a highest value of 9.3 K for pure niobium. However, Onnes found that the superconductivity is destroyed by the application of magnetic field beyond a certain value called the critical magnetic field, H_c.

This critical field in pure metals is quite low, of the order of a few hundred Oersteds. Further, the superconductivity can also be destroyed by passing a current through the superconductor just above a critical value, called critical current, I_c. Thus a superconductor is bound by three critical parameters, T_c, H_c and I_c. Superconductivity is destroyed if any of these parameters is exceeded.

In 1986, a class of new compounds showing superconducting transition temperature well beyond the previous limit of 23 K in Nb_3Ge sputtered films was discovered. Superconductivity was reported at 90 K in $Y_1Ba_2Cu_3O_{7-\delta}$ compound. This was the first time that the T_c shifted from liquid helium regime to liquid nitrogen regime.

When the magnetic field is applied to a superconductor, some of the field can pierce it through the generation of magnetic vortices, each of which contains some magnetic flux. Although the superconducting state of the material outside each vortex is maintained (destroyed within each vortex) the interaction of vortices with a current passing through the material can cause them to move, dissipating energy and thereby a source of resistance. A limit on technological applications of HTSC comes from the intrinsically low critical current density J_c and from the ease with which magnetic vortices can move. It was shown by Nelsen and Vinokur [1] that the vortices in highly layered high T_c superconductors like the bismuth and thallium family of compounds can be pinned effectively by columnar defects, produced by Swift Heavy Ions (SHI), as had already been established experimentally for the first time by Budhani et al. [2]. Due to their well defined cylindrical shape, columnar tracks provide efficient pinning centres. Every ion, if its electronic energy loss S_e exceeds a material dependent threshold S_{th} (typically of the order of 1.2 keV/A for HTSC), creates one latent track along its path. Introduction of columnar defects improves J_c, the most crucial property of cuprate superconductors for device applications. Introduction of columnar defects is expected to improve the properties of cuprate superconductors for device applications. For the fabrication of superconducting microwave devices such as circulators where the films are subjected to magnetic fields, it is essential to be able to control dissipation due to the motion of vortices. This can be done by locking the vortices to the columnar defects.

Engineering of the HTSC: Intra- and Inter-granular Modifications

SHI irradiation is one of the widely used methods to produce controlled defects inside HTSC materials [3]. A study of transient behaviour of resistivity during irradiation of films was carried out by Mohanty et al. [4] with an aim to understand the mechanism of creation and annealing of defects in cuprate superconductors. Thin films (~3000 Å) of $YBa_2Cu_3O_{7-\delta}$ (YBCO) deposited on $LaAlO_3$ were studied with 100 MeV O ions. The resistivity was measured continuously during the irradiation and also after turning the beam off. Studies were also carried out on bulk Bi(Pb)-2223 superconductor irradiated with 100 MeV oxygen ions. Positron lifetime technique was employed to probe the defect structure. Changes observed in resistivity and the critical temperature, T_c were interpreted by taking into account the microstructural changes, the defect structure and effect of SHI on the grain boundaries in sintered materials as well as in thick films [5]. It was observed that S_e leads to various different types of effects depending on the nature of the target. SHI can partially anneal defects originally present in the material, can create phase transformation (amorphization or crystallization), anisotropic growth of the material etc. Atomic size point defects can be produced if S_e is below the threshold value S_{th} and the microstructure can be modified.

In situ resistance measurements by Agnihotri et al. [6] during 100 MeV O ion irradiation of Bi-Pb-Sr-Ca-Cu-O bulk superconductors show that the oxygen stoichiometry of the unirradiated sample plays a critical role in its response to ion irradiation. The ion induced defects are associated with the displacement of the excess oxygen atoms. The study led to the conclusion that due to the mobile nature of these defects, they are redistributed leading to a reduction in resistance below certain fluence ~ 3×10^{12} ions cm^{-2} for the 100 MeV oxygen ion and a concomitant increase of the transport critical current.

The grain boundaries and oxygen deficient regions in the grains are known to act as weak links to superconduction in sintered cuprate superconductors. Proper oxygen-stoichiometry is crucial to all cuprate superconductors. For example, the perfect $Y_1Ba_2Cu_3O_7$ superconductor should have O_7. Oxygen deficient YBCO turns in to a semiconductor when O is below 6.5. Further oxygen deficiency causes $Y_1Ba_2Cu_3O_6$ (with O_6) to become an insulator. SHI irradiation has been found useful for improving oxygen stoichiometry in these materials. Sudarsan et al. [7] indeed observed displacement of oxygen from the columnar tracks to the oxygen deficient regions of Bi-system when irradiated with 250 MeV Ag ions. This eliminates weak links and improves the superconducting behaviour of such cuprates. Ag residing at the grain boundaries in the composite films has a strong affinity for oxygen [8]. The oxygen ions, which are liberated from the ion tracks during irradiation, can thus migrate over a much larger distance (compared to the ion track radius) leading to improved oxygen stoichiometry in oxygen deficient regions.

Kaplan et al. [9] compared the far infrared (FIR) transmission spectra in the spectral range ~2 to 80 cm^{-1} of $YBa_2Cu_3O_{7-\delta}$ thin films with columnar defects produced by 270 MeV Ag ions. These measurements performed in the normal as well as the superconducting state in a magnetic field as high as 9 Tesla show significant increase in FIR transmission of samples with columnar defects. However, the magnetic field does not change the transmission appreciably which could otherwise be related to the vortex physics. The changes in transmission produced by irradiation has been modelled in the framework of Maxwell-Garnett effective medium theory which treats the amorphous track as insulating regions dispersed in a metallic background. The fraction of the insulating phase required to fit the data agrees with the density of columnar defects.

The irradiation experiments [10] of cuprate superconductors by High Energy Light Ions (HELI) show that the response of the superconductor to ion irradiation is governed by the pre-radiation defect structure in the films. The experiments by Mishra et al. [10] on irradiation response of two types of YBCO films: YBCO/Ag composite and Ag-free YBCO film indicate that YBCO/Ag composite films do not show any change in the normal state resistivity when irradiated with 140 MeV Si ions upto a fluence of 8×10^{14} ions cm^{-2}. At the same time the T_c shows hardly a scatter of about 0.4 K. The YBCO/Ag composite film thus appears to be insensitive to ion irradiation. The pure YBCO film, on the other hand, shows a large increase in the normal state resistivity and a sharp drop in T_c as the fluence rate is increased.

HELI irradiation induces only point defects in sintered granular YBCO thick films. These defects cause a faster degradation of the grain boundaries as compared to the grain. The superconducting phase coherence across the grain boundaries is therefore suppressed and the samples show semiconducting behaviour. In addition to this general kind of irradiation effect in a granular medium, an interesting aspect of irradiation in inducing improvement in the sample quality with improved grain alignment has been observed [11].

In comparison to pure YBCO, the YBCO/Ag composite thick films have much suppressed weak link resistivity. Response of this system to SHI irradiation is therefore mostly intragranular. The evolution of this system with fluence when irradiated with 250 MeV Ag ions with $S_e > S_{eth}$, should be governed by the volume fraction of the amorphized latent tracks and T_c suppression should occur only when current percolation is inhibited by these tracks. For a fluence of 5×10^{10} ions cm^{-2}, the damage area is calculated to be ~4% of the area exposed to ion beam. With such a low fraction of the damage area, the percolation conduction in the thin film cannot be eliminated. The decrease of T_c at this fluence therefore is not expected from track. Even for the fluence of 5×10^{11} ions cm^{-2}, tracks occupy 40% of the surface area, which is well below the percolation threshold of 60 % in two dimensions. Against this expectation, there is T_c suppression by about 10 K even when the volume fraction of the ion tracks is an order of magnitude less than that required to bring about this change. This result points to an enhanced interaction cross section of the SHI in the medium which was explained in terms of frozen point defects created by SHI induced secondary electrons [12, 13]. The effect of ion irradiation in inducing materials modification arises not only through creation of amorphized latent tracks along the ion path, but also through creation of atomic disorder in the oxygen sublattice in the Cu-O chains of YBCO by the secondary electrons. These electrons are emitted radially from the tracks during the passage of the SHI [12]. In situ XRD experiments during 200 MeV Ag ion irradiation of YBCO thin film reveals [14] the picture that SHI induces three different types of defects in YBCO at low temperature ~89 K: (i) amorphous ion tracks provided $S_e > S_{eth}$, (ii) strained region around ion track, and (iii) oxygen disorder point defects in the Cu-O basal plane. The track diameter is ~1.9 nm and one defected zone is developed around the track in the range ~97 nm. This observation is explained on the basis of SHI induced secondary electrons which create oxygen disorder by an electron capture process in the Cu-O chains of a fully oxygenated YBCO structure. This could also be due to the transiently heated annular region surrounding the ion track. In the case of NBCO the compositional variation of spatial density of defects and the concomitant vortex mobility was shown [15].

Effect of Irradiation on Excess Conductivity in HTSC

High temperature superconductors are characterized by high T_c, small coherence length and have layered structure of the conducting CuO_2 planes. As a result these materials exhibit excess conductivity or paracoherent regions which extend at temperature well above T_c. Fluctuation effects are thus expected to dominate at such temperatures in these superconductors. Effect of disorder on the critical exponent of the electrical conductivity in the paracoherent region of the high temperature superconductor YBCO was studied by irradiating the material with 100 MeV $^{16}O^{7+}$ ions at liquid nitrogen temperature. The critical exponent was found to change from a value of about 2 for the pristine sample to a value of 1.62 in the irradiated sample [16, 17]. An analysis of the excess conductivity has been made within the framework of Aslamazov-Larkin (AL) [18] and Lawrence-Doniach (LD) theories [19]. 2D to 3D transition is observed with increasing fluence. The coherence length and the Josephson coupling can be estimated from LD theory [20].

Modifications of Defect Structures

A pronounced peak is observed in the microwave surface resistance in thin films of YBCO in magnetic fields, which is associated with "peak effect" [21]. Introduction of artificial pinning centres like columnar defects as a result of irradiation with 200 MeV Ag ion leads to the suppression of the peak in films exhibiting the 'peak-effect'. The peak effect (PE) in surface resistance is observed in thin HTSC films at subcritical currents at 4.88 and 9.55 GHz microwave frequencies in dc magnetic fields in the range 0.2–0.8 T. The PE phenomenon observed in different samples shows that the nature of the peak depends on the density of pinning centres, pinning potential, and hence, on the defect structure in the samples [21].

Amorphous Columnar Tracks and Flux Pinning

The irradiation induced enhanced flux pinning of the high T_c material suppresses the flux motion induced noise in the vicinity of T_c. Effect of 250 MeV ^{107}Ag ion irradiation induced columnar defects on the noise properties of YBCO was investigated [22]. The noise performance in the vicinity of the superconducting transition as well as in the normal state is found to improve by an order of magnitude with these ions. Influence of columnar defects produced by 250 MeV Ag ions on magneto-transport and magnetization properties of BSSCO single crystal was studied by Pradhan et al. [23]. It was shown that the columnar defects produced by the ions affect significantly the magnetotransport properties of $Bi_2Sr_2CaCu_2O_{8+y}$ single crystals. The generally observed long tail in the magneto-resistance curve is drastically reduced and magnetic phase diagram is modified in an irradiated BSCCO crystal displaying the suppression of thermal fluctuations and enhanced interlayer coupling. SHI irradiation induced effects on the magnetic properties of MgB_2 thin films have been studied with films of thickness 300-400 nm and 200 MeV Au ions, along with the change in surface morphology [24].

Frequency Dependence of Vortex Dynamics in SHI Irradiated High T_C Cuprates

Microwave absorption in YBCO thin films with columnar defects was studied [25] by measuring the angular dependence of microwave magneto-absorption of both pristine and 270 MeV Ag ion irradiated YBCO C-axis oriented films. Budhani et al. [2] had seen using dc measurements that columnar defects are introduced at these energies. These measurements effectively measure the depth of the pinning potential. Lofland et al. [26] obtained similar results in the microwave frequency regime. The detailed study at 10 and 32 GHz frequency indicated that the observed magneto-absorption arise from two contributions, namely, a small Lorentz-force-free component and another, the loss due to viscous damping of the vortices. It was found by Lofland et al. [27] that at 10 GHz there is an increased pinning but it reduces at higher frequencies. Patnaik et al. [28] made a detailed study of the mixed state properties of highly c-axis oriented Bi-2223 superconductors using a high frequency oscillator technique. These measurements of rf penetration depth in the unirradiated and SHI irradiated pellets of this superconductor unambiguously establish granularity in these composite conductors. They also carried out a quantitative estimation of the pinning force constant and its enhancement by SHI. The localizing effect of columnar defect sites as a function of temperature, defect density, anisotropy and magnetic field strength have been studied through a precision angle dependent measurement of the radio frequency response. The SHI irradiated samples show cusp-like feature in the rf response when the external magnetic field is brought into alignment with the defect sites. Vortex dynamics in SHI irradiated epitaxial thin films of $Tl_2Ba_2CaCu_2O_8$ superconductor has been studied by Sahoo et al. [29] using a micro-Hall probe based ac susceptometer, using 250 MeV Ag ions.

Observations on the vortex locked-in state in YBCO thin films with columnar defects produced by 270 MeV Ag ions showed that the effective pinning along the column increases with increasing fluence, and is a function of applied field [27]. Shifting of irreversibility line to higher magnetic field and temperature in Bi-2212 single crystals has been observed by Pradhan et al. [30].

Enhancement of Critical Current Density

Practical applications of high temperature superconductors demands that the material should have high critical current density (J_c). As the cuprates are granular in nature, the high density of grain boundaries impedes the flow of current. It is generally believed that grain boundaries together with the poor flux pinning are the main factors for low J_c in HTSC. Irradiation with lighter ions like 120 MeV S, which does not create amorphized latent tracks ($S_e < S_{th}$) is seen to cause grain alignment in sintered YBCO thick films as revealed from the enhancement of current percolation factor with ion fluence [10] and grain aligned sample show better conduction.

The weak link problem between the grains, responsible for poor J_c in these materials, has attracted the attention of many researchers. Attempts have been made to prepare samples of small misorientation angle between the grains and by reducing grain boundaries. The layered structure of HTSC with complex crystal chemistry and the inherent granularity lead to strong structural disorder at the microscopic and the mesoscopic levels respectively. The inhomogeneities occur at widely varying length scales in single crystals, grain aligned films, melt-textured and sintered samples. The mesoscopic inhomogeneities like grain boundaries, cracks, voids etc. have much larger length scale than coherence length. However, the best grain boundary J_c value is much smaller than the J_c in grains and much less for practical applications especially in high magnetic field.

Swift Heavy Ion (SHI) Irradiation technique is a versatile technique for creating controlled columnar defects in cuprate superconductors [31]. These defects pin the flux lines along entire path of the beam rather than occasional pinning at randomly distributed point defects. These controlled columnar defects serve as efficient pinning centres. Pinning centres can also be achieved in nanostructured materials. Several recent studies have reported improved flux pinning in nanostructured HTSC materials prepared by a variety of routes [32]. The extremely local nature of energy deposition leads to creation of nanosized hillocks or nanodots under normal incidence. Akcoltekin et al. have shown that multiple nanodots can be created if a single high energy Xenon ion strikes the surface at a grazing angle by SHI and it will produce latent tracks [32].

SHI irradiation is of interest in increasing critical current density in presence of moderate magnetic field and producing a shift in the irreversibility line towards higher temperatures [31-34]. Enhancement of flux pinning in cuprate superconductors have been observed in HTSC [35] by irradiating the material with 200 MeV Ag ions. An enhancement of J_c(H) in La-2125 thin films has been reported [35] with increasing Ag^{+15} ion irradiation upto an ion fluence of 1×10^{11} ions cm^{-2}. Interestingly, the maximum enhancement in J_c(H) is observed to be five-fold higher (1.6×10^7 $A.cm^{-2}$) compared to J_c(H) of pristine thin films. This large enhancement in J_c(H) of irradiated thin film is attributed to augmented flux pinning caused by the columnar defects.

Haugan et al. [36] have shown that high density pinning centres can be achieved in YBCO in form of ~8 nm nano particles which will increase J_c by a factor of two or three at high magnetic field [36].

The interaction of the ions with the material modification (craters, hillock, tracks etc.) is still not exactly understood. The thermal spike model describes the energy transport out of the electronically heated region whereas Coulomb explosion model couples the electronic excitation due to repulsive forces [37]. Both the models show the material modifications though in different ways [38]. It has been reported that columnar defects generated by irradiating HTSC materials with heavy ions significantly enhance the infield critical current density [39].

It has been found that J_c in MgB_2 films increases after irradiation, the enhancement being more pronounced for Au ion irradiation, which is attributed to higher flux-pinning efficiency in the irradiated samples [40, 41]. Kaushik et al. [42] have ascertained that extended defects alter the band properties of this two-gap superconductor MgB_2 more effectively than the point defects when irradiated with SHI. They find that while all defects alter the σ band scattering mechanism, extended defects along the c-axis preferentially affect the π band. The improvement in upper critical field and critical current density is intricately related to the type and density of created defects. $NdBa_2Cu_3O_{7-\delta}$ (NBCO) thin films (2000 Å) grown by excimer KrF laser, irradiated with Au^{+13} (200 MeV) ions, show an almost four-fold increase in J_c at the optimum ion dose, indicating a robust pinning of vortices [43].

These experiments demonstrate that the SHI irradiation is a very potent technique to control the defect structure in a HTSC and increase the J_c value very significantly. Ion tracks are of significant importance for its role in flux pinning of vortices in the high T_c superconducting thin films.

6.8 SPECIAL OXIDES (FERRITES, MULTIFERROICS AND LCMO)

Ferrites

Ferrites are ferrimagnetic ceramic compounds with the chemical formula AB_2O_4 (where A and B represent various cations). They are usually derived from iron oxide such as hematite (Fe_2O_3) and oxide of other metals. Due to their low coercivity, these are widely used as magnetic cores in transformers. There are variation in ferrites by replacing fraction of cations at A site by other ions, to achieve variations in electrical and magnetic properties.

Irradiation of nanocrystalline thin film of $Li_{0.25}Mg_{0.5}Mn_{0.1}Fe_{2.15}O_4$, on Si(1 0 0) substrate by 190 MeV Au^{14+} ions results in drastic change in resistivity (reduction by three orders of magnitude) as revealed by in situ measurement of electrical resistance using two-probe method [1]. On-line ERDA measurement shows that there is no loss of oxygen content within the film with the increase in ion fluence. Transformation from ferrimagnetic to superparamagnetic behaviour occurs in nanocrystalline Li-Mg ferrite under 190 MeV Au ion irradiation beyond certain fluence [2].

The study of the pristine and SHI irradiated single crystal plates of ferrimagnetic yttrium iron garnet (111)-YIG:Si ($Y_3Fe_{4.94}Si_{0.06}O_{12}$) and barium hexaferrite (00.1)-BaM ($BaFe_{12}O_{19}$) or (00.1)-BaM:Co,Ti ($BaFe_{9.1}Co_{1.4}Ti_{1.5}O_{19}$) in the electronic slowing down regime, above the S_e threshold of formation of ion track, reveals that magnetomechanical effect generated by the stress field induced by the amorphous tracks flips the magnetization along the track-axis direction [3]. This shows that the stress created by SHI can be used in tilting the plane of magnetization.

Multiferroics

Multiferroics are the materials having more than one ferroic order in a single phase. These ferroic order parameters are ferromagnetism, ferroelectricity and ferroelasticity. The ferroelectric material undergoes spontaneous electrical polarizaion under electric field and thus ferroelectricity is an electrical anologue of ferromgnetism. The ferroelastic material undergoes phase transition by application of stress and thus ferroelasticity is the mechanical anologue of ferromagnetism. Another, fourth ferroic order parameter, which is under debate is the ferrotoroidicity pertaining to magnetic vortices [4].

Shukla et al., showed that a muliferroic film of $BiMnO_5$ (synthesized by pulsed laser deposition) undergoes a transition from antiferririmagnetic to ferrimagnetic by 200 MeV Ag ion irradiation at low fluence due to evolution of Mn^{2+} [5]. It demonstrates the capability of the SHI irradiation to tailor the properties of oxide multiferroics. The multiferroic $BiFeO_3$ thin film (existing in rhombohedral phase) undergoes complete amorphization under 200 MeV Ag ion irradiation, indicating that S_e deposited in the film is larger than the threshold for the creation of ion track [6].

LCMO

Transport properties of doped pervoskite of type $A_{1-x}B_xMnO_{3+y}$ (where A is La, Nd and Y, and B is Ca, Ba and Sr) have attracted attention due to their large magnetoresistance referred to as Colossal Magneto Resistance (CMR) [7]. Basically in the parent compound of $RMnO_3$ (where R is rare earth element), R is partially substituted by divalent cations such as Ca, Sr etc, to achieve high CMR. Similarly when the rare earth is partially subsituted by tetra valent Ce, a high value of CMR is observed. The difference being that divalent doping results in hole-doped system whereas Ce doping results in electron-doped system.

The resistivity transition temperature and magnetic transition temperature in thin LCMO ($La_{0.25}Ca_{0.75}MnO_3$) film deposited by PLD can be tuned by 90 MeV O ion irradiation at different fluences [8]. LCMO thin film with composition $La_{0.7}Ca_{0.3}MnO_3$ on irradiation with 250 MeV Ag ions results in the formation of ion tracks, increase in resistivity by an order of magnitude and decrease in Curie temperature [9, 10]. High values of CMR and high temperature coefficient of resistance (TCR) in preferred range of temperature and magnetic field are desirable. Irradiation of $La_{0.7}Ca_{0.3}MnO_3$ thin films by 200 MeV Ag ions can tune the TCR value and magnetic sensitivity in different temperature regime by ion fluence [11]. A study on 250 MeV Ag ion irradiation of $La_{0.7}Ca_{0.25}MnO_3$ thin films reveal that the metal insulator transition temperature and resistivity can be varied by the ion fluence [12].

These examples of SHI irradiation of special oxides such as ferrites, multiferroics and LCMO show that the properties of these special oxides can be tuned by SHI irradiation.

6.9 RESISTIVE RANDOM ACCESS MEMORY (RRAM) DEVICES BASED ON OXIDES

Introduction

Modern semiconductor nonvolatile memories, such as flash memory, have been scaled down to achieve large capacity memories through state of art photolithography technology. The conventional memory scaling, however, is expected to meet technical and physical limits in the near future. The new materials such as ferroelectric random access memory (FeRAM), in which the polarization of a ferroelectric material is reversed, magneto resistive RAM (MRAM), which uses magnetic tunnel junctions, and phase-change RAM (PRAM), which uses the change in resistance between crystalline and amorphous states of a chalcogenide compound, have attracted attention of researchers and technologists for use as next-generation nonvolatile memories. Recently, a new candidate—resistance random access memory (RRAM)—based on resistance switching has emerged [1, 2]. It is based on metal oxides [3] and organic compounds [4, 5], which exhibit a resistive switching phenomenon. The RRAM memory cell has a capacitor-like structure, which consist of insulating or semiconducting layer sandwiched between two metal electrodes. A recent study has shown that the switching speed can be in nanoseconds [6]. In 1962, Hickmott [7] first reported hysteretic current–voltage (I–V) characteristics in metal insulator-metal (MIM) structures of $Al/Al_2O_3/Al$, indicating that resistive switching occurs as a result of applied electric fields. Resistive switching has subsequently been reported in a wide variety of MIM structures composed of binary metal oxides, such as SiO [8] and NiO [9]. Some models for the driving mechanism of resistive switching have been proposed, e.g., the charge trap model [8] and conductive filament model [9]. Earlier studies on binary metal oxides have been reviewed [10]. In the 1990s, complex transition metal oxides, such as perovskite-type manganites and titanates, became the focus because of a report of resistive switching in $Pr_{0.7}Ca_{0.3}MnO_3$ (PCMO) by Asamitsu et al. [11].

Classification of Resistive Switching Behaviour

The resistive switching phenomenon has been observed in a wide variety of transition metal oxides, such as PCMO [12], NiO [3, 9], TiO_2 [13] and $SrTiO_3$ [14, 15]. However, the observed switching behaviour seems to differ depending on the material. On the basis of I-V characteristics, the switching behaviours can be classified into two types: unipolar (nonpolar) and bipolar. In unipolar resistive switching, the switching direction depends on the amplitude of the applied voltage and not on the polarity. An as-prepared memory cell is in a highly resistive state and is put into a low-resistance state (LRS) by applying a high voltage stress. This is called the 'forming process'. After the forming process, the cell in a LRS is switched to a high-resistance state (HRS) by applying a threshold voltage ('reset process'). In the set process, the current is limited by adding a series resistor. Bipolar resistive switching shows directional resistive switching depending on the polarity of the applied voltage. This type of resistive switching behaviour occurs with many semiconducting oxides, complex perovskite oxides [12].

Resistive Switching Mechanisms

The resistive switching can be categorized in terms of the type of conducting path. One class shows a filamentary conducting path, in which the resistive switching originates from the formation and rupture of conductive filaments in an insulating matrix. The driving mechanism for unipolar resistance switching is attributed to filament-type resistive switching. Here, during the forming process, filamentary conducting paths form as a soft breakdown in the dielectric material. Rupture of the filaments takes place during the reset process, and filament formation during the set process. There are indications that the filamentary paths are formed in grain boundaries. Thermal redox and/or anodization near the interface between the metal electrode and the oxide is widely considered to be the mechanism behind the formation and rupture of the filaments [13, 14]. In contrast, in bipolar-type switching, electrochemical migration of oxygen ions is regarded as the driving mechanism [15].

The Role of Swift Heavy Ion irradiation (SHI) in RRAM Research

Since the polycrystalline films of the oxide compounds show better resistive switching properties than their epitaxial counterpart, it favours the defect state model for the retentive behaviour. SHI is an ideal (and only) tool for selectively creating the defect states in a controlled manner in the epitaxial films to examine the defect state model for resistive switching in detail. Combination of experiments such as fluence dependence of resistance switching phenomena along with on-line elastic recoil detection analysis (ERDA) is a unique tool to examine the loss of surface oxygen or the electrochemical migration of oxygen atoms. The irradiation effects of binary oxides (such as Li-doped NiO and anatase TiO_2 thin films) by 100 MeV Ag^{+14} ions results in hysteresis curves, giving evidence of the resistance switching (RS) behaviour induced by SHI irradiation [16].

The RS ratio (R_{high}/R_{low}) of nearly 230% was estimated at an applied voltage of ± 2.4 V for the sample irradiated with the fluence of 1×10^{13} ions cm^{-2}. The RS is highly symmetric and depends on the ion fluence. The switching ratio increases non-linearly with ion fluence. Further, the RS becomes nearly constant for ion fluence doses of 2×10^{12} and beyond. This suggests that the switching is induced by the bulk defects and/or the oxygen vacancy in the vicinity of the Ag/NiO interface, created by swift heavy ion irradiation. Interestingly, upon SHI irradiation, the current values were found to increase by more than two orders of magnitude as compared to that of the pristine sample.

The observed resistance switching in SHI irradiated NiO films reveals several features, such as, independent of bias polarity, high degree of symmetry, enhanced on to off resistance ratio and increased current values compared to the pristine sample. The primary effect of SHI irradiation is to induce gross defects viz., point, planner and/or columnar, to create interstitials, vacancies as well as dislocations in the band gap in oxide system, which results in novel electrical transport properties of it. The unipolar switching in polycrystalline NiO is chiefly due to the formation and rupture of metal filaments which causes, the low resistance state (LRS) and high resistance state (HRS), respectively [3, 7]. However, in most cases the filamentary model is treated as local phenomena, close to the metal/TMO interface. Heavy ion irradiation yields number of gross defect states within the oxide band gap and not only at the metal/TMO interface. SHI irradiation is known not only to create and/or change the defect density, but also the type of defects in the system. The formation of defects depend upon the type of ion, its energy and fluence. The gross defects, the large area filament density, i.e., number of filaments per unit area away from the metal/

TMO interface, may be responsible for the observed RS switching in two dimensional nanostructured Li:NiO thin films [16].

In search for another mechanism for RRAM devices, particularly the bipolar type RS observed in perovskite oxides, recent studies have indicated that the electrochemical migration of oxygen vacancies in the vicinity of the interface drives resistive switching [17]. On-line Elastic recoil detection analysis (ERDA) employed to monitor the compositional changes as a result of irradiation with 100 MeV Ag^{+14} ions [18-20], indicated that there is insignificant change, in the amount of oxygen. However, preferential loss of oxygen very close to the surface and at the metal/oxide interface could not be ruled out [21]. Li-doped NiO is a p-type oxide semiconductor in which the oxygen vacancies are considered to be an acceptor scavenger. Therefore, even a small preferential oxygen loss can result in the interfacial oxygen vacancies upon SHI irradiation that, in turn, cause the depletion layer to become narrower, resulting in a decrease in the observed film resistivity. Combined with the SHI irradiation and fluence dependence results, the observed ERD analysis suggested that the RS mechanism of Ag/Li:NiO/Ag planner structure can be attributed chiefly to the gross defect formation and possible change in the conducting filament density as compared to the interfacial oxygen vacancies.

RRAM based on resistive switching is potential candidate for memory device. SHI irradiation is shown to enhance the resistive switching behaviour as well as a possibility to resolve the basic questions on mechanism of switching.

6.10 QUASICRYSTALS

Introduction

A crystal is defined as a three-dimensional periodic arrangement of atoms with translational periodicity along its three principal axes. Thus it is possible to obtain an infinitely extended crystal structure by aligning building blocks called unit-cells until the space is filled up. Quasicrystals are structural forms that are ordered but nonperiodic. They form patterns that fill all the space but lack translational symmetry. Normally "quasicrystal" refers to a class of binary and higher order metallic alloys that show translational quasiperiodicity and are distinguished by their rotational symmetry of an icosahedron that are forbidden in crystalline structures. Most of the known quasicrystalline alloys contain 60% to 70% of Aluminium (Al). Among the most well-studied system in literature, Al-Cu-Fe and Al-Pd-Mn are well-established icosahedral phases.

These specialized structural forms which can be classified intermediates between crystal and glass are called "quasicrystals". Just like crystals, quasicrystals produce modified Bragg diffraction, but where crystals have a simple repeating structure, quasicrystals are more complex.

The basic question—'where are the atoms in quasicrystals?'—has been essentially a challenge for crystallographers. The theory of aperiodic crystals was given by Levine and Steinhardt [1]. Schetman et al. [2] reported the bright diffraction spots on transmission electron diffraction pattern taken for $Al_{86}Mn_{14}$. Quasicrystalline materials offer a new challenge to material scientists and to production engineers. The main areas that arise from the premises of technological use of quasicrystals and also from the point of fundamentals in physics and chemistry are the potential applications of quasicrystalline coatings where surface properties of these quasicrystals like low friction, reduced apparent surface energy, corrosion and oxidation resistance and hardness are exploited. The unique property of "very low coefficient of friction" particularly against steel in vacuum has special relevance to aerospace applications and vacuum technology. In this regard the icosahedral quasicrystals will play an important role in making of the heavy load bearings and latches of the space vehicles. Quasicrystals are promising candidates for coatings, hydrogen storage materials, thermal barrier, infra red sensors and many others.

It has been pointed out that the reduced density of states at the Fermi energy which is a characteristic of icosahedral and related crystalline material, plays an important role and affects two very important surface properties of quasicrystals. The surface energy of these alloys is low, ~500 mJ/meter square.

Al-Pd-Mn or Al-Cu-Fe icosahedral phases are well-studied systems and their atomic and electronic structures are also well-characterized. The study of the influence of electronic structure on adhesion and friction on this class of intermetallic alloys, and study of electron-phonon coupling on these materials are relevant. Since electronic excitations by swift heavy ions and friction within a contact area (a macroscopic phenomenon) are both expected to cause heat-spikes and strain in the material, the two studies are correlated. Dubois et al. have shown [3-4] that the DOS at E_F plays an important role in deciding important surface properties (friction etc.) that have a bearing on potential applications of quasicrystals.

Structural Modifications with Swift Heavy Ions

The electronic excitation and ionization arising from the energy loss of SHI induce structural changes in insulators and semiconductors. It was believed [5] that no effect of this kind could exist in metallic alloys, due to the ability of the free electrons to rapidly and efficiently smear out the perturbation caused by the SHI bombardment. Later, it was shown [6] that structural changes due to S_e also occur in metallic alloys above a critical electronic stopping power threshold. It was pointed out that for metallic alloys to be affected by the SHI requires efficient electron–phonon coupling, and thus a large density of states at the Fermi level. Quasicrystals have a very low density of states at the Fermi level, thus it was interesting to study the effect of SHI irradiation on metallic alloys with a quasicrystalline structure. Chatterjee et al. [7] investigated the effects of SHI irradiation on quasicrystals.

Investigations have been done on the interaction of swift heavy ions (SHI) with quasicrystalline metallic alloys focussed on the Al-Cu-Fe system [8-10]. These quasicrystals irradiated with high energy projectiles (up to 900 MeV Pb with ~40 keV/nm and 780 MeV Xe ions with ~25 keV/nm energy deposition) in the electronic energy deposition regime showed only minor structural modifications (production of defects). The phase transitions in Al-Cu-Fe can be governed by (phason) atomic jumps. Phason is a quasiparticle existing in quasicrystals due to their specific, quasiperiodic lattice structure. Phason is associated with atomic motion, similar to phonon. However, whereas phonons are related to translation of atoms, phasons are associated with atomic rearrangements. As a result of these rearrangements, waves, describing the position of atoms in crystal, change phase, thus the term "phason". Such jumps have been observed experimentally at or near phase transition temperatures [9-10]. These jumps can create a structural defect that is typical of quasicrystals. However, from these studies it was learnt that the principal effect of irradiation in electronic excitation/electronic energy deposition regime was a minute global broadening of the line shape in Bragg diffraction pattern only.

Quasicrystals are remarkably stable under irradiation in this electronic energy deposition regime. It was established firmly that these broadening of peaks do not depend on the phason elastic constants and the defects created by irradiations are not phason defects.

In 2006, Mechler et al. [11] demonstrated that by choosing a metastable quasicrystal which is presumably more sensitive to amorphization, e.g. $Zr_{64.5}Ti_{11.4}Ni_{13.8}Cu_{10.3}$, deposition of ~38-42 keV/nm energy in the electronic excitation regime could induce complete amorphization of the alloy after a fluence of 1×10^{13} ions cm^{-2}. Single amorphous ion tracks after irradiation up to lower fluences were also observed.

Present efforts are to understand the influence of the electronic structure on adhesion and friction. Efforts are in the direction to study thin-films of Al-Pd-Mn based quasicrystals and study the relevance of electron phonon coupling in these films by using swift heavy ions to get electronic excitations in the system. Such studies are expected to provide a wide range of applications for the quasicrystals.

6.11 ALKALI HALIDES

Alkali halides are ionic crystals having large band gap and are electronic insulators. The impurities, excess constituents, defects etc. give rise to different types of colour centres. Some of the experiments with SHI are briefly described here.

SHI irradiation of LIF crystals produces F centres in a large halo of 5 to 30 nm around ion path [1]. At S_e above 10 keV/nm, complex defect clusters are formed in a narrow cylindrical region of ~2 nm [2].

Pure calcium fluoride (CaF_2) single crystals irradiated with 100 MeV Ni^{7+} ions at fluences in the range $5 \times 10^{11} - 2.5 \times 10^{13}$ ions cm^{-2} show a broad absorption band with a peak at ~556 nm and a weak one at ~220 nm, whereas Ytterbium (Yb) doped CaF_2 crystals show two strong absorption bands at ~300 and 550 nm under similar irradiation. TL measurements of irradiated pure CaF_2 samples show a strong TL peak at ~510 K and Yb-doped crystals show two TL peaks at ~406 and 496 K [3]. Detailed investigation of SHI irradiation of CaF_2 single crystal by different ions from ^{32}S to ^{209}Bi at energies between 1 and 12 MeV/amu suggest that out of plane swelling occurs which increases linearly with fluence [4]. It is seen that SHI irradiation induces swelling at the surface in some other materials also like quartz. The irradiation of CaF_2 by 4 MeV/amu Pb ions result in reduction in grain size and amorphization [5]. The ion tracks are created in LiF and CaF_2 single crystals as examined by scanning electron microscope [6].

A detailed study on SHI irradiation of NaCl and LiF crystals for different electronic energy depositions at room temperature and at 8 K reveal that anion interstitials (I and H centres) are created on irradiation at 8 K only, which disappear at 100 K [7].

A large number of irradiation experiments of alkali halide crystals reveal that various effects like chemical etchability, surface hillock creation, swelling, sputtering etc., have different S_e thresholds [8]. A general phenomenon in the irradiated halides is the creation of mechanical stress arising due to ion tracks [8].

The SHI irradiation induces structural and optical modifications in polycrystalline LiF thin films deposited on glass [9] and silica substrates [10]. The irradiations are performerd at RT and at 77 K (liquid nitogen temperature) with different ion species (Au, Ag and Ni) in the fluence range of 5×10^{10}–1×10^{13} ions cm^{-2}. The average grain size decreases systematically from 46.3 nm for the pristine sample to 18.3 nm for the sample irradiated at a fluence of 3×10^{12} ions cm^{-2} [9]. Thereafter at higher fluences, it remains constant. The optical absorption studies of the irradiated LiF thin films show dominant absorption bands of F_3 (385 nm) and F_2 and/or F_3^+ (445 nm) colour centres in case of glass substrate whereas the films on silica substrates show three bands at 260, 380 and 445 nm corresponding to F, F_3 and F_2 colour centres, respectively. The PL spectra of irradiated samples show well resolved bands at 534 and 665 nm corresponding to F_3^+ and F_2 colour centres [9-12]. A study on comparison of the PL intensity behaviour of two bands, (a) F_2 and (b) F_3^+, as a function of fluence for different ions and irradiation temperature [12] reveal that the luminescence intensity ratio of F_3^+ and F_2 colour centres increases with irradiation fluence. At higher fluence, the luminescence intensity of F_3^+ colour centre is nearly twice as strong as the F_2 colour centre intensity. The irradiation at liquid nitrogen temperature results in decrease of the intensity of CCs and increase in the ratio of F_3^+ and F_2 CCs with irradiation fluence. The concentration ratio between F_3^+ and F_2 can be enhanced by irradiation at low temperature even at lower ion fluence. Thus, by controlling the ion fluence and irradiation temperature, one can have controlled creation of F_2 and/or F_3^+ CCs.

The lamellae structures were observed in LiF thin films irradiated with 120 MeV Ag ions at liquid nitrogen temperature at an angle of 10° for a fluence of 5×10^{13} ions cm^{-2} [13]. However, lower fluence irradiation at 1×10^{12} ions cm^{-2} did not show any lamellae formation.

SHI irradiation of LiF crystal have two different zones of damage consisting of different type of colour centres. The intensity of colour centres can be tuned by SHI irradiation in alkali halides.

6.12 PLASMONIC MATERIALS

Introduction

Plasmon is a collective excitation of the electrons in a piece of conducting material. When the electrons are displaced, the positive charge left behind exerts an attractive force on the electrons, trying to pull them back to their original positions. This interaction makes the electrons to oscillate once collectively. These Coulomb interactions with the positive background core also compel the excited plasmons to obey certain boundary conditions at the interface of nanoparticle surface and host material. If the surface of metal itself has nanoscale patterns, due to extremely small surface area the plasmons undergo high-electric field variations and they can have discrete allowed plasmonic modes strongly influenced by existing nanopatterns. However large structures (e.g., thin metal films) can have propagating plasmon modes over a broad range of wavelengths.

Noble metal nanoparticles are of enormous interest due to their fascinating property of absorbing visible light around a narrow band of wavelength, due to localized surface plasmon resonance (LSPR). The noble metal-dielectric nanocomposites (noble metal nanoparticles embedded in an insulating matrix) can support various electromagnetically coupled plasmon modes, open the new avenues for manipulating light and sensing different inorganic-organic molecules. These plasmonic nanomaterials have potential applications in photonics, telecommunications, and biomedical engineering due to their excellent optical and electrical properties in nanoscale region [1]. The LSPR and hence optical properties of these noble metal dielectric nanocomposites depend on their size, shape, particle-distribution and the environment in which they are embedded. Therefore, for the actual applications of noble metal-dielectric nanocomposites, a careful control of the size, shape and their distribution of NPs is needed to achieve the desired tailoring of the properties for functional applications. The ion beams have unique role in controlling the size and shape of the noble metal particles as discussed in Chapter 5. Especially the shape of nanoparticles embedded in a matrix is difficult to control by any other means than ion beams.

A change in the local refractive index of the surrounding medium causes a significant blue or red shift in the SPR wavelength of NPs depending upon whether refractive index of surrounding medium decreases or increases. This property makes the SPR as a potential tool in biosensors.

Elongated nanoparticles exhibit different plasmon resonant absorptions with respect to the transverse and longitudinal polarization axes [2]. These interesting aspects of metal-dielectric nanocomposites lead to major challenges for their synthesis. There are several methods for the synthesis of NPs embedded in different dielectric matrices and although nanostructures with different sizes and shapes have been synthesized, their controlled synthesis in nanometric dimensions (from 5 nm to 20 nm) remains a challenge [1-5].

Synthesis of Plasmonic Nanocomposites

Chemical synthesis technique has been widely used to produce NPs with different sizes and morphology. Various types of physical and biological routes have been tried for their synthesis [6]. Gold and silver nanostructures embedded in different dielectric matrices can be synthesized by atom beam (energy 1.5 keV) co-sputtering, a novel technique [7]. Synthesis of nanocomposites using this technique results in relatively narrow particle size distribution. Gold-silicon core shell nanostructures with tunable localized surface plasmon resonance (LSPR) have been synthesized by atom beam co-sputtering and subsequent annealing [8]. The Au NPs at the surface of quartz have been synthesized by electron beam evaporation and subsequent annealing in Argon environment [9, 10]. Gold-silicon core shell nanostructures and Au-ZnO nanocomposite with tunable LSPR have been synthesized [8, 11]. Fullerene thin films of identical thickness were deposited on Au-ZnO nanocomposite and Si substrate. It was shown that the Raman scattering from fullerene was enhanced when it was on the nanocomposite [11]. It was attributed to electric field enhancement in the vicinity of noble metal particle (due to LSPR) which enhances the polarization of fullerene molecule, resulting in surface enhanced Raman scattering (SERS).

Tuning of SPR of Noble Metal-fullerene Nanocomposites

The nanocomposite of noble metal NPs with carbon is of interest due to wide range of the value of the refractive index (1.5 to 2.3 from amorphous carbon to fullerenes) among different allotropes of carbon which provides the possibility to tune the SPR wavelength of NPs according to the requirement in various applications. These have been studied in detail by synthesizing Au and Ag nanoparticles (NPs) in fullerene (C_{60} and C_{70} separately) matrix by thermal co-evaporation technique. The Ag-C_{60}, Ag-C_{70}, Au-C_{60}, and Au-C_{70} nanocomposite thin films on glass substrates and TEM grids were irradiated with a beam of 120 MeV Ag ions at different fluences from 1×10^{12} to 3×10^{13} ions cm^{-2}. After ion irradiation at a fluence of 3×10^{13} ions cm^{-2}, nanocomposite thin films showed two interesting features: (i) a regular and large blue shift in SPR wavelength and (ii) growth of metal NPs, with increasing ion fluence. In the case of Ag-C_{60} and Ag-C_{70} nanocomposites, the blue shift was 49 nm and 100 nm respectively after ion irradiation at a fluence of 3×10^{13} ions cm^{-2} [12, 13], whereas in the case of Au-C_{60} and Au-C_{70} nanocomposites, the blue shift after ion irradiation at a fluence of 3×10^{13} ions cm^{-2} was 119 and 135 nm. Raman spectroscopy on these samples confirmed the complete transformation of fullerene matrix (C_{60} and C_{70}) into amorphous carbon at a fluence of 3×10^{13} ions cm^{-2} evident by appearance of D and G peaks in irradiated fullerene films [14, 15]. Figure 6.14 shows the absorbance of the Ag-C_{70} nanocomposite at different fluences of 100 MeV Ag ions. Since the refractive index of amorphous carbon (1.6) is less than that of fullerene (2.1 for C_{60} and 2.3 for C_{70}), blue shift in SPR wavelength is expected by ion irradiation of these nanocomposites. Blue shift is higher when NPs are embedded in C_{70} compared to C_{60}, which may be due to the higher refractive index of C_{70} than C_{60}. Ion beam irradiation of these nanocomposites provides a unique way to spatially tune the SPR wavelength at desired value in visible region. The SPR frequency can be tuned, controllably and systematically, by simply varying the ion fluence.

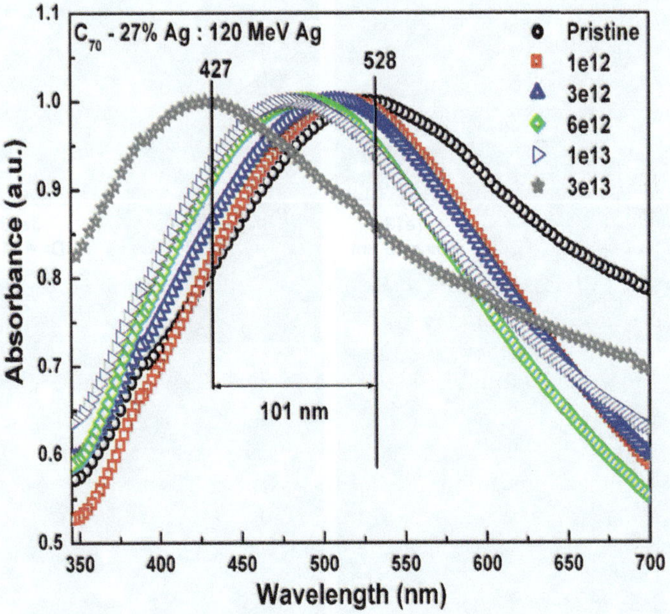

Figure 6.14: The absorbance of Ag-C_{70} nanocomposite thin film at different fluences of 100 MeV Ag ions. [Reprinted from R. Singhal, D.C. Agarwal, Y.K. Mishra, F. Singh, J.C. Pivin, R. Chandra and D.K. Avasthi, *J. Phys. D: Appl. Phys.*, **42** (2009) 155103 with permission from Institute of Physics Publishing Ltd.]

With ion irradiation, NPs grow in size. A growth of Ag NPs from 6 to 10 nm for Ag-C_{60} and 7 to 11 nm for Ag-C_{70} was observed after ion irradiation at a fluence of 3×10^{13} ions cm^{-2}, whereas in the case of Au-C_{60} and Au-C_{70}, a growth from 6.3 to 9.5 and 5.7 to 8.6 was observed respectively. Figure 6.15 gives the TEM pictures of pristine and 120 MeV Ag ion irradiated Ag-C_{70} nanocomposite thin films at the fluence of 10^{13} and 3×10^{13} ions cm^{-2}. The growth of NPs with irradiation can be understood in the frame of thermal spike model. According to the Mie theory calculations, a growth of NPs by 3–4 nm induces a red shift in SPR wavelength but only by few nm, whereas the decrease in refractive index of fullerene matrix, from ~ (2.3 to 1.6) for C_{70} and from ~ (2.1 to 1.6) for C_{60}, gives a large blue shift to SPR wavelength. So the large blue shift (induced by the transformation of fullerene into amorphous carbon with a smaller refractive index) dominates the small red shift due to increase in the particle size.

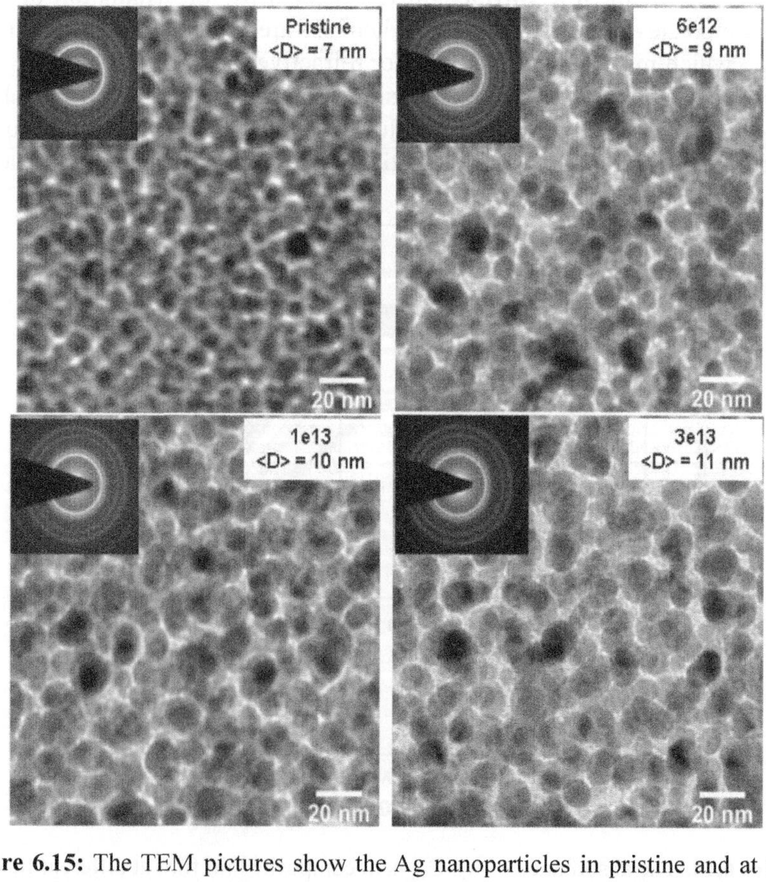

Figure 6.15: The TEM pictures show the Ag nanoparticles in pristine and at three different fluences, showing the growth of nanoparticles with ion fluence. [Reprinted from R. Singhal, D.C. Agarwal, Y.K. Mishra, F. Singh, J.C. Pivin, R. Chandra and D.K. Avasthi, *J. Phys. D: Appl. Phys.*, **42** (2009) 155103 with permission from Institute of Physics Publishing Ltd.]

Optical Properties of SHI Irradiated Metal-polymer Nanocomposite

Ag-PET nanocomposites synthesized by atom beam co-sputtering [16] and irradiated by 120 MeV Ni ions were studied [17]. The pristine Ag-PET nanocopomposite thin film show an interesting property of transmission of UV light at 320 nm suitable for UV filter application. The irradiation with 120 MeV Ni ion results in reduction of few nm in transmittance (further narrowing the fwhm of transmittance of UV filter), depending on the ion fluence [18]. SHI irradiation can control these properties by choosing ion beams characteristics.

The Ag nanoparticles (embedded in soda lime glass) aligned along the ion beam direction exhibit the SPR response as expected in elongated particle giving indication of coupling of plasmonic mode in adjacent particles. The SPR frequency of such an array of Ag nanostructures embedded in glass for transverse and longitudinal modes were ~1.5 ev apart [18]. Similar type of linear arrays of Ag nanostructures (obtained by ion beam irradiation) studied by Z scan measurements show saturable absorption [19]. In the linear arrays of Au nanoparticles obtained by Au ion implantation followed by annealing and 10 MeV Si ion irradiation, splitting of SPR is observed depending on the polarization angle of incident light [20]. The study of the optical properties of elongated Cu, Ag and Au nanoparticles show that the frequencies of transverse and longitudinal modes depend on aspect ratio of anisotropic nanoparticles as well as on the end shape [21].

The ability of ion beam irradiation to control the plasmonic properties are highlighted. The SPR depends on the particle size, its shape and surrounding matrix. The ion beam has potential to alter the size which can alter the SPR by a small amount, whereas in certain specific cases it is possible to control the shape anisotropy by ion beam, which results in significant changes in the SPR due to the two modes (longitudinal and transverse) of the collective oscillations of electrons. Especially in case of metal fullerene nanocomposite (Ag or Au nanoparticles embedded in C_{60} or C_{70} matrix), the tuning of SPR is demonstrated in large range, due to the transformation of fullerene to a:C, thereby changing the surrounding matrix of the metal particles by ion beams. The shape anisotropy in noble metal particles induced by SHI irradiation is capable of tuning the SPR.

6.13 MATERIALS FOR ENERGY

The development of materials for energy is of great importance due to depleting conventional resources of energy. Ion beams are providing ways to engineer properties of materials required for this purpose. A brief account is given here of the role of ion beam in the materials required for (i) Thermoelectrics, (ii) H by splitting of water, (iii) H storage and (iv) Solar energy.

Thermoelectric Materials

Thermoelectric material can convert solar heat or waste heat directly into electricity as well as electricity into thermal energy for cooling or heating. The efficiency of thermoelectric material is measured by its Figure of Merit (*ZT*) and Power Factor (P_F) defined as:

$$ZT = S^2 \sigma T/k; \; P_F = S^2 \sigma$$

where *S* is thermopower, σ is electrical conductivity, *k* is thermal conductivity and *T* is the temperature in degrees absolute at which the properties are measured. It is normally difficult to increase the Figure of Merit because *S*, σ and *k* are interdependent. For reduced dimensionality materials the thermal conductivity reduces significantly without affecting the power factor. All these properties can be independently controlled using ion beams. Ordered nanostructures created by swift heavy ion bombardment improve the thermoelectric properties of the materials. It is shown that the properties of thermoeletcric materials are improved by irradiation with 5 MeV Si ions at different fluences [1, 2]. The band gap can be tuned by appropriate ion fluence. A smaller band gap leads to increase of electrical conductivity. Ion irradiation also causes reduction in thermal conductivity due to interface scattering and absorption of phonons. Budak et al. [1] found that the Figure of Merit (*ZT*) of Bi_2Te_3/Sb_2Te_3 multilayers (50 layers of 5 nm thickness) is enhanced from 0.0049 to 1.3 as a result of irradiation by 5 MeV Si ions at 1×10^{14} ions cm^{-2} fluence, beyond which it decreases until 0.042 at 5×10^{15} ions cm^{-2} [1]. $Si_{1-x}Ge_x/Si$ multilayers (70 layers with 10 nm period) with ion bombardment of 5 MeV Si ions enhancing the *ZT* from ~0.008 to 1 at 1×10^{14} ions cm^{-2} fluence [2]. Grain boundaries of nano clusters formed by ion bombardment increase phonon scattering, limits the phonon mean free path, reducing the thermal conductivity. Electrical conductivity also increases due to increase in the density of states by ion irradiation [2]. The grain growth causes the enhancement in electrical conductivity and thermo power [3]. Swift heavy ions can engineer the grain size to achieve the enhancement. It is shown that 100 MeV Au ion irradiation increases the grain size of nano crystalline cadmium sulfide from 3.2 nm to 69.2 nm [3]. Thermoelectric generator has been fabricated using ion track lithography by ~2.3 GeV Pb ion irradiation at fluences of 5×10^8 to 5×10^9 ions cm^{-2} giving improved thermoelectric properties [4].

Hydrogen Evolving Photoelectrochemical Material

Photoelectrochemical systems are advance alternative to the conventional photovoltaic cells for solar energy utilization offering just one step process of converting solar energy into directly usable clean fuel i.e., hydrogen. A photoelectrochemical (PEC) hydrogen production cell consists of a semiconductor electrode, metal counter electrode and a reference electrode immersed in an aqueous electrolyte. Main requirements for suitable semiconductor photoelectrodes are: (a) sufficient light absorption, (b) high chemical stability, (c) favourable energetic positions of the band edges with respect to water oxidation potential, (d) fast transport of photo-generated charge carriers and (e) low cost. Nanostructured metal oxide semiconductors are being investigated as potential materials to develop an efficient photo-electrochemical device for solar hydrogen generation. Nanostructured semiconductor can be used to obtain large specific surface area for solar energy absorption and ultrafine nanoparticles can be explored to modify the local electron polarization and spatial separation of photogenerated electrons and holes.

Swift heavy ions (SHI) provide ways to modify nanomaterials for various applications. Such modifications, occurring at atomic and grain level, lead to alteration in the properties viz., band gap, resistance, additional states, which can modify the PEC response of the material. The role of SHI irradiation in modification of semiconductor materials and their effect on PEC water splitting for hydrogen production have been investigated [5-8]. Results related to these studies are summarized here.

Fe_2O_3 Thin Films

Iron oxide thin films (prepared using spray pyrolysis) irradiated with 170 MeV Au^{13+} ions at the fluence ranging from 10^{11} to 10^{13} ions cm^{-2} show enhanced photocurrent upon irradiation at 10^{12} ions cm^{-2} fluence, attributed to the formation of tubular structures in films [5]. However, at fluence of 10^{13} ions cm^{-2} marginal decrease in the photocurrent was observed probably due to the collapse of the initially formed tubular structures, resulting in the formation of large number of discontinuities/dislocations, which act as recombination centres for photo-generated charge carriers. The X-ray diffraction analysis exhibited diffused nature, indicating highly disordered material at the fluence of 10^{13} ions cm^{-2}. Elastic recoil detection analysis exhibited no change occurring in stochiometry of thin films throughout irradiation.

The films irradiated by 120 MeV Ag ions show an additional peak at 656 cm^{-1} as analyzed by Raman spectroscopy which suggests that partial transition of phase occurs from α-Fe_2O_3 to Fe_3O_4. The irradiation of Ag^{9+} ions on α-Fe_2O_3 improves PEC response. The photocurrent density increases with fluence. Sample irradiated at 1×10^{13} ions cm^{-2} fluence exhibits maximum photocurrent density, five times larger than from the unirradiated sample [6].

ZnO Thin Films

A study on ZnO films irradiation by 120 MeV Ag^{9+} ions at different fluences revealed that the effects of SHI irradiation on resistivity are more pronounced in ZnO films sintered at 500°C and 600°C. SHI irradiation at low fluence results in increase in the crystallinity, but at higher fluences crystallinity decreases. The highest photo response is recorded with sintered films at 500°C irradiated at fluence 3×10^{12} ions cm^{-2} [7].

TiO_2 Thin Films

TiO_2 thin films deposited on conducting glass substrate (SnO_2:In), prepared by the sol-gel spin coating method, irradiated at room temperature with 120 MeV Ag^{9+} ions with fluence values of 5×10^{11}, 1×10^{12}, 5×10^{12} and 1×10^{13} ions cm^{-2} show a significant decrease in the average grain size (from 23 to 11 nm). The PEC cell having thin films irradiated at fluence 5×10^{11} ions cm^{-2} exhibits a photocurrent of 0.76 mA cm^{-2} at zero bias conditions, significantly better than that of the unirradiated sample [8].

The swift heavy ion irradiation of these materials results in 2 to 5 times enhancement in photoelectrochemical current [5-8]. For most of the oxides, increased photocurrent has been obtained at low fluences of 10^{11} to 10^{12} ions cm^{-2} of ion irradiation.

H-Storage Materials

Ion beams play an important role in quantifying H as well as in increasing the H-storage capacity of the materials. They provide method for standardizing the contents of H in samples. For example, SIMS is a powerful technique but the quantification of H by SIMS requires a standard sample containing known amount of H and such standard measurements are performed by ion beam techniques like ERDA and NRA, discussed in Chapter 3.

A significant increase (about 18 %) in hydrogenation properties is achieved by SHI (120 MeV Ag ion) irradiation of Pd capped Pr layers (Pd-Pr layers) [9]. Ion induced defects surrounding the ion path (nanodimension track) in the Pd-Pr layer enhance the H diffusion and improve the hydrogenation. The samples are cooled to liquid nitrogen temperature so that H does not escape from the sample during ERDA. The H loading of the irradiated samples is performed in-situ by introducing H gas in the irradiation chamber at 760 torr.

Solar Energy Materials

There are various approaches for harnessing solar energies like Si based photovoltaic cell, dye sensitized solar cell, polymer based solar cell etc. The ion beams play a role in improving the properties of the solar cell constituent materials and in testing the solar cell at extreme conditions. Transparent conducting oxide (TCO) is used in most of the solar cell configurations. The most common TCO for this purpose is indium tin oxide (ITO). The desirable property for TCO as electrode in solar cell is high transmittance and high conductivity.

Effect of swift heavy ion irradiation on the structural, optical and electrical properties of ITO to study the effect on photoanode of Dye-Sensitized Solar Cell (DSSC) was investigated with a view to examine the stability of ITO in space irradiations [10]. The DSSC photoanode consists of: transparent conducting oxide (TCO) as electrodes at top and bottom, TiO_2 or other porous oxide material with wide band gap, and suitable dye. RF sputtered ITO film on corning glass substrate were irradiated with 110 MeV Ni^{8+} ions at different fluences ranging from 3.0×10^{11} ions cm^{-2} to 1.0×10^{14} ions cm^{-2}. XRD patterns indicated that at low fluences of up to 3.0×10^{12} ions cm^{-2} there is strain in film and beyond this fluence strain is relaxed. A significant improvement in transmittance from 75% for pristine to 84% was observed for irradiation at the fluence 3×10^{11} ions cm^{-2}, revealing an interesting result in the context of DSSC electrode study. Although the resistivity increased marginally but it remained in acceptable limits.

In another experiment Deshpande et al. [11] reported a study on the effect of 100 MeV Au ion irradiation of ITO samples, prepared by Spray Pyrolysis (SP) method. A decrease of about 30% in optical transmittance and also in bulk resistance was observed [11]. The ITO synthesized by two different techniques produces different effects under SHI irradiation.

SHI irradiation of a DSSC results in improvement of the efficiency of dye sensitized solar cell [12]. The DSSC cell consisting of F-doped tin oxide (FTO) as electrode and TiO_2 as blocking layer and N719 dye, irradiated by 100 MeV O ion irradiation improves the energy conversion efficiency from 3.8% to 5.5%.

There are definite improvements in different type of energy materials under ion impingement. SHI irradiation also has decisive role in testing the energy materials under extreme conditions.

6.14 NUCLEAR MATERIALS

Ion beams help in the development of nuclear materials by simulating the effects produced in reactor environment. Low energy ion irradiation having dominant elastic collisions in energy loss processes are used to simulate the defects produced by the neutrons in reactor materials. The density of defects produced by ion beams is very large compared to that produced by the neutrons and thus the damage equivalent to that produced by the neutrons in several years is equivalent to that produced by ion beam in a short time, with the achievable fluences. High energy heavy ions are useful in testing the inert matrix fuel (IMF) by simulating the reactor core environment, so that an appropriate choice out of vast possibilities among IMF can be made. Inert matrix, as suggested by it's name, does not lead to the formation of any fissile material after irradiation. It is a neutron transparent matrix used for diluting the fissile phase to the volumetric considerations required by reactor control considerations [1]. Inert matrix fuels are of interest for futuristic Accelerator Driven Subcritical (ADS) reactors. One of the criteria for the selection of these materials is their radiation stability at elevated temperatures in the radiation environment of a reactor. Energetic ion beams (SHI as well as low energy ions) can experimentally simulate the same damage that a material would experience in a radiation environment in the reactor. Ions and the energies of SHIs are chosen so that they are equivalent to fission fragments.

Materials in a reactor undergo structural modifications due to radiation damage. The main sources of radiation being plutonium and minor actinides (MA's; Am, Np Cm etc.) produced by α-particles, neutrons (>1 MeV) and fission fragments (A = Ga-Dy and energy 70 to 120 MeV). Accumulation and release of fission gases (He, Xe, Kr) pose other problems. The recoils and primary knock-on atoms (PKA) created by neutrons cause damage in materials by depositing energy via elastic collisions. Fission fragments deposit energy by interacting with the electronic sub-system of the target atoms by inelastic collisions. Radiation causes point defects and defect clusters which then evolve in formation of dislocation loops thus leading to various kinds of damage. Moreover, high temperatures existing in a reactor, influence the damage recovery and structural modifications of materials. All this occurs inside the reactor in a radioactive environment. Simulation of radiation damage outside a reactor can be done using energetic ion beams.

Potential materials for IMF are, pyrochlores ($A_2B_2O_7$) of various r_a/r_b (where r_a and r_b are the radii of A and B cations) CeO_2, ThO_2, spinels such as $MgAl_2O_4$, etc. The pyrochlore structure consists of AO_8 and BO_6 polyhedra and most importantly A and B occupy specific positions forming a superstructure lattice with a = 10 Å [2]. Pyrochlores are considered as potential candidates for disposal of nuclear waste [3]. This is due to their ability for accommodating MAs and stability against radiation damage in the nuclear stopping regime. The radiation stability of pyrochlores is very well studied in the nuclear stopping regime and it is seen to decrease with an increase in r_a/r_b [4, 5]. It is shown [6, 7] that the radiation stability of pyrochlores obeys the same rule even in the electronic stopping regime (Fig. 6.16), envisaging their use also as potential IMF. Three different pyrochlores [$Gd_2Zr_2O_7$ (r_a/r_b = 1.46), $Nd_2Zr_2O_7$ (r_a/r_b = 1.54) and $Gd_2Ti_2O_7$ (r_a/r_b = 1.75)] of varying r_a/r_b were studied with different SHIs of varying S_e (equivalent to some FF's like 90 MeV I). The XRD in Fig. 6.16 summarizes the results. $Gd_2Zr_2O_7$ and $Nd_2Zr_2O_7$ transform to anion deficient fluorite and the former does not amorphize even with ions of highest S_e (typical of a FF) and fluence. $Nd_2Zr_2O_7$ amorphizes at S_e = 16 keV/nm and $Gd_2Ti_2O_7$ amorphizes at even lower S_e and fluence. Thus pyrochlore with lowest r_a/r_b is most radiation resistant. This criterion for radiation stability is due to the energy required for formation of cation antisites (randomization of A and B cations from their specific locations) which decreases with decrease in r_a/r_b. The pyrochlore with lowest r_a/r_b prefers to undergo disordering to fluorite structure rather than amorphizing. There have been other experiments on SHI irradiation of pyrochlores or other potential candidates for IMF [8-10].

SHI irradiation of pyrochlores (one of the potential candidates of IMF) reveal that pyrochlores with smallest value of r_a/r_b are most stable, whereas those with largest value of r_a/r_b are prone to radiation damage. This observation is similar to what is observed for low energy ions where nuclear energy loss is dominant.

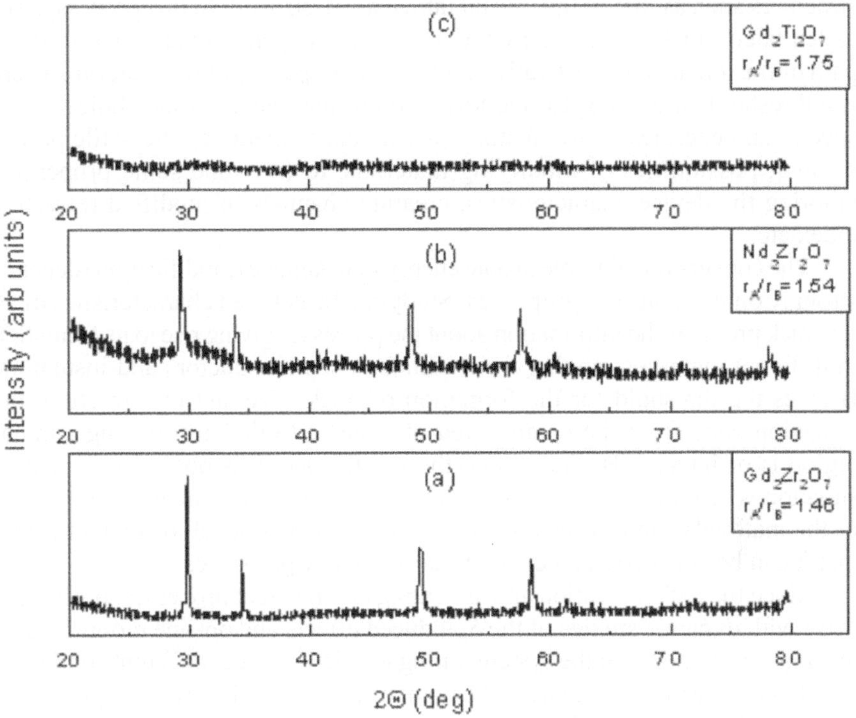

Figure 6.16: The XRD spectra of different pyrochlores after irradiation by 90 MeV I ions at a fluence of 10^{13} ions cm^{-2}. To begin with all the pyrochlores were perfectly crystalline. [Reprinted from M.K. Patel, V. Vijayakumar, D.K. Avasthi, S. Kailas, J.C. Pivin, V. Grover, B.P. Mandal and A.K. Tyagi, *Nucl. Instr. and Meth.*, **B 266** (2008) 2898 with permission from Elsevier.]

6.15 SUMMARY

Swift heavy ion irradiation of a wide variety of materials has been used to study the modifications of their properties by ion matter interaction. One of the important goals in materials science is to be able to design materials with specific properties. Phenomenal developments in ion beam technology are providing ways to tailor the properties in the desired direction to create the materials required for specific applications.

When SHI penetrates a solid, it loses its energy predominantly through inelastic scattering with electrons (Electronic energy loss, S_e) and some through elastic scattering with atoms (Nuclear energy loss, S_n). For thin film with the thickness smaller than the range of the ion, there is negligible loss due to elastic collisions leading to smooth transfer of energy from the ion to the solid, which is locked into electron excitations. This then gets coupled to the lattice which results in the modification of material along the trajectory of the ion. Atoms are pushed out of their normal position, molecules are split and ordered structures such as that of the crystal lattice are destroyed. In this process, a so-called latent track is created by the ion. What happens to the material in the path of the ion depends critically on the basic properties of the material. There is a threshold value of S_e for the formation of the track. A wide choice of ion species and energies allows features which makes them very versatile probes opening possibilities of modifying materials, to fine tune some properties, tailoring the device characteristics, creating channels of modified refractive index etc.

The conversion of the electronic energy into damage (modification) depends strongly on the materials properties. Studying the detailed characteristics of an ion track provides the information about the processes giving rise to its formation in different materials ranging from metals to semiconductors and insulators. Besides the threshold for the formation of track, size and the structure, the transition zone from the central area of damage to the surrounding annular region is of interest. By increasing the ion fluence, it is possible to create a situation where the tracks overlap, then the physical and chemical properties of the materials can also be altered on a macroscopic level to such an extent that it can be considered a new material with new properties.

Since the defect creation and annealing are transient processes, probing of these with dynamic studies of the S_e induced effects and on-line monitoring of the transient response of the system during irradiation under different conditions are of immense importance from the point of view of engineering the properties of materials with the ion beams.

Research so far is mostly aimed at a better understanding of the changes produced in materials by ion beams. It is now gradually getting focussed on the desired modifications in the properties of the materials to make them functional for specific applications. Ultimately it will be possible to direct efforts towards the development of improved materials and possibly completely new materials.

Present and potential application of ion beam for materials engineering needs detailed understanding of ion matter interaction and the material in nano region. Various types of non-equilibrium and non-linear conditions are produced by ion matter interaction which can be exploited for producing new types of materials. A deeper understanding of the nanoscale pattern formation could lead to many useful applications.

SHI has unique feature of depositing high energy density along its ion path which creates a narrow cylindrical zone of modified material significantly different from its surrounding. This ion track is an amorphous zone in case of most of insulators like polymers and ceramics, may have colour centres in ionic crystals and small nanoclusters in Si-based gels and polymers, etc. These nanometric size ion tracks have interesting properties of tremendous interest because these are not achievable so efficiently by any other route.

Challenges in tailoring polymers and their composites for applications are opening possibilities of using ion beam for producing low band gap polymer, surface engineering of fillers in polymer nanocomposites, encapsulation of thermoelectric devices, tailoring of biodegradable polymers, creating nanocomposites for biomedical applications, fabricating functional co-polymers, tailoring structural, optical and electrical properties of MWCNT nanocomposites etc. Ion tracks (latent as well as etched) in polymers are of importance for grafting of monomers of interest to achieve desirable properties.

Ion tracks in high Tc superconductor materials are crucial for flux pinning, enhancing the current carrying capabilities at higher magnetic fields. Ion tracks in fullerene, DLC films and Si-based gels and polymers such as MTES, MPS etc., are basically conducting cylindrical zones because these are either filled up with conducting C clusters in case of MTES or they are conducting zones due to transformation of material to conducting form of C. The field emission properties of such conducting nanowires are of immense importance. In the field emission properties the aspect ratio of conducting wire, inter-separation of the nanowires and their parallelness are of importance. These parameters are well controllable by ion beam parameters. For example, the aspect ratio can be controlled by choosing the film thickness and the S_e of the ion beam, where film thickness defines the length of the nanowire and S_e will dictate the diameter of the nanowire. The inter separation of nanowire is controlled by ion fluence. Ion tracks in silica can be used to control the size and shape of embedded nanoparticles in nanocomposite thin films. Experiments have shown that the size of embedded nanoparticles can be increased, if the inter particle separation is small and can be reduced if the inter particle separation is large. The shape of the embedded particles can be engineered as the aspect ratio can be changed by the deposited electronic energy ($S_e \times \phi$). More experiments are required to understand these transformations and their dependence on the ion beam parameters. The plasmonic properties of the noble metal dielectric nano-composites can be tuned by changing the shape of particles and modifying the matrix.

Perpendicular magnetic anisotropy (PMA) in magnetic materials has been of wide interest. For the systems with magnetic nanoparticles embedded in silica matrix, the SHI irradiation induces perpendicular magnetic anisotropy (PMA). There are open questions related to the aspects of elongation of magnetic particles and its correlation with PMA, which needs to be addressed in the SHI irradiation experiments. In GMR multilayers, there are possibilities of engineering the interface in terms of smoothness, roughness and mixed layer, which can control the PMA behaviour, by varying the electronic energy deposition. The capability of SHI of dissolving the small magnetic metal clusters in the semiconducting matrix is of interest from DMS point of view.

Large electronic energy density deposition in metals (except for some exception like Fe) does not cause any effect due to high thermal and electrical conductivity of metals resulting in fast smearing of electronic energy in the system. However, it does effect metals if it is in nanodimensional form. For example, bulk Au shows small sputtering under the impact of SHI whereas thin Au film (~ 20 nm thickness) shows sputtering which is two orders of magnitude larger than expected in bulk Au. The materials which are immiscible and do not form compound by ion beam mixing can become miscible by ion beam mixing if the individual components are nanosize grains. The interaction of SHI with nanodimensional materials is different from that of bulk which needs to be investigated in detail. The experiments on electronic sputtering and ion beam mixing for nanodimensional system are of great interest.

SHI can cause various phase transformations. The transformation from crystalline to amorphous phase due to damage within the ion track is well known but apart from this the phase transformation from crystalline to crystalline (for example anatase TiO_2 to rutile TiO_2), amorphous to crystalline and formation of high pressure phases have been observed in and around ion track. Thus apart from damaging action, SHI can cause crystallization (ion beam induced epitaxial crystallization) and crystalline to crystalline phase transformation (in some materials) can be achieved simply by tuning the ion beam parameters. The ion beam parameters optimized for a given system for a process do not hold good for another system due to difference in dissipation of deposited electronic energy, which is due to differences in physical parameters such as thermal conductivity, specific heat and electron phonon coupling.

The unique feature of high energy density deposition by SHI is of interest to test the new materials under extreme conditions. SHI irradiation, on one hand, provides conditions to simulate the conditions existing in reactor and on the other hand high energy light ions [HELI] can provide conditions to simulate the space radiation. The electronic chips and circuits prone to radiation damage can be tested at the accelerator for the fail safe flight of satellite. In medical field apart from synthesis of useful short-lived isotopes (by a few MeV light ions) required in medical diagnostic, SHIs up to carbon and oxygen have been shown to be useful in ion beam cancer therapy.

SHI is emerging as a very powerful tool for providing unique features to the materials and for tailoring them for possible applications. At present it may appear that these interesting possibilities will not be feasible for practical applications when considered from their commercial viability. Development of table-top accelerators for SHI will solve this problem. There are presently very specific applications in sectors like space, defense and health care, which are gradually increasing. Applications of swift heavy ions in various fields are leading towards the research and developments in multidisciplinary science and various technologies. Ion beams have unique role in the nanoworld. The physics in this domain and the ion matter interaction have to be understood to enable us to engineer the nanomaterials, including the bio-materials at the cellular level.

REFERENCES

6.1 Carbon

1. H.W. Kroto, J.R. Heath, S.C. O'Brien, R.F. Curl and R.E. Smalley, *Nature* (London), **318** (1985) 162.
2. S. Iijima, *Nature*, **354** (1991) 56.

3. S. Iijima and T. Ichihashi, *Nature*, **363** (1993) 603.
4. B.W. Smith, M. Monthioux and D.E. Luzzi, *Nature*, **396** (1998) 323.
5. S. Bandow, M. Takizawa, K. Hirahara, M. Yudasaka and S. Lijima, *Chem. Phys. Lett.*, **337** (2001) 48.
6. K.S. Novoselov, A.K. Geim, S.V. Morozov, D. Jiang, Y. Zhang, S.V. Dubonos, I.V. Grigorieva and A.A. Firsov, *Science*, 3**06** (2004) 666.
7. K.S. Novoselov, A.K. Geim, S.V. Morozov, D. Jiang, M.I. Katsnelson, I.V. Grigorieva, S.V. Dubonos and A.A. Firsov, *Nature* (London), **438** (2005) 197.
8. J. Krauser, A.–K. Nix, H.-G. Gehrke, H. Hofsass, C. Trautmann, A. Weidinger, F. Wunsch and J. Bruns, *J. Vacuum Science Techn.*, **B 26** (2008) 2468.
9. Jens-Hendrik Zollondz and A. Weidinger, *Nucl. Instr. and Meth.*, **B 225** (2004) 178.
10. J. Krauser, J.-H. Jollondz, A. Wedeinger and C. Trautmann, *J. Appl. Phys.*, **94** (2003) 1959.
11. A. Tripathi, Amit Kumar, D. Kabiraj, S.A. Khan, V. Baranwal and D.K. Avasthi, *Nucl. Instr. and Meth.*, **B 244** (2006) 15.
12. Amit Kumar, D.K. Avasthi, A. Tripathi, D. Kabiraj, F. Singh and J.C. Pivin, *J. Appl. Phys.*, **101** (2007) 014308.
13. Amit Kumar, D.K. Avasthi, A. Tripathi, L.D. Filip, J.D. Carey and J.C. Pivin, *J. Appl. Phys.*, **102** (2007) 044305.
14. S.K. Srivastava, D.K. Avasthi and E. Pippel, *Nanotechnology*, **17** (2006) 2518.
15. Amit Kumar, F. Singh, A. Tripathi J. Pernot, J.C. Pivin and D.K. Avasthi, *J. Phys. D: Appl. Phys.*, **41** (2008) 95304.
16. Amit Kumar, F. Singh, J.C. Pivin and D.K. Avasthi, *J Phys. D: Appl. Phy.*, **40** (2007) 2083.
17. Amit Kumar, F. Singh, D.K. Avasthi and J.C. Pivin, *Nucl. Instr. and Meth.*, **B 244** (2006) 221.
18. Amit Kumar, D.K. Avasthi and J.C. Pivin, *Appl. Phys. Express*, **1** (2008) 125002.
19. Amit Kumar, D.K. Avasthi, J.C. Pivin, R.M. Papaleo, A. Tripathi, F. Singh and I. Sulania, *Journal of Nanoscience and Technology*, **7** (2007) 2201.
20. Amit Kumar, D.K. Avasthi, J.C. Pivin, A. Tripathi and F. Singh, *Phys. Rev.*, **B 74** (2006) 153409.
21. T.L. Makarova, B. Sundqvist, R. Höhne, P. Esquinaz, Y. Kopelevich, P. Scharff, V.A. Davydov, L.S. Kashevarova and A.V. Rakhmanina, *Nature*, **413** (2001) 716.
22. P. Esquinazi, D. Spemann, R. Hohne, A. Setzer, K.H. Han and T. Butz, *Phys. Rev. Lett.*, **91** (2003) 227201.
23. S. Talapatra, P.G. Ganesan, T. Kim, R. Vajtai, M. Huang, M. Shima, G. Ramanath, D. Srivastava, S.C. Deevi and P.M. Ajayan, *Phys. Rev. Lett.*, **95** (2005) 97201.
24. W.L. Wang, O.V. Yazyev, S. Meng and E. Kaxiras, *Phys. Rev. Lett.*, **102** (2009) 157201.
25. O.V. Yazyev and L. Helm, *Phys. Rev.*, **B 75** (2007) 125408.
26. Amit Kumar, D.K. Avasthi, J.C. Pivin, P.M. Koinkar, *Appl. Phys. Lett.*, **92** (2008) 221904.
27. Kiran Jeet, V.K. Jindal, L.M. Bharadwaj, D.K. Avasthi and K. Dharamvir, *J. Appl. Phys.*, **108** (2010) 034302.
28. K. Dharamvir, K. Jeet, N. Pan, C. Du and V.K. Jindal, *J. Nano Research*, **10** (2010) 1.
29. P.M. Koinkar, R.S. Khairnar, S.A. Khan, R.P. Gupta, D.K. Avasthi and M.A. More, *Nucl. Instr. and Meth.*, **B 244** (2006) 217.

30. P.T. Pandey, G.L. Sharma, D.K. Avasthi and V.D. Vankar, *Vacuum*, **72** (2003) 297.
31. D.K. Avasthi, Amit Kumar, Rahul Singhal, Ambuj Tripathi and D.S. Misra, *Journal of Nanoscience and Technology*, **10** (2010) 3767.
32. Huang-Chin Chen, U. Palnitkar, B. Sundaravel, I.-N. Lin, A.P. Singh and Ravi Kumar Ranade, *Diamond and Related Materials*, **18** (2009) 164.
33. Abha Misra, Pawan K. Tyagi, Padmnabh Rai, D.S. Misra, Jay Ghatak, P.V. Satyam and D.K. Avasthi, *Appl. Phys. Lett.*, **89** (2006) 091907.
34. Abha Misra, Pawan K. Tyagi, Padmnabh Rai, Dipti Ranjan Mahopatra, Jay Ghatak, P.V. Satyam, D.K. Avasthi and D.S. Misra, *Phys. Rev.*, **B 76** (2007) 014108.
35. A.V. Krasheninnikov and K.Nordlund, *J. Appl. Phys.*, **107** (2010) 071301.
36. N. Bajwa, A. Ingale, D.K. Avasthi, R. Kumar, A. Tripathi, K. Dharamvir and V.K. Jindal, *J. Appl. Phys.*, **94** (2003) 326.
37. N. Bajwa, A. Ingale, D.K. Avasthi, R. Kumar, A. Tripathi, K. Dharamvir and V.K. Jindal, *Radiation Measurements*, **36** (2003) 737.
38. N. Bajwa, A. Ingale, D.K. Avasthi, R. Kumar, A. Tripathi, K. Dharamvir and V.K. Jindal. *J. Appl. Phys.*, **104** (2008) 054306.
39. S. Lotha, A. Ingale, D.K. Avasthi, V.K. Mittal, S. Mishra, K.C. Rustagi, A. Gupta, V.N. Kulkarni and D.T. Khathing, *Solid. Stat. Comm.*, **111** (1999) 55.
40. M. Terrones, F. Behnhart, N. Grobert, J.-C. Charlier, H. Terrones and P.M. Ajayan, *Phys. Rev. Lett.*, **89** (2002) 075505.
41. A.V. Krasheninnikov, K. Nordlund and J. Keinonen, *Phys. Rev.*, **B 66** (2002) 245403.
42. Amit Kumar, F. Singh, R. Kumar, A. Tripathi, J.C. Pivin, and D.K. Avasthi, *Sol. Stat. Comm.*, **138** (2006) 448.
43. Amit Kumar, A. Podhorodecki, J. Misiewicz, D.K. Avasthi and J.C. Pivin, *J. App. Phys.*, **105** (2009) 024314.
44. S. Prawer, K.W. Nugent, S. Biggs, D.G. McCulloch, W.H. Leong, A. Hoffman and R. Kalish, *Phys. Rev.*, **B 52** (1995) 841.
45. R. Singhal, Amit Kumar, Y.K. Mishra, S. Mahopatra, J.C. Pivin and D.K. Avasthi, *Nucl. Instr. and Meth.*, **B 266** (2008) 3257.
46. R. Singhal, F. Singh, A. Tripathi and D.K. Avasthi, *Rad. Eff. and def. in Sol.*, **164** (2009) 38.
47. H.C. Barshilia, S. Sah, B.R. Mehta, V.D. Vankar, D.K. Avasthi, Jaipal and G.K. Mehta, *Thin Solid Films*, **258** (1995) 123.
48. P. Wesolowski, Y. Lyutovich, F. Banhart, H.D. Carstanjen and H. Kronmuller, *Appl. Phys. Lett.*, **71** (1997) 1948.
49. J. Liu, R. Neumann, C. Trautmann and C. Muller, *Phys. Rev.*, **B 64** (2001) 184115.
50. L.P. Biro, B. Szabo, G.I. Mark, J. Gyulai, K. Havancsak, J. Kurti, A. Dunlop, L. Frey and H. Ryssel, *Nucl. Instr. and Meth.*, **B 148** (1999) 1102.
51. L.P. Biro, G.I. Mark, J. Gyulai, K. Havancsak, S. Lipp, Ch. Lehere, L. Frey and H. Ryssel, *Nucl. Instr. and Meth.*, **B 147** (1999) 142.
52. A. Tripathi, S.A. Khan, M. Kumar, V. Baranwal, R. Krishna and A.C. Pandey, *Nucl. Instr. and Meth.*, **B 244** (2006) 225.
53. D.K. Avasthi, *Vacuum*, **47** (1996) 1249.
54. H. Hofsass (Invited talk), Swift heavy ion in material engineering & characterization (SHIMEC) 6-9 October, 2010, New Delhi.

55. D. Borka, D.J. Mowbray, Z.L. Miskovic, S. Petrovic and N. Neskovic, *J. Phys: Conference Ser.*, **133** (2008) 012015.
56. A.V. Krasheninnikov and K. Nordlund, *Phys. Rev.*, **B 71** (2005) 245408.

6.2 Polymers

1. E. Balanzat, N. Betz and S. Bufford, *Nucl. Instr. and Meth.*, **B 105** (1995) 46.
2. D. Fink (Ed.), *Fundamentals of Ion–irradiated Polymers*, Springer (2004).
3. P.K. Diwan, V. Sharma, S. Kumar, V.K. Mittal, S.A. Khan and D.K. Avasthi, *Nucl. Instr. and Meth.*, **B 258** (2007) 293.
4. M. Singh and L. Singh, *Radiat. Eff. & Defects Solids*, **163** (2008) 605.
5. Neetu, Pratibha, V. Sharma, P.K. Diwan and S. Kumar, *Rad. Meas.*, **44** (2009) 363.
6. P.K. Diwan, V. Sharma, Pratibha, S. Kumar, S.A. Khan and D.K. Avasthi, *Nucl. Instr. and Meth.*, **B 266** (2008) 4738.
7. P.K. Diwan and S. Kumar, *Nucl. Instr. and Meth.*, **B 215** (2004) 27.
8. A. Adla, V. Buschmann, H. Fuess and C. Trautmann, *Nucl. Instr. and Meth.*, **B 185** (2001) 210.
9. C. Trautmann, *Nucl. Instr. and Meth.*, **B 105** (1995) 81.
10. P. Apel, A. Schultz, R. Spohr, C. Trautmann and V. Vutsadakis, *Nucl. Instr. and Meth.,* **B 131** (1997) 55.
11. A. Biswas, D.K. Avasthi, B.K.Singh, S. Lotha, J.P. Singh, D. Fink, B.K. Yadav, B. Bhattacharya and S.K. Bose, *Nucl. Instr. and Meth.*, **B 151** (1999) 84.
12. D. Fink, L.T. Chadderton, K. Hoppe, W.R. Fahrner, A. Chandra and A. Kiv, *Nucl. Instr. and Meth.*, **B 261** (2007) 727.
13. D. Fink, A. Chandra, W.R. Fahrner, K. Hoppe, H. Winkelmann, A. Saad, P. Alegaonkar, A. Berdinsky, D. Grasser and R. Lorenz, *Vacuum*, **82** (2008) 900.
14. D. Fink, A. Saad, S. Damodaran, A. Chandra, W.R. Fahrner, K. Hoppe and L.T. Chadderton, *Rad. Meas.*, **43** (2008) S546.
15. K. Hoppe, W.R. Fahrner, D. Fink, S. Damodaran, A. Petrov, A. Chandra, A. Saad, F. Faupel, V.S.K. Ckakravadhanula and V. Zaporotchenko, *Nucl. Instr. and Meth.*, **B 266** (2008) 1642.
16. M. Saroch, S. Srivastava, D. Fink and Amita Chandra, *Radiat. Eff. Defects Solids*, **163** (2008) 645.
17. M. Saroch, S. Srivastava, D. Fink and A. Chandra, *Sensors*, **8** (2008) 6355.
18. Y.K. Vijay, N.K. Acharya, S. Wate and D.K. Avasthi, *Int. J. of Hydrogen Energy,* **28** (2003) 1015.
19. V. Rao, J.V. Amar, D.K. Avasthi and R.N. Charyulu, *Rad. Meas.*, **36** (2003) 585.
20. N. Reber, H. Omichi, R. Spohr, M. Tamada, A. Wolf and M. Yoshida, *Nucl. Instr. and Meth.*, **B 105** (1995) 275.
21. V.K. Mittal, S. Lotha and D.K. Avasthi, *Radiat. Eff. Defects Solids*, **147** (1999) 199.
22. D.K. Avasthi, J.P. Singh, A. Biswas and S.K. Bose, *Nucl. Instr. and Meth.*, **B 146** (1998) 504.
23. D. Ila, A.L. Evelyn and G.M. Jenkins, *Nucl. Instr. and Meth.*, **B 91** (1994) 580.
24. H.K. Upfer and G.K. Wolf, *Nucl. Instr. and Meth.*, **B 166-167** (2000) 722.
25. J. Chen, F. Zhu, H. Pan, J. Cao, D. Zhu, H. Xu, Q. Cai, J. Shen, L. Chen and Z. He, *Nucl. Instr. and Meth.*, **B 169** (2000) 26.

26. D.M. Ruck, S. Brunner, K. Tinschert and W.F.X. Frank, *Nucl. Instr. and Meth.*, **B 106** (1995) 447.
27. A. Qureshi, N.L. Singh, S. Shah, F. Singh and D.K. Avasthi, *J. Macromolecular Science*, **A 45** (2008) 265.
28. J. Kaur, S.K. Chakarvarti, D. Kanjilal and S. Singh, *Pramana*, **72** (2009) 759.
29. E. Ferain and R. Legras, *Nucl. Instr. and Meth.*, **B 174** (2001) 116.
30. A. Srivastava, V. Singh, C. Dhand, M. Kaur, T. Singh, K. Witte and U.W. Scherer, *Sensors*, **6** (2006) 262.
31. N.L. Singh, V. Shrinet, N.R. Pandya, A. Sharma, N.V. Patel and D.K. Avasthi, *Nucl. Instr. and Meth.*, **B 156** (1999) 191.
32. K. Friese, V. Placek, R. Mehnert, N. Angert, R. Spohr and Ch. Trautmann, *Nucl. Instr. and Meth.*, **B 105** (1995) 139.
33. K. Schmidt, K. Friese, N. Angert and Ch. Trautmann, *Nucl. Instr. and Meth.*, **B 107** (1996) 381.
34. S. Chawla, A.K. Ghosh, D.K. Avasthi, P.K. Kulriya and S. Ahmad, *J. Appl. Poly.*, **105** (2007) 3578.
35. R. Mazzei, G. Garcýia Bermúdez, N. Betz and E. Cabanillas, *Nucl. Instr. and Meth.*, **B 226** (2004) 575.
36. M.C. Porté-Durrieu, C. Aymes-Chodur, N. Betz and Ch. Baquey, *J. Biomed Mater. Res.*, **52** (2000) 119.
37. P. Mazzoldi and G.W. Arnold (Eds.), *Ion Beam Modification of Insulators*, vol. 2. Elsevier, Amsterdam (1987).
38. T. Venkatesan, L. Calcagno, B.S. Elman and G. Foti, *Ion Beam Modifications of Insulators,* P. Mazzoldi and G.W. Arnold (Eds.), Elsevier, Amsterdam (1987) 301.
39. A.M.P. Hussain, A. Kumar, F. Singh and D.K. Avasthi, *J. Phys. D: Appl. Phys.*, **39** (2006) 750.
40. A. Kumar and A.M.P. Hussain, *Nucl. Instr. and Meth.*, **B 251** (2006) 451.
41. G.B.V.S. Lakshmi, Azher M. Siddiqui and M. Zulfequar, *International Journal of Polymeric Materials*, **59** (2010) 970.
42. R.C. Ramola, Subhash Chandra, J.M.S. Rana, Raksha Sharma, S. Annapoorni, R.G. Sonkawade, F. Singh and D.K. Avasthi, *Current Science*, **97** (2009) 1453.
43. Amarjeet Kaur, Anju Dhillon and D.K. Avasthi, *J. of Appl. Phys.*, **106** (2009) 073715.
44. Anju Dhillon, Amarjeet Kaur and D.K. Avasthi, *Thin Solid Films*, **519** (2010) 998.
45. P. Sonar, A.L. Sharma, A. Chandra, K. Muellen and A. Srivastava, *Current Applied Physics*, **3** (2003) 247.
46. A.M.P. Hussain, D. Saikia, F. Singh, D.K. Avasthi and A. Kumar, *Nucl. Instr. and Meth.*, **B 240** (2005) 834.
47. A.M.P. Hussain, A. Kumar, D. Saikia, F. Singh and D.K. Avasthi, *Nucl. Instr. and Meth.*, **B 240** (2005) 871.
48. G.B.V.S. Lakshmi, Azher M. Siddiqui, Vazid Ali, Pawan K. Kulriya, M. Zulfequar, *Nucl. Instr. and Meth.*, **B 266** (2008) 1685.
49. D. Saikia, A. Kumar, F. Singh and D.K. Avasthi, *J. Phys. D: Appl. Phys.*, **39** (2006) 4208.
50. D. Saikia, A. Kumar, F. Singh, D.K. Avasthi and N.C. Mishra, *J. Appl. Phys.*, **98** (2005) 043514.

6.3 Semiconductors

1. J. Vetter, R. Scholz and N. Angert, *Nucl. Instr. and Meth.*, **B 91** (1994) 129.
2. O. Herre, W. Wesch, E. Wendler, P.I. Gaiduk, F.F. Komarov, S. Klaumünzer and P. Meier, *Phys. Rev.*, **B 58** (1998) 4832.
3. G. Szenes, Z.E. Horvath, B. Pecz, F. Paszti and L. Toth, *Phys. Rev.*, **B 65** (2002) 45206.
4. F.F. Komarov, P.I. Gaiduk, L.A. Vlasukova, A.J. Dydik and V.N. Yuvchenko, *Vacuum*, **70** (2003) 75.
5. P.I. Gaiduk, A. Nylandsted Larsen, C. Trautmann and M. Toulemonde, *Phys. Rev.*, **B 66** (2002) 045316.
6. C.S. Schnohr, P. Kluth, R. Giulian, D.J. Llewellyn, A.P. Byrne, D.J. Cookson and M.C. Ridgway, *Phys. Rev.*, **B 81** (2010) 075201.
7. A. Kamarou, W. Wesch, E. Wendler, A. Undisz and M. Rettenmayr, *Phys. Rev.*, **B 73** (2006) 184107.
8. W. Wesch, C.S. Schnohr, P. Kluth, Z.S. Hussain, L.L. Araujo, R. Giulian, D.J. Sprouster, A.P. Byrne and M.C. Ridgway, *J. Phys. D: Appl. Phys.*, **42** (2009) 115402.
9. S.T. Chavan, P.S. Bhave, V.N. Bhoraskar and D.K. Avasthi, *J. Appl. Phys.*, **78** (1995) 2328.
10. S.T. Chavan, S.D. Dhole, V.N. Bhoraskar, D. Kanjilal and G.K. Mehta, *J. Appl. Phys.*, **82** (1997) 4805.
11. A. Narsale, Y.P. Ali, U. Bhambhani, V.P. Salvi, B.M. Arora, D. Kanjilal and G.K. Mehta, *J. Appl. Phys.*, **82** (1997) 4228.
12. Shiva S. Hullavarad, T.A. Railkar, S.V. Bhoraskar, P. Madukumar, A.S. Gokarna, V.N. Bhoraskar, S. Badrinarayanan and N.R. Pawaskar, *J. Appl. Phys.*, **83** (1998) 1962.
13. P.S. Bhave and V.N. Bhoraskar, *Nucl. Instr. and Meth.*, **127/128** (1997) 383.
14. P.C. Srivastava, S.P. Pandey, D.K. Avasthi and K. Asokan, *Vacuum*, **48** (1997) 965.
15. S. Arulkumaran, J. Arokiaraj, M. Udhayasankar, J. Kumar and D. Kanjilal, *Nucl, Instr. and Meth.*, **B 117** (1996) 243.
16. Yousuf Pyar Ali, A.M. Narsale, Uma Bhambhani, Arun Damle, V.P. Salvi, B.M. Arora, A.P. Shah and D. Kanjilal, *Nuc. Instr. and Meth.*, **B 117** (1996) 129.
17. K.G. Bhole, B.A. Kamalapurkar, S.K. Dube, A.D. Yadav, T.K. Gundu Rao, T. Mohanti and D. Kanjilal, *Nucl. Inst. and Meth.*, **212** (2003) 525.
18. S. Kumar, Y.S. Katharria, Y. Batra and D. Kanjilal, *J. Phys.D: Appl. Phys.*, **40** (2007) 6892.
19. R. Singh, S.K. Arora and D. Kanjilal, *Mater. Sci. Semicond. Process.*, **4** (2001) 425.
20. S. Kumar, Y.S. Katharria, S. Kumar and D. Kanjilal, *J. Appl. Phys.*, **100** (2006) 113723.
21. L. Vines, E. Monakhov, B.G. Svensson, J. Jensen, A. Hallen and A.Y. Kuznetsov, *Phys. Rev.*, **B 73** (2006) 085312.
22. S. Kumar, Y.S. Katharria and D. Kanjilal, *J. Appl. Phys.*, **103** (2008) 044504.
23. S. Kumar, Y.S. Katharria and D. Kanjilal, *J. Phys. D: Appl. Phys.*, **41** (2008) 105105.
24. B.R. Mehta, M.K. Sahay, L.K. Malhotra, D.K. Avasthi and R.K. Soni, *Thin Solid Films*, **289** (1996) 95.

25. Tejashree M. Bhave, S.V. Bhoraskar, Prabhat Singh and V.N. Bhoraskar, *Nucl. Instr. and Meth.*, **132** (1997) 409.
26. Tejashree M. Bhave, S.S. Hullavarad, S.V. Bhoraskar, S.G. Hegde and D. Kanjilal, *Nucl. Instr. & Meth.*, **156** (1999) 121.
27. V.V. Ison, A.R. Rao, V. Dutta, P.K. Kulriya, D.K. Avasthi and S.K. Tripathi, *J. Appl. Phys.*, **106** (2009) 023508.
28. V.V. Ison, A.R. Rao, V. Dutta, P.K. Kulriya, D.K. Avasthi and S.K. Tripathi, *J. Phys. D: Appl. Phys.*, **41** (2008) 105113.
29. V.V. Ison, A.R. Rao, V. Dutta, P.K. Kulriya and D.K. Avasthi, *Nucl. Instr. and Meth.*, **B 267** (2009) 2480.
30. W. Wesch, A. Kamarou and E. Wendler, *Nucl. Instr. and Meth.*, **B 225** (2004) 111.
31. V.V. Ison, A.R. Rao and V. Dutta, *Sol. Energy Mater. Sol. Cells*, **93** (2009) 1507.
32. S.V.S. Nageswara Rao, Anand P. Pathak, Azher M. Siddiqui, D.K. Avasthi, Claudiu Muntele, Daryush Ila, B.N. Dev, R. Muralidharan, F. Eichhorn, R. Groetzschel and A. Turos, *Nucl. Instr. and Meth.*, **B 212** (2003) 442.
33. S.V.S. Nageswara Rao, D.K. Avasthi, E.T. Subramanayam, Kundan Singh, G.B.V.S. Lakshmi, S.A. Khan, Azher M. Siddiqui, A. Tripathi, S.K. Srivastava, Sarvesh Kumar, T. Srinivasan, Umesh Tiwari, S.K. Mehta, R. Muralidharan, R.K. Jain and Anand P. Pathak, *AIP Conf. Series*, **680** (2003) 94.
34. S.V.S. Nageswara Rao, A.K. Rajam, Azher M. Siddiqui, D.K. Avasthi, T. Srinivasan, Umesh Tiwari, S.K. Mehta, R. Muralidharan, R.K. Jain and Anand P. Pathak, *Nucl. Instr. and Meth.*, **B 212** (2003) 473.
35. A.P. Pathak, S.V.S. Nageswara Rao, A.M. Siddiqui, G.B.V.S. Lakshmi, S.K. Srivastava, S. Ghosh, D. Bhattacharya, D.K. Avasthi, D.K. Goswami, P. Satyam, B.N. Dev and A. Turos, *Nucl. Inst. and Meth.*, **B 193** (2002) 319.
36. S.V.S. Nageswara Rao, A.P. Pathak, D.K. Avasthi, R. Muralidharan, Umesh Tiwari, T. Srinivasan, S.K. Mehta, R.K. Jain, F. Eichhorn, R. Groetzschel and N. Schell, Physics at Surface and Interface, Ed. B.N. Dev, *World Scientific* (2003) 158.
37. Anand P. Pathak, S.V.S. Nageswara Rao, D.K. Avasthi, Azher M. Siddiqui, S.K. Srivastava, F. Eichhorn, R. Groetzschel, N. Schell and A. Turos, *AIP Conf. Series*, **680** (2003) 593.
38. S.V.S. Nageswara Rao et al., *Solid State Physics* (India), **44** (2001) 227.
39. S. Dhamodaran, A.P. Pathak, A. Turos, G. Sai Saravanan, S.A. Khan, D.K. Avasthi and B.M. Arora, *Nucl. Instr. and Meth.*, **B 266** (2008) 583.
40. S. Dhamodaran, G. Devaraju, A.P. Pathak, A. Turos, D.K. Avasthi, R Kesavamoorthy and B.M. Arora, *Nucl. Instr. and Meth.*, **B 266** (2008) 1810.
41. G. Devaraju, S. Dhamodaran, A.P. Pathak, G. Sai Saravanan, J. Gaca, M. Wojcik, A. Turos, S.A. Khan, D.K. Avasthi and B.M. Arora, *Nucl. Instr. and Meth.*, **B 266** (2008) 3552.
42. A.P. Pathak, N. Sathish, S. Dhamodaran and D. Emfietzoglou, *Vaccum*, **82** (2008) 911.
43. N. Sathish, S. Dhamodaran, A.P. Pathak, S.A. Khan, M. Ghanashyam Krishna, S.A. Khan, D.K. Avasthi, A. Pandey, R. Muralidharan, G. Li and C. Jagadish, *Nucl. Instr. and Meth.*, **B 256** (2007) 281.
44. N. Sathish, S. Dhamodaran, A.P. Pathak, C. Muntele, D. Ila, S.A. Khan and D.K. Avasthi, Proceeding on Conference on Applications of Accelerators in Research and Industry (CAARI 2008).

45. N. Sathish, S. Dhamodaran, A.P. Pathak, B. Sundravel, K.G.M. Nair, S.A. Khan, D.K. Avasthi, M. Bazzan, E. Trave and P. Mazzoldi, Proceeding on Conference on Applications of Accelerators in Research and Industry (CAARI 2008).
46. A.P. Pathak, G. Devaraju, N. Sathish and I. Kyriakou, *Vaccum*, **84** (2010) 1049.
47. G. Devaraju, N. Sathish, A.P. Pathak, A. Turos, M. Bazzan, E. Trave, P. Mazzoldi and B.M. Arora, *Nucl. Instr. and Meth.*, **B 268** (2010) 3001.
48. R. Sreekumar, R. Jayakrishnan, C. Sudhakartha, K.P. Vijayakumar, S.A. Khan and D.K. Avasthi, *J. Appl. Phys.*, **103** (2008) 023709.
49. S. Chandramohan, R. Sathyamoorthy, P. Sudhagar, D. Kanjilal, D. Kabiraj and K. Asokan, *Thin Solid Films*, **516** (2008) 5508.
50. R.R. Ahire, A.A. Sagade, S.D. Chavhan, V. Huse, Y.G. Gudage, F. Singh, D.K. Avasthi and R. Sharma, *Current Applied Phys.*, **9** (2009) 374.
51. R. Sharma, A.A. Sagade, S.R. Gosavi, Y.G. Gudage, A. Ghosh, P.K. Kulariya, I. Sulaniya, R.S. Mane and Sung-Hwan Han, *J. of Non Cryst. Solids*, **355** (2009) 1653.

6.4 Transparent Conducting Oxides (TCOs)

1. M. Hagerott, H. Jean, A.V. Nurmikko, W. Xie, D.C. Grillo, M. Kobayashi and R.L. Gunshor, *Appl. Phys. Lett.*, **60** (1992) 2825.
2. P. Nunes, E. Fortunato, P. Tonello, F.B. Fortunato, P. Vilarinho and R. Martins, *Vacuum*, **64** (2002) 281.
3. Cathleen A. Hoel, Thomas O. Mason, Jean-Franc-ois Gaillard and Kenneth R. Poeppelmeier, *Chem. Mater.*, **22-12** (2010) 3569.
4. Z.L. Wang and Z.C. Kang, Functional and Smart Materials. Plenum, New York (1998).
5. T. Makino, C.H. Chia, N.T. Tuan, Y. Segawa, M. Kawasaki, A. Ohtomo, K. Tamura and H. Koinuma, *Appl. Phys. Lett.*, **76** (2000) 3549.
6. J. Hu, T.W. Odam and C.M. Lieber, *Acc. Chem. Res.*, **32** (1999) 435.
7. J. Zhou, N. Xu and Z.L. Wang, *Adv. Mater.*, **18** (2006) 2432.
8. D.C. Agarwal, F. Singh, D. Kabiraj, S. Sen, P.K. Kulriya, I. Sulania, S. Nozaki, R.S. Chauhan and D.K. Avasthi, *J. Phys. D: Appl. Phys.*, **41** (2008) 045305.
9. D.C. Agarwal, D.K. Avasthi, F. Singh, D. Kabiraj, P.K. Kulariya, I. Sulania, J.C. Pivin and R.S. Chauhan, *Surf. and Coat. Tech.*, **203** (2009) 2427.
10. D.C. Agarwal, Amit Kumar, S.A. Khan, D. Kabiraj, F. Singh, A. Tripathi, J.C. Pivin, R.S. Chauhan and D.K. Avasthi, *Nucl. Instr. and Meth.*, **B 244** (2006) 136.
11. D.C. Agarwal, R.S. Chauhan, D.K. Avasthi, S.A. Khan, D. Kabiraj and I. Sulania, *J. Appl. Phys.*, **104** (2008) 24304.
12. N. Fujimura, T. Nishihara, S. Goto and J. Xua, *J. Cryst. Growth*, **130** (1993) 269.
13. Y.J. Kim, Y.T. Kin, H.J. Yang, J.C. Park, J.I. Han, Y.E. Lee and H.J. Kim, *J. Vac. Sci. and Technol.*, **A 15** (1997) 1103.
14. P.M. Ratheesh Kumar, C. Sudha Kartha, K.P. Vijayakumar, F. Singh, D.K. Avasthi, T. Abe, Y. Kashiwaba, G.S. Okram, M. Kumar and Sarvesh Kumar, *J. Appl. Phys.*, **97** (2005) 013509.

15. Sanju Rani, Somnath C. Roy, N.K. Puri, M.C. Bhatnagar and D. Kanjilal, *Nanosensor Technology*, **2008** (2008) Article ID 395490.
16. S. Rani, S.C. Roy, M.C. Bhatnagar and D. Kanjilal, *Sensors 2008 IEEE*, 886.

6.5 Transition Metal Oxides (TMO)

1. E. Fujii, A. Tomozawa, H. Torii and R. Takayama, *Jpn. J. Appl. Phys.*, **35** (1996) L 328.
2. K. Yoshimura, T. Miki and S. Tanemura, *Jpn. J. Appl. Phys.*, **34** (1995) 2440.
3. U.S. Joshi, S.J. Trivedi, K.H. Bhavsar, U.N. Trivedi, S.A. Khan and D.K. Avasthi, *J. Appl. Phys.*, **105** (2009) 73704.
4. P. Mallick, D.C. Agarwal, C. Rath, R. Biswal, D. Behera, D.K. Avasthi, D. Kanjilal, P.V. Satyam and N.C. Mishra, *Nucl. Instr. and Meth.*, **B 266** (2008) 3332.
5. P. Mallick, Chandana Rath, S. Majumdar, R. Biswal, D.C. Agarwal, Shikha Varma, D.K. Avasthi, P.V. Satyam and N.C. Mishra, *Appl. Surf. Sci.*, **256** (2009) 521.
6. P. Mallick, R. Biswal, Chandana Rath, D.C. Agarwal, A. Tripathi, D.K. Avasthi, D. Kanjilal, P.V. Satyam and N.C. Mishra, *Nucl. Instr. and Meth.*, **B 268** (2010) 470.
7. A. Benyagoub, *Nucl. Instr. and Meth.*, **B 245** (2006) 225; A. Benyagoub, *Phys. Rev.*, **B 72** (2005) 094114; A. Benyagoub and F. Levesque, *Europhys. Lett.*, **60** (2002) 580.
8. B. Schattat, W. Bolse, S. Klaumunzer, I. Zizak and R. Scholz, *Appl. Phys. Lett.*, **87** (2005) 173110.
9. P. Mallick, Chandana Rath, Jai Prakash, D.K. Mishra, R.J. Choudhary, D.M. Phase, A. Tripathi, D.K. Avasthi, D. Kanjilal and N.C. Mishra, *Nucl. Instr. and Meth.*, **B 268** (2010) 1613.
10. C. Garapon, C. Champeaux, J. Mugnier, G. Panczer, P. Marchet, A. Catherinot and B. Jacquier, *Appl. Surf. Sci.*, **96-98** (1996) 836.
11. L. Castaneda, A. Maldonado and M. dela L. Olvera, *Sens. Actuators*, **B 133** (2008) 687.
12. T. Yuji, H. Akatsuka, N. Mungkung, B.W. Park and Y.M. Sung, *Vacuum*, **83** (2009) 124.
13. M. Kadoshima, M. Hiratani, Y. Shimamoto, K. Torii, H. Miki, S. Kimura and T. Nabatame, *Thin Solid Films*, **424** (2003) 224.
14. R. Jourdani, A. Outzourhit, A. Oueriagli, D. Aitelhabti, E.L. Ameziane, S. Barazzouk and S. Hotchandani, *Act. Passive Electron. Compon.*, **27** (2004) 125.
15. C.-M. Wang, S.-Y. Lin and Y.-C. Chen, *J. Phys. Chem. Solids*, **69** (2008) 451.
16. H. Rath, P. Dash, T. Som, P.V. Satyam, U.P. Singh, P.K. Kulriya, D. Kanjilal, D.K. Avasthi and N.C. Mishra, *J Appl. Phys.*, **105** (2009) 074311.
17. T. Mohanty, A. Pradhan, S. Gupta and D. Kanjilal, *Nanotechnology*, **15** (2004) 1620.
18. N. Koshizaki, A. Narazaki and T. Sasaki, *Appl. Surf. Sci.*, **197-198** (2002) 624.
19. M. Toulemonde, Ch. Dufour, A. Meftah and E. Paumier, *Nucl. Instr. and Meth.*, **B 166-167** (2000) 903.
20. D.K. Avasthi, W. Assmann, H. Nolte, H.D. Mieskes, S. Ghosh and N.C. Mishra, *Nucl. Instr. and Meth.*, **B 166-167** (2000) 345.

21. Madhavi Thakurdesai, A. Mahadkar, D. Kanjilal and Varsha Bhattacharyya, *Vacuum*, **82** (2008) 639.
22. Madhavi Thakurdesai, A. Mahadkar, P.K. Kulriya, D. Kanjilal and Varsha Bhattacharyya, *Nucl. Instr. and Meth.*, **B 266** (2008) 1343.
23. Madhavi Thakurdesai, D. Kanjilal and V. Bhattacharyya, *Appl. Surf. Sci.*, **254** (2008) 4695.
24. Madhavi Thakurdesai, T. Mohanty, J. John, T.K. Gundu Rao, R. Pratap, V. Bhattacharyya and D. Kanjilal, *J. Nanosci. and Nanotech.*, **8** (2008) 4231.
25. Madhavi Thakurdesai, T. Mohanty, D. Kanjilal, Pratap Raychaudhuri and Varsha Bhattacharyya, *Appl. Surf. Sci.*, **255** (2009) 8935.
26. Madhavi Thakurdesai, D. Kanjilal and Varsha Bhattacharyya, *Semicond. Sci. Tech.*, **24** (2009) 085023.
27. E.J.W. Verwey, *Nature*, **144** (1939) 327.
28. S.A. Wolf, D.D. Awschalom, R.A. Buhrman, J.M. Daughton, S. von Molnár, M.L. Roukes, A.Y. Chtchelkanova, and D.M. Treger, *Science*, **294** (2001) 1488.
29. Ravi Kumar, M. Wasi Khan, J.P. Srivastava, S.K. Arora, R.G.S. Sofin, R.J. Choudhary, and I.V. Shvets, *J. Appl. Phys.*, **100** (2006) 033703.
30. Ram Prakash, R.J. Choudhary, Shailja Tiwari, D.M. Phase and Ravi Kumar, *Nucl. Instr. and Meth.*, **B 266** (2008) 1242.

6.6 Diluted Magnetic Semiconductors (DMS) and Other Magnetic Materials

1. S.B. Ogale, *Advanced Materials*, **22** (2010) 3125.
2. T. Minami, *Semicond. Sci. Technol.*, **20** (2005) S35.
3. S.J. Pearton, C.R. Abernathy, M.E. Overberg, G.T. Thaler, D.P. Norton, N. Theodoropoulou, A.F. Hebard, Y.D. Park, F. Ren, J. Kim and L.A. Boatner, *J. Appl. Phys.*, **93** (2003) 1.
4. K. Ando, *Solid State Sciences: Magneto-Optics* (2000), Springer, N.Y., 128, 211.
5. K. Sato and H. Katayama-Yoshida, *Jap. J. Appl. Phys.*, **39** (2000) L555.
6. D.A. Schwartz, K.R. Kittilstved, D.R. Gamelin, *Appl. Phys. Lett.*, **85** (2004) 1395.
7. H. Saeki, H. Matsui, T. Kawai, H. Tabata, *J. Phys.: Cond. Matter*, **16** (2004) S5533.
8. T. Wakano, N. Fujimura, Y. Morinaga, N. Abe, A. Ashida and T. Ito, *Physica E*, **10** (2001) 260.
9. Z. Yin, N. Chen, F. Yang, S. Song, C. Chai, J. Zhong, H. Qian and K. Ibrahim, *Sol. Stat. Comm.*, **135** (2005) 430.
10. S. Zhou, K. Potzger, K. Kuepper, J. Grenzer, M. Helm, J. Fassbender, E. Arenholz and J.D. Denlinger, *J. Appl. Phys.*, **103** (2008) 043901.
11. W. Yu, L.H. Yang, X.Y. Teng, J.C. Zhang, Z.C. Zhang, L. Zhang and G.S. Fu, *J. Appl. Phys.*, **103** (2008) 093901.
12. N. Dave, B.G. Pautler, S.S. Farvid and P.V. Radovanovic, *Nanotechnology*, **21** (2010) 134023.
13. B. Angadi, Y.S. Jung, W.K. Choi, R. Kumar, K. Jeong, S.W. Shin, J.H. Lee, J.H. Song, M.W. Khan and J.P. Srivastava, *Appl. Phys. Lett.*, **88** (2006) 142502.
14. Ravi Kumar, Abhinav Pratap Singh, P. Thakur, K.H. Chae, W.K. Choi, Basavaraj Angadi, S.D. Kaushik and S. Patnaik, *J. Phys. D: Appl. Phys.*, **41** (2008) 155002.

15. M. Julliere, Phys. Letters, **54A** (1975) 225.
16. T. Banerjee, T. Som, D. Kanjilal and J.S. Moodera, *Eur. Phy. J. Appl. Phys.*, **32** (2005) 115.
17. Y. Conraux. J.P. Nozieres, V. da Costa, M. Toulemonde and K. Ounadjela, *J. Appl. Phys.*, **93** (2003) 7301.
18. P.C. Srivastava and J.K. Tripathi, *J. Phys. D: Appl. Phys.*, **39** (2006) 1465.
19. J. Narayan, S. Nori, D.K. Pandya, D.K. Avasthi and A.I. Smirnov, *Appl. Phys. Lett.*, **93** (2008) 82507.
20. J. Juraszek, A. Fnidiki and M. Toulemonde, *J. Appl. Phys.*, **89** (2001) 3151.
21. A. Gupta, A. Paul, R. Gupta, D.K. Avasthi and G. Prinicipi, *J. Phys.: Condens. Matter Phys.*, **10** (1998) 9669.
22. R. Dubey, A. Gupta, P. Sharma, M. Darbowski and G. Schumacher, *Int. J. Mag. Mat.*, **310** (2007) 2491.
23. A. Fnidiki, J. Jurasjek, J. Teillet and F. Studer, *Appl. Phys. Lett.*, **75** (1999) 1296.

6.7 High Temperature Superconductors

1. D.R. Nelson and V.M. Vinokur, *Phys. Rev.*, **B 48** (1993) 13060.
2. R.C. Budhani, W.L. Holstein and M. Suenaga, *Phys. Rev. Lett.*, **72** (1994) 566.
3. T. Banerjee, R.N. Viswanath, D. Kanjilal, R. Kumar and S. Ramasamy, *Solid State Communications*, **114** (2000) 655.
4. T. Mohanty, A.K. Mohapatra, P.K. Samal, B.N. Dash, D.B. Behera, H.P. Mahapatra, S. Mishra, N.C. Mishra, K. Patnaik, L. Senapati, D. Kanjilal and G.K. Mehta, *Indian J. Phys.*, **69A** (1995) 29.
5. G.K. Mehta, *Nucl. Instr. and Meth.*, **B 212** (2003) 8.
6. S.A. Agnihotry, R.H. Bhawalkar, S. Chandra, L. Senapati, D. Kanjilal and G.K. Mehta, *Supercond. Sci. Technol.*, **5** (1992) 582.
7. Y. Sudershan, Amit Rastogi, S.V. Bhat, A.K. Grover, Y. Yamaguchi, K. Oka, Y. Nishihara, L. Senapati and D. Kanjilal, *Appl. Phys. Lett.*, **71** (1997) 1576.
8. D. Kumar, M. Sharon, R. Pinto, P.R. Apte, S.P. Pai, S.C.Purandare, L.C. Gupta and R. Vijayaraghavan, *Appl. Phys. Lett.*, **62** (1993) 3522.
9. S.G. Kaplan, M. Chen, H.D. Drew, M. Rajeshwari, R. Liu, T. Venkatesan, D. Kanjilal, L. Senapati, G.K. Mehta, *Physica*, **C 232** (1994) 174.
10. N.C. Mishra, D. Behera, T. Mohanty, D. Mohanta, D. Kanjilal, G.K. Mehta and R. Pinto, *Nucl. Instr. and Meth.*, **B 156** (1999) 30.
11. D. Behera, K. Patnaik and N.C. Mishra, *Modern Physics Letters*, **B 15** (2001) 69.
12. D. Behera, T. Mohanty, S.K. Dash, T. Banerjee, D. Kanjilal and N.C. Mishra, *Radiation Measurement*, **36** (2003) 125.
13. R. Biswal, J. John, D. Behera, P. Mallick, Sandeep Kumar, D. Kanjilal T. Mohanty, P.Raychaudhuri and N.C. Mishra, *Supercond. Sci. Technol.*, **21** (2008) 085016.
14. R. Biswal, J. John, P. Mallick, B.N. Dash, P.K. Kulriya, D.K. Avasthi, D. Kanjilal, D. Behera, T. Mohanty, P. Raychaudhuri and N.C. Mishra, *J. Appl. Phys.*, **106** (2009) 053912.
15. R.C. Budhani, Studies on High Temperature Superconductors. A. Narlikar (ed.) Nova Science, New York (2000) 116.
16. Manmeet Kaur Marhas, K. Balakrishnan, P. Saravanan, V. Ganesan, R. Srinivasan, D. Kanjilal, G.K. Mehta, M. Vedwyas, S.B. Ogale, S.P. Pai, M.S. Ramachandra

Rao, R. Pinto, G. Mohan Rao, S. Senthilnathan and S. Mohan, *Vacuum*, **48** (1997) 983.

17. M.K. Marhas, K. Balakrishnan, V. Ganeshan, R. Srinivasan, D. Kanjilal, G.K. Mehta, G.K. Murlidhar, G. Mohan Rao, S. Nathan and S. Mohan, *Bullet. Mater. Sci.*, **17** (1994) 585.
18. Aslamasov and A. Larkin, *Phys. Lett.*, **26A** (1968) 238.
19. J. Lawrence and S. Doniach, *In:* Proc. 12th Conf. Low-temp. Phys. Kyoto. E. Kanda (ed.) Keigaku, Tokyo (1970) 361.
20. T. Banerjee, Ravi Kumar, D. Kanjilal and S. Ramasamy, *Physica C*, **341** (2000) 1185.
21. Tamalika Banerjee, Avinash Bhangale, D. Kanjilal, S.P. Pai and R. Pinto, *Pramana*, **58** (2002) 949.
22. S.K. Arora, Ravi Kumar, D. Kanjilal, G.K. Mehta, S. Khatua, R. Pinto, Vijay Kumar and A.K. Gupta, *Bulletin of Material Science*, **22** (1999) 251.
23. A.K. Pradhan, B. Chen, G. Pindoria, K. Nakao, N. Koshizuka, D. Kanjilal, A.J.S. Chowdhary and B.M. Wanklyn, *Phys. Rev.*, **B 55** (1997) 11129.
24. Ravindra Kumar, H.M. Agrawal, R.P.S. Kushwaha and D. Kanjilal, *Nucl. Instr. and Meth.*, **B 263** (2007) 414.
25. G.K. Mehta, *Nucl. Instr. and Meth.*, **A 382** (1996) 335.
26. S.E. Lofland, S.D. Tyagi, S.M. Bhagat, M. Rajeshwari, T. Venketeshan, D. Kanjilal and G.K. Mehta, *Physica*, **C 267** (1996) 79.
27. S.E. Lofland, S.M. Bhagat, M. Rajeshwari, T. Venketesan, D. Kanjilal, L. Senapati and G.K. Mehta, *Phys. Rev.*, **B 51** (1995) 8489.
28. S. Patnaik, R.C. Budhani, M. Konczykowski, Y.L. Yang and M. Suenaga, *Physica*, **C 349** (2001) 155.
29. L.K. Sahoo, R.C. Budhani, D. Kanjilal and G.K. Mehta, *Physica*, **C 383** (2002) 8.
30. A.K. Pradhan, S.B. Roy, P. Chaddah, D. Kanjilal, C. Chen and B.M. Wanklyn, *Phys. Rev.*, **B 53** (1996) 2269.
31. D. Kanjilal, *Current Science*, **80** (2001) 1560.
32. E. Akcoltekin, T. Peters, R. Meyer, A. Duvenbeck, M. Klusmann, I. Monnet, H. Lebius and M. Schleberger, *Nature Nanotechnology*, **2** (2007) 290.
33. D. Kanjilal, *Vacuum*, **48** (1997) 979.
34. R.C. Budhani, Y. Zhu and M. Suenaga, *Appl. Phys. Lett.*, **61** (1992) 985.
35. K.R. Mavani, D.S. Rana, S. Rayaprol, R.N. Parmar, D.G. Kuberkar, Ravi Kumar, M. Tonouchi, J. John and R. Nagarajan, *Solid State Communications*, **142** (2007) 462.
36. T. Haugan, P.N. Barnes, R. Wheeler, R. Meisenkothen and M. Sumption, *Nature*, **430** (2004) 867.
37. R.L. Fleischer, P.B. Price and R.M. Walker, *J. Appl. Phys.*, **36** (1965) 3645.
38. S. Klaumunzer, *Nucl. Instr. and Meth.*, **B 225** (2004) 136.
39. L. Civale, A.D. Marwick, T. Worthington, M. Kirk, J.R. Thompson, L. Krusin Elbaum, Y. Sun, J.R. Clem and F. Holzberg, *Phys. Rev. Lett.*, **67** (1991) 648.
40. Himanshu Narayan, Ravindra K. Bhatt, H.M. Agrawal, R.P.S. Kushwaha and H. Kishan, *J. Phys.: Condens. Matter*, **19** (2007)136209.
41. Ravindra Kumar, H.M. Agrawal, R.P.S. Kushwaha and D. Kanjilal, *Supercond. Sci. Technol.*, **19** (2006) 1047.

42. S.D. Kaushik, Ravi Kumar, P.K. Mishra, J.E. Giencke, D.M. Kim, C.B. Eom and S. Patnaik, *Physica*, **C 442** (2006) 73.
43. Sanghamitra Khatua, P.K. Mishra, Ravi Kumar, V.C. Sahni and R. Pinto, *Supercond. Sci. Technol.*, **15** (2002) 324.

6.8 Special Oxides (Ferrites, Multiferroics and LCMO)

1. Sanjukta Ghosh, P. Ayyub, N. Kumar, S.A. Khan and D. Banerjee, *Nucl. Instr. and Meth.*, **B 212** (2003) 510.
2. S. Ghosh, A. Gupta, P. Ayyub, N. Kumar, S.A. Khan, D. Banerjee and R. Bhattacharya, *Nucl. Instr. and Meth.*, **B 225** (2004) 310.
3. Jean-Marc Costantini, Francis Studer and Jean-Claude Peuzin, *J. Appl. Phys.*, **90** (2001) 126.
4. Bas B. Vas Aken, Jean Pierre Rivera, Hans Schmid and Manfred Fiebig, *Nature*, **449** (2007) 702.
5. D.K. Shukla, Ravi Kumar, S. Mollah, R.J. Choudhary, P. Thakur, S.K. Sharma, N.B. Brookes and M. Knobel, *J. Appl. Phys.*, **107** (2010) 09D903.
6. B.N. Dash, Priyadarshini Dash, Haripriya Rath, P. Mallick, R. Biswal, P.K. Kulriya and N.C. Mishra, *Ind J. of Physics*, **84** (2010) 1315.
7. A.P. Ramierez, *J. Phys.: Condens. Matter*, **9** (1997) 8171.
8. R. Bathe, S.K. Date, S.R. Shinde, L.V. Saraf, S.B. Ogale, S.I. Patil, Ravi Kumar, S.K. Arora and G.K. Mehta, *J. Appl. Phys.*, **83** (1998) 7174.
9. S.B. Ogale, K. Ghosh, J.Y. Gu, R. Shreekala, S.R. Shinde, M. Downes, M. Rajeswari, R.P. Sharma, R.L. Greene, T. Venkatesan, R. Ramesh, Ravi Bathe, S.I. Patil, R. Ravikumar, S.K. Arora and G.K. Mehta, *J. Appl. Phys.*, **84** (1998) 6255.
10. S.B. Ogale, Y.H. Li, M. Rajeswari, L. Salamanca Riba, R. Ramesh, T. Venkatesan, A.J. Millis, Ravi Kumar, G.K. Mehta, Ravi Bathe and S.I. Patil, *J. Appl. Phys.*, **87** (2000) 4210.
11. R.J. Choudhary, Ravi Kumar, S.I. Patil, Shahid Husain, J.P. Srivastava and S.K. Malik et al., *Appl. Phys. Letts.*, **86** (2005) 222501.
12. S.K. Arora, Ravi Kumar, Rajendra Singh, D. Kanjilal, G.K. Mehta, Ravi Bathe, S.I. Patil and S.B. Ogale, *J. Appl. Phys.*, **86** (1999) 4452.

6.9 Resistive Random Access Memory (RRAM) Devices Based on Oxides

1. A. Sawa, *Materials Today*, **11** (2008) 28.
2. R. Wasser and M. Aono, *Nature Materials*, **6** (2007) 833.
3. I.G. Baek et al., *Tech. Dig. IEDM* (2004) 587.
4. R. Müller, R. Naulaerts, J. Billen, J. Genoe, and P. Heremans, *Appl. Phys. Lett.*, **90** (2007) 063503.
5. Liping Ma, Seungmoon Pyo, Jianyong Ouyang, Qianfei Xu and Yang Yang, *Appl. Phys. Lett.*, **82** (2003) 1419.
6. Chikako Yoshida, Kohji Tsunoda, Hideyuki Noshiro and Yoshihiro Sugiyama, *Appl. Phys. Lett.*, **91** (2007) 223510.
7. T.W. Hickmott, *J. Appl. Phys.*, **33** (1962) 2669.
8. J.G. Simmons, R.R. Verderber, *Proc. R. Soc. London, Ser.*, **A 301** (1967) 77.

9. J.F. Gibbons and W.E. Beadle, *Solid-State Electron.*, **7** (1964) 785.
10. G. Dearnaley, A.M. Stoneham, D.V. Morgan, *Rep. Prog. Phys.*, **33** (1970) 1129.
11. A. Asamitsu, Y. Tomioka, H. Kuwahara and Y. Tokura, *Nature*, **388** (1997) 50.
12. S.Q. Liu, N.J. Wu and A. Ignatiev, *Appl. Phys. Lett.*, **76** (2000) 2749.
13. Y. Watanabe, J.G. Bednorz, A. Bietsch, Ch. Gerber, D. Widmer, A. Beck and C.J. Wind, *Appl. Phys. Lett.*, **78** (2001) 3738.
14. B.J. Choi, D.S. Jeong, S.K. Kim, C. Rohde, S. Choi, J.H. Oh, H.J. Kim and C.S. Hwang, *J. Appl. Phys.*, **98** (2005) 033715.
15. Krzysztof Szot, Wolfgang Speier, Gustav Bihlmayer and Rainer Waser, *Nat. Mater.*, **5** (2006) 312.
16. U.S. Joshi, S.J. Trivedi, K.H. Bhavsar, U.N. Trivedi, S.A. Khan and D.K. Avasthi, *J. Appl. Phys.*, **105** (2009) 73704.
17. Y.B. Nian, J. Strozier, N.J. Wu, X. Chen and A. Ignatiev, *Phys. Rev. Lett.*, **98** (2007) 146403.
18. D.K. Avasthi and W. Assmann, *Curr. Sci.*, **80** (2001) 1532.
19. D.K. Avasthi, S. Ghosh, S.K. Srivastava and W. Assmann, *Nucl. Instr. and Meth.*, **B 219-220** (2004) 206.
20. S.A. Khan, Manvendra Kumar and D.K. Avasthi, *Nucl. Instr. and Meth.*, **B 266** (2008) 1912.
21. W.M. Arnoldbik, N. Tomozeiu and F.H.P.M. Habraken, *Nucl. Instr. and Meth.*, **B 219-220** (2004) 312.

6.10 Quasicrystals

1. D. Levine and P.J. Steinhardt, *Phys. Rev. Lett.*, **53** (1984) 2477.
2. D. Schetman, I. Blech, D. Gratias and J.W. Cahn, *Phys. Rev. Lett.*, **53** (1984) 1951.
3. J.M. Dubois, P. Plaindoux, E. Belin-Ferre, N. Tamura and D.J. Sordelet, *In:* Quasicrystals. S. Takeuchi and T. Fujiwara (eds.) World Scientific, Singapore (1998) 733.
4. J.M. Dubois, *J. Non Cryst. Sol.*, **334-335** (2004) 481 (Proceedings of 8th Int. Conf. on Quasicrystals).
5. A. Dunlop, D. Lesueur and J. Dural, *Nucl. Instr. and Meth.*, **B 42** (1989) 182.
6. A. Dunlop, D. Lesueur, J. Marillo, J. Dural, R. Spohr and J. Vetter, *Nucl. Instr. Methods Phys. Res.*, **B 48** (1990) 419.
7. Ratnamala Chatterjee, D. Kanjilal and R.A. Dunlap, *Nucl. Instr. and Meth.*, **B 156** (1999) 201.
8. R. Chatterjee, A. Kanjilal, A. Dunlop and J.M. Ramillon, *Sol. Stat. Commun.*, **120** (2001) 289.
9. R. Chatterjee and A. Kanjilal, *J. Non-Cryst. Solids*, **334-335** (2004) 431.
10. G. Coddens, A. Dunlop, H. Dammak, R. Chatterjee, Y. Calvayrac, M. Quiquandon, E. Elkaim, M. Gailhanou and S. Rouziere, *Nucl. Instr. and Meth.*, **B 211** (2003) 122.
11. S. Mechler, C. Abromeit, N. Wanderka, M.P. Macht, T. Zumkley, G. Schumacher and S. Klamunzer, *NIM*, **B 245** (2006) 133.

6.11 Alkali Halides

1. C. Trautmann, M. Toulemonde, K. Schwartz, J.M. Costantini and A. Muller, *Nucl. Instr. and Meth.*, **B 164-165** (2000) 365.
2. K. Schwartz, C. Trautmann, T. Steckenreiter, O. Geib and M. Kramer, *Phys. Rev.*, **B 58** (1998) 11232.
3. C. Pandurangappa, B.N. Lakshminarasappa, F. Singh and K.R. Nagabhushana, *Nucl. Instr. and Meth.*, **B 269** (2011) 185.
4. M. Boccanfuso, A. Benyagoub, K. Schwartz, M. Toulemonde and C. Trautmann, *Progress in Nucl. Energy*, **38** (2001) 271.
5. M. Boccanfuso, A. Benyagub, M. Toulemonde, C. Trautmann, K. Schwartz and Ch. Dufour, *Nucl. Instr. and Meth.*, **B 175** (2001) 590.
6. C. Muller, M. Cranney, A. El-Said, N. Ishikawa, A. Iwase, M. Lang and R. Neumann, *Nucl. Instr. and Meth.*, **B 191** (2002) 246.
7. K. Schwartz, A.E. Volkov, M.V. Sorokin, C. Trautmann, K.-O. Voss, R. Neumann and M. Lang, *Phys. Rev.*, **B 78** (2008) 024120.
8. K. Schwartz, C. Trautmann and R. Neumann, *Nucl. Instr. and Meth.*, **B 209** (2003) 73.
9. M. Kumar, F. Singh, S.A. Khan, V. Baranwal, S. Kumar, D.C. Agarwal, A.M. Siddiqui, A. Tripathi, A. Gupta, D.K. Avasthi and A.C. Pandey, *J. Phys. D: Appl. Phys.*, **38** (2005) 637.
10. M. Kumar, F. Singh, S.A. Khan, A. Tripathi, D.K. Avasthi and A.C. Pandey, *J. Phys. D: Appl. Phys.*, **39** (2006) 2935.
11. M. Kumar, F. Singh, S.A. Khan, A. Tripathi, D.K. Avasthi and A.C. Pandey, *Nucl. Instr. and Meth.*, **B 244** (2006) 183.
12. M. Kumar, S.A. Khan, F. Singh, A. Tripathi, D.K. Avasthi and A.C. Pandey, *J. Lumin.*, **127** (2007) 302.
13. M. Kumar, F. Singh, S.A. Khan, A. Tripathi, D.K. Avasthi and A.C. Pandey, *Phys. Stat. Solid*, **C 4** (2007) 1075.

6.12 Plasmonic Materials

1. Polman, *Science*, **322** (2008) 868.
2. K.L. Kelly, E. Coronado, L.L. Zhao and G.C. Schatz, *J. Phys. Chem.*, **B 107** (2003) 668.
3. S.I. Bozhevolnyi, V.S. Volkov, E. Devaux and T.W. Ebbesen, *Phys. Rev. Lett.*, **95** (2005) 046802.
4. A.V. Zayats, I.I. Smolyaninov and Alexei A. Maradudin, *Phys. Rep.*, **408** (2005) 131.
5. V. Ramaswamy, T.E. Haynes, C.W. White, W.J. Moberlychen, S. Roorda and M.J. Aziz, *Nano Lett.*, **5** (2005) 373.
6. Y. Sun and Y. Xia, *Science*, **298** (2002) 2176.
7. D. Kabiraj, S.R. Abhilash, L. Vanmarcke, N. Cinausero, J.C. Pivin and D.K. Avasthi, *Nucl. Instr. and Meth.*, **B 244** (2006) 100.
8. S. Mohapatra, Y.K. Mishra, D. Kabiraj, D.K. Avasthi, J. Ghatak and S. Verma, *Appl. Phys. Lett.*, **92** (2008) 103105.
9. Y.K. Mishra, D. Kabiraj, I. Sulania, J.C. Pivin and D.K. Avasthi, *J. Nanosci. Nanotechnol.*, 7 (2007) 1878.

10. Y.K. Mishra, S. Mohapatra, D. Kabiraj, A. Tripathi, J.C. Pivin and D.K. Avasthi, *J. Opt. A: Pure Appl. Opt.*, **9** (2007) S410.
11. Y.K. Mishra, S. Mohapatra, R. Singhal, D.K. Avasthi, D.C. Agarwal and S.B. Ogale, *Appl. Phys. Lett.*, **92** (2008) 43107.
12. R. Singhal, D.C. Agarwal, Y.K. Mishra, S. Mohapatra, D.K. Avasthi, A.K. Chawla, R. Chandra and J.C. Pivin, *Nucl. Instr. and Meth.*, **B 267** (2009) 1349.
13. R. Singhal, D.C. Agarwal, Y.K. Mishra, F. Singh, J.C. Pivin, R. Chandra and D.K. Avasthi, *J. Phys. D: Appl. Phys.*, **42** (2009) 155103.
14. R. Singhal, A. Kumar, Y.K. Mishra, S. Mohapatra, J.C. Pivin and D.K. Avasthi, *Nucl. Instr. and Meth.*, **B 266** (2008) 3257.
15. R. Singhal, F. Singh, A. Tripathi and D.K. Avasthi, *Radiation Effects and Defects in Solids*, **164** (2009) 38.
16. D.K. Avasthi, Y.K. Mishra, D. Kabiraj, N.P. Lalla and J.C. Pivin, *Nanotechnology*, **18** (2007) 125604.
17. Y.K. Mishra, S. Mohapatra, V.S. Chakrvadhaluna, N.P. Lalla, V. Zeprotchenko and D.K. Avasthi, *J. Nanoscience and Nanotechnology*, **10** (2010) 2833.
18. L.A. Stweatlock, S.A. Maier, H.A. Atwater, J.J. Penninkhof and A. Polman, *Phys. Rev.*, **B 71** (2005) 235408.
19. R. Rangel-Rojo, J. McCarthy, H.T. Bookey, A.K. Kar, L. Rodriguez-Fernandez, J.C. Cheang-Wong, A. Crespo-Sosa, A. Lopez-Suarez, A. Oliver, V. Rodriguez-Iglesias and H.G. Silva-Pereyra, *Optics Communications*, **282** (2009) 1909.
20. V. Rodguez-Iglesias, O. Pea-Rodrguez, H.G. Silva-Pereyra, L. Rodrguez-Fernndez, G. Kellermann, J.C. Cheang-Wong, A. Crespo-Sosa and A. Oliver, *J. Phys. Chem.*, **C 114** (2010) 746.
21. A.L. González, J.A. Reyes-Esqueda and Cecilia Noguez, *J. Phys. Chem.*, **C 112** (2008) 7356.

6.13 Materials for Energy

1. S. Budak, C.I. Muntele, R.A. Minamisawa, B. Chhay and D. Ila, *Nucl. Instr. and Meth.*, **B 261** (2007) 608.
2. B. Zheng, Z. Xiao, B. Chhay, R. Zimmerman and D. Ila, *Nucl. Instr. and Meth.*, **B 266** (2008) 73.
3. R.R. Ahire, Abhay A. Sagade, N.G. Deshpande, S.D. Chavhan, Ramphal Sharma and F. Singh, *J. Phys. D: Appl. Phys.*, **40** (2007) 4850.
4. E. Koukharenko, X. Li, I. Nandhakumar, N. Frety, S.P. Beeby, D. Cox, M.J. Tudor, B. Schiedt, C. Trautmann, A. Bertsch and N.M. White, *J. Micromech. Microengg.*, **18** (2008) 104015.
5. Yatendra S. Chaudhary, Saif A. Khan, Rohit Shrivastav, Vibha R. Satsangi, Satya Prakash, Umesh K. Tiwari, D.K. Avasthi, Navendu Goswami and Sahab Dass, *Thin Solid Films*, **492** (2005) 332.
6. Aadesh P. Singh, Saroj Kumari, Rohit Shrivastav, Sahab Dass and Vibha R. Satsangi, *J. Phys. D: Appl. Phys.*, **42** (2009) 085303.
7. Monika Gupta, Jaya Shrivastava, Vidhika Sharma, Anjana Solanki, Ananad Pal Singh, V.R. Satsangi, S. Dass and Rohit Shrivastav, *Advanced Materials Research*, **67** (2009) 95.
8. Aadesh P. Singh, Saroj Kumari, A. Tripathi, F. Singh, Karen J. Gaskell, Rohit Shrivastav, Sahab Dass, S.H. Ehrman and Vibha R. Satsangi, *J. Phys. Chem.*, **C 114** (2010) 622.

9. Shubhra Kala, B.R. Mehta, S.A. Khan, and D.K. Avasthi, *Appl. Phys. Lett.*, **90** (2007) 151321.
10. H.K. Singh, D.C. Agarwal, P.M. Chavhan, R.M. Mehra, S. Agarwal, P.K. Kulariya, A. Tripathi and D.K. Avasthi, *Nucl. Instr. and Meth.*, **B 268** (2010) 3223.
11. N.G. Deshpande, A.A. Sagade, S.D. Chavhan, J.C. Vyas, F. Singh, A.K. Tripathi, D.K. Avasthi and Ramphal Sharma, *Vacuum*, **82** (2008) 39.
12. P. Sudhagar, K. Asokan, J.H. Jung, Y.G. Lee, S. Park and Y.S. Kang, *Nanoscale Research Letters* (2010) DOI: 10.1007/s 11671-010-9763-2.

6.14 Nuclear Materials

1. IAEA Report IAEA-TECDOC-1516.
2. M.A. Subramanian, G. Aravamudan and G.V. Subba Rao, *Prog. Solid State Chem.*, **15** (1983) 55.
3. Rodney C. Ewing, William J. Weber and Jie Lian, *J. Appl. Phys.*, **95** (2004) 5949.
4. K.E. Sickafus, R.W. Grimes, J.A. Valdez, A. Cleave, M. Tang, M. Ishimaru, S.M. Corish, C.R. Stanek and B.P. Uberuaga, *Nature Materials*, **6** (2007) 217.
5. K.E. Sickafus, L. Minervini, R.W. Grimes, J.A. Valdez, M. Ishimaru, F. Li, K.J. McClellan and T. Harimann, *Science*, **289** (2000) 748.
6. M.K. Patel, V. Vijayakumar, D.K. Avasthi, S. Kailas, J.C. Pivin, V. Grover, B.P. Mandal and A.K. Tyagi, *Nucl. Instru. Meth. Phys. Res.*, **B 266** (2008) 2898.
7. M.K. Patel, V. Vijayakumar, S. Kailas, D.K. Avasthi, J.C. Pivin and A.K. Tyagi, *J. Nuclear Materials*, **380** (2008) 93.
8. G. Sattonnay, S. Moll, L. Thomé, C. Legros, M. Herbst-Ghysel, F. Garrido, J.-M. Costantini and C. Trautmann, *Nucl. Instr. and Meth.*, **B 266** (2008) 3043.
9. M. Lang, J. Lian, J. Zhang, F. Zhang, W.J. Weber, C. Trautmann and R.C. Ewing, *Phys. Rev.*, **B 79** (2009) 224105.
10. T. Wiss and Hj. Matzke, *Rad. Measurements*, **31** (1999) 507.

Appendix 1

Epitaxial Crystallization

Ion implantation process provided the ways to tailor the conducting properties of semiconductors and thereby initiated the transistor revolution in electronics. During implantation of the desired element in a semiconductor, ion-induced collisions produce athermal atomic movements at and around the surface or interface, which can be controlled by varying the temperature and ion-beam characteristics, guiding the system between non-equilibrium and quasi-equilibrium states. Subject to the ion implantation process, high degree of damage formation due to collision cascades can lead to amorphization in semiconductors like Si and Ge. At temperatures where defects are mobile and interact, irradiation can lead to layer-by-layer amorphization, whereas at higher temperatures irradiation can lead to the recrystallization of previously amorphized layers.

Silicon-on-insulator (SOI) structure is used in the integrated circuits. SOI refers to the use of a layered silicon-insulator-silicon substrate in place of conventional silicon substrates in semiconductor manufacturing, especially microelectronics, to reduce parasitic device capacitance and thereby improving performance. SOI-based devices differ from conventional silicon-built devices in that the silicon junction is above an electrical insulator, typically silicon dioxide.

SOI structures are typically synthesized by implantation of oxygen ions in silicon (at elevated temperatures), which gives rise to amorphization, introduces roughness, and produces defects. Various annealing methods like furnace, laser and electron-beam have been employed to either recover the damage or to recrystallize the amorphous layers. In these processes annealing temperatures greater than 1000°C are generally needed. With the reduction in the size of the devices to submicron scales it has become a challenge to regain the lattice structure at lower temperatures to avoid undesired diffusion of the dopants [1]. It can be achieved in silicon at lower temperatures (only a few hundred °C) by simultaneous irradiation with energetic ions [2]. This recrystallization process, achieved by ion-atom collision, is termed as ion beam induced epitaxial crystallization (IBIEC).

Ion beam induced epitaxial crystallization (IBIEC) has been shown to take place in silicon and other materials at considerably lower target temperatures than necessary for thermal annealing when performed under irradiation. Skorupa et al. showed that amorphous Si layer deposited by chemical vapour deposition on silicon substrate can be epitaxially recrystallized by IBIEC at 400°C by implanting Si ions at 330 keV with a dose of $\sim 1 \times 10^{17}$ cm^{-2} [3].

IBIEC provides a way not only to anneal the defects at low processing temperature but has unique characteristics such as layer-by-layer and selective area crystallization, and dynamic defect annealing. Ion energy is chosen such that their projected range is well beyond the original amorphous/crystalline interface [4].

Heera et al. [5] showed that the ion irradiation substantially reduces the onset temperature of both the epitaxial layer regrowth and the random nucleation of crystalline grains.

Low-to-medium energy IBIEC has been mainly ascribed to the migration and recombination of defects (at the amorphous/crystalline (a/c) interface) created by the elastic collisions between ions and target atoms. The role of inelastic scattering processes on IBIEC was first pointed out by Nakata [6]. The inelastic electronic scattering of ^{84}Kr at energies of 0.5-5 MeV and $^{131,132}Xe$ ions at energies of 1-5 MeV were used for IBIEC of amorphous Si layers on crystalline Si substrate at 310-450°C. It was found that the crystallization rate per unit vacancy (normalized crystallization rate) created by the elastic nuclear scattering of the incident ion beam at the amorphous-crystalline (a/c) interface is increased 40-50% by increasing the inelastic electronic scattering three to four fold while maintaining the same elastic nuclear scattering conditions at the a/c interface [7]. Sahoo et al. [8] comprehensively showed 100 MeV Ag ion induced recrystallizataion of Si, where inelastic scattering process plays a major role.

Swift Heavy Ion Beam Induced Epitaxial Crystallization (SHIBIEC)

Epitaxial recrystallization of 200 nm amorphous Si layers by swift heavy ions (50 and 100 MeV Au^{8+}) was investigated by Rutherford backscattering spectrometry and Micro-Raman spectroscopy [9]. Good epitaxial recrystallization was observed in the range of 473-673 K, which is a lower temperature regime as compared to the one needed for conventional solid phase epitaxial growth of Si. The self ion (50 MeV Si) induced SHIBIEC showed that the regrowth rate is higher as compared to 50 MeV or 100 MeV Au ions. The enhancement in the regrowth rates shows a systematic dependence on the S_e/S_n ratio. For 50 MeV Si ions, for which the S_e/S_n ratio is about 3000, an enhancement of an order of magnitude in the normalized regrowth rate is seen. The mechanism of layer-by-layer growth of the amorphous layer involves creation and migration of vacancies towards interface causing creation of large vacant spaces around the interface. These vacant spaces make the thermal vibrations of Si atoms around the interface more free, as a consequence of which, the redistribution and recrystallization of Si atoms occurs resulting in the layer-by-layer epitaxial growth of the amorphous region. For swift heavy ions a hot region around the ion track is created due to the very large electronic energy loss. Vacancies created in this hot region migrate, increase the vacant spaces at the a/c interface, and enhance the regrowth rate dramatically.

SHI induced recrystallization of materials has been studied extensively by Som et al. [10-12]. SHIBIEC of a buried Si_3N_4 layer was observed at temperatures as low as 373 K, at energies where the projectile ions lose their energy mainly by inelastic collision processes [13, 14]. Complete recrystallization of silicon nitride layer, having good quality interfaces with the top- and the substrate-Si, can be obtained this way at significantly lower temperatures of 373, 423 and 473 K for O, Si and Ag ions, respectively.

Recent advances in the scaling down approach of microelectronic devices have given rise to the possibility of using a combination of high-k dielectric materials with high-mobility substrates. Germanium has higher carrier mobilities than silicon (3800 versus 1900 $cm^2\ V^{-1}\ s^{-1}$ for electrons and 1820 versus 500 $cm^2\ V^{-1}\ s^{-1}$ for holes) and is thus attractive material.

Amorphous regions in Ge, produced by ion implantation, regrow epitaxially in the solid phase well below its melting point. The regrowth related defects are very stable ones in high-energy implants in Ge and can be removed only by annealing at temperatures as high as 1123 K [15].

Benyagoub et al. investigated SHIBIEC by irradiating SiC with low and high energy ions and also successively with both types of ions. Sequential irradiations revealed that the damage formed by the low energy ion irradiation can be readily removed by electronic excitations generated by SHI [16].

Som et al. [17] made extensive measurements to get the experimental evidence of intense electronic excitation induced athermal crystallization of *a*-Ge grown on crystalline Ge substrate by using 100 MeV silver ions to the fluence of 1×10^{14} ions cm^{-2}. High-resolution transmission electron microscopic (HRTEM) studies showed complete recrystallization of the *a*-Ge layer induced by the Ag ions at room temperature. Cross-sectional TEM (XTEM) images collected from various parts of the sample showed uniform morphology, which is single crystalline in nature. This was further confirmed by the selected area electron diffraction (SAED) patterns recorded at different regions in the sample. The observed recrystallization results from the local transient melting due to intense electronic excitation along the ion trajectory and is not because of bulk heating of the sample by the ions. However, no signature of recrystallization in *a*-Ge was observed when the samples were irradiated by 70 MeV Si and 100 MeV O ions. This indicates to a possible existence of a threshold S_e value for SHI induced recrystallization to take place in *a*-Ge. Thus, it is clear that room temperature recrystallization of Group IV semiconducting materials is possible only by using swift heavy ions where high electronic excitation induced processes dominate. The irradiation by 150 MeV Ag ions induce recrystallization in buried Si_3N_4 and Si overlayer [18].

Results of various experiments in different conditions reveal that although the phenomenon of SHIBIEC resembles the process of IBIEC, it differs from it significantly. High recrystallization rate at low temperature can be accounted for by a mechanism [19] based on the melting of the amorphous zoned through a thermal spike process followed by an epitaxial recrystallization. The conventional IBIEC is triggered by the atomic displacements generated by nuclear collisions, while the effects in SHIIEC is related to the energy deposited by the incoming ions into the target electrons leading to thermal spike process.

REFERENCES

1. J. Linnros, G. Holman and B. Svensson, *Phys. Rev.*, **B 32** (1985) 2770.
2. J.S. Williams, R.G. Elliman, W.L. Brown and T.E. Seidel, *Phys. Rev. Lett.*, **55** (1985) 1482.
3. W. Skorupa, M. Voelskow, J. Mathai and P. Knothe, *Electronics Letters*, **24** (1988) 875.
4. A. Dunlop, G. Jaskierowicz, G. Rizza and M. Kopcewicz, *Phys. Rev. Lett.*, **90** (2003) 015503.
5. V. Heera, R. Kogler, W. Skorupa and J. Stoemenos, *Appl. Phys. Lett.*, **67** (1995) 1999.
6. J. Nakata, *Phys. Rev.*, **B 43** (1991) 14643.
7. J. Nakata, *J. Appl. Phys.*, **79** (1996) 682.
8. P.K. Sahoo, T. Som, D. Kanjilal and V.N. Kulkarni, *Nucl. Instr. Meth. Phys. Res.*, **B 240** (2005) 239.
9. P.K. Sahoo, T. Mohanty, D. Kanjilal, A. Pradhan and V.N. Kulkarni, *Nucl. Instr. and Meth.*, **B 257** (2007) 244.
10. T. Som, B. Satpati, O.P. Sinha and D. Kanjilal, *J. Appl. Phys.*, **98** (2005) 013532.
11. T. Som, B. Satpati, O.P. Sinha, D.K. Avasthi and D. Kanjilal, *Nucl. Instr. and Meth. Phys. Res.*, **B 245** (2006) 255.
12. T. Som, O.P. Sinha, J. Ghatak, B. Satpati and D. Kanjilal, *J. Appl. Phys.*, **101** (2007) 034912.
13. T. Som, B. Satpati, O.P. Sinha, N.C. Mishra and D. Kanjilal, *Nucl. Instr. Meth. Phys. Res.*, **B 244** (2006) 213.
14. T. Som, O.P. Sinha, J. Ghatak, B. Satpati and D. Kanjilal, *Defence Sci. J.*, **59** (2009) 351.
15. D.P. Hickey, Z.L. Bryan, K.S. Jones, R.G. Elliman and E.E. Haller, *Appl. Phys. Lett.*, **90** (2007) 132114.
16. A. Benyagoub, A. Audren, L. Thome and F. Garrido, *Appl. Phys. Lett.*, **89** (2006) 241914.
17. T. Som, J. Ghatak, O.P. Sinha, R. Sivakumar and D. Kanjilal, *J. Appl. Phys.*, **103** (2008) 123532.
18. G.S. Virdi, B.C. Pathak, D.K. Avasthi and D. Kanjilal, *Nucl. Instr. and Meth.*, **B 187** (2002) 189.
19. A. Benyagoub and A. Audren, *J. Appl. Phys.*, **106** (2009) 83516.

Appendix 2

Sputtering

An energetic ion penetrating a solid causes electronic excitations and nuclear collisions. Atoms are emitted from the surface if the energy transfer in such collisions is enough to overcome surface binding energy. This emission process of surface atoms by energetic ions is called sputtering. The average number of atoms ejected from the target per incident ion is called the sputter yield which depends on the ion incident angle, the energy of the ion, the masses of the ion and target atoms, and the surface binding energy of atoms in the target. For a crystalline target the orientation of the crystal axes with respect to the target surface are relevant. Depending on projectile energy loss, different scenario of sputtering occur, that include materials removal due to atomic collision cascade due to nuclear energy loss, so-called "Nuclear Sputtering" and "Electronic Sputtering" governed by the electronic energy loss process. Nuclear sputtering and electronic sputtering are completely different processes. Some ion-induced phenomena depend on the internal (potential) projectile energy, particularly if this potential energy greatly exceeds the kinetic projectile energy. The stored potential energy in a highly charged ion (HCI) can be quite high and produce sputtering in insulators, which is known as the "Potential Sputtering".

Nuclear Sputtering

The nuclear sputtering is essentially kinetic sputtering, which is due to elastic energy transfer from the ion to solid atom [1]. Any ion impinging on the solid matrix knocks out atoms due to a series of collisions in its journey. Knocked atoms may also remove other atoms from their lattice site due to secondary collisions. Thus a collision cascade develops inside the lattice matrix. When such cascades recoil and reach the target surface with an energy above the surface binding energy, an atom can be ejected. If the target is thin on an atomic scale the collision cascade can reach the back side of the target and atoms can escape the surface providing sputtering 'in transmission' mode, if the energy exceeds the surface binding energy.

Nuclear sputtering has a lot of applications in different fields such as: surface cleaning and thinning, thin film deposition, micromachining etc. The controlled removal of material on an atomic scale from the surface by sputtering is the basis of many analytical techniques which provide techniques to determine the sample composition as a function of depth.

Potential Sputtering

Some ion-induced phenomena depend on the internal (potential) projectile energy, particularly if this potential energy greatly exceeds the kinetic energy of the projectile. The stored potential energy in a highly charged ion can be quite high. The stored potential energy is equal to the energy spent in removing a part, say q, of their Z electrons (Z being the projectiles nuclear charge). This potential energy becomes very large for high values of q (the ion charge state). Upon surface impact this potential energy induces various inelastic processes while the ion regains its missing q electrons to again become fully neutralized. The ion deposits its potential energy in a short time (typically about 100 fs) within a small area (typically less than 1 nm^2). In the course of the neutralization of the highly charged ion (HCI) at the surface a multiply-excited neutral particle with empty inner shells is formed, which is known as the "Hollow Atom".

Hollow atoms are short-lived multiply-excited neutral atoms which carry a large part of their Z electrons (Z is projectile nuclear charge) in high-n levels while inner shells remain (transiently) empty. This population inversion exists for typically 100 femtoseconds during the interaction of a slow highly charged ion (HCI) with a solid surface. For impact on insulator surfaces the potential energy contained by hollow atom may also cause the release of target atoms and ions via potential sputtering [2-5] which can cause the formation of nanostructures on a surface. For metal surfaces, even rather sudden perturbations of the electronic structure can occur without inducing any structural modification while the excitation energy is being rapidly dissipated in the target material.

The extent to which the electronic relaxation of hollow atom takes place above or below the surface is closely related to the way it dissipates its large potential energy. Emission of electrons and X-ray photons carries away a fraction of the total potential energy originally stored in the highly charged ion. The remaining part is deposited into the solid and converted into electronic excitation of a small surface region. This electronic excitation causes modifications in the material if it is insulating. For metal surfaces the excitation energy gets rapidly dissipated in the target material without inducing any structural modification. On certain insulator surfaces a quite dramatic increase of the yields for total sputtering and secondary ion emission with increasing q has been observed.

Electronic Sputtering

The ejection of atoms due to the electronic energy loss is referred as the electronic sputtering. It is caused by very high-energy or highly charged heavy ions which lose energy to the solid mostly by electronic stopping power. Simple and complex molecules can be ejected intact into the vapour phase when a material is electronically excited by incident particles, which provides an excellent probe to study the behaviour of condensed matter at high excitation densities and has applications in fields as diverse as astrophysics and biomolecular mass spectrometry [6].

Dependence of Electronic Sputtering on the Film Thickness

Electronically mediated sputtering in thin gold films with 200 MeV Ag ions was studied with ex-situ thickness measurements of the film using X-ray reflectivity technique. The sputter yield was observed to be a few orders of magnitude higher as compared to that normally encountered in the regime of elastic collisions and depends upon the film thickness as shown in Fig. A2.1 [7]. The increased number of oscillations in the irradiated sample is indication of decreased thickness as compared to the unirradiated sample. Sputtering is accompanied by a significant smoothening of the film surface and smearing of the boundaries between the grains. A systematic decrease in sputtering yield of carbon with increase in film (C_{60}/silicon) thickness was observed [8].

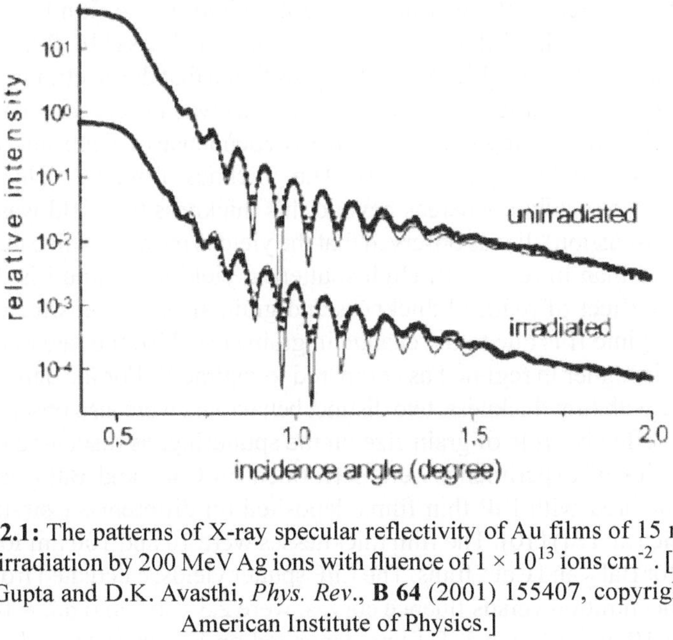

Figure A2.1: The patterns of X-ray specular reflectivity of Au films of 15 nm before and after irradiation by 200 MeV Ag ions with fluence of 1×10^{13} ions cm^{-2}. [Reprinted from A. Gupta and D.K. Avasthi, *Phys. Rev.*, **B 64** (2001) 155407, copyright (2009), American Institute of Physics.]

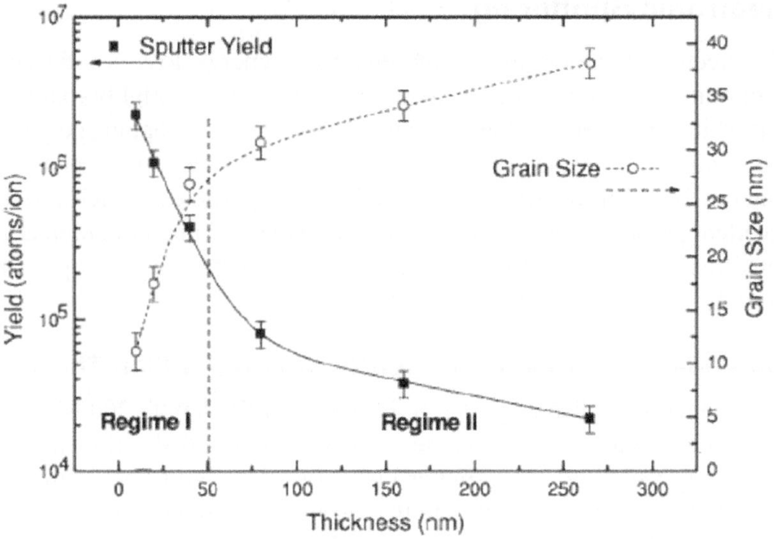

Figure A2.2: Two different regimes of sputtering as well as grain size vs film thickness are shown. Sputtering yield decreases and grain size of pristine film increases with increasing film thickness. The sputtering is enhanced about one order of magnitude for film <100 nm. [Reprinted from Manvendra Kumar, S.A. Khan, Parasmani Rajput, F. Singh, A. Tripathi, D.K. Avasthi and A.C. Pandey, *J. Appl. Phys.*, **102** (2007) 083510, copyright (2009), American Institute of Physics.]

The dependence of electronic sputtering yield on the film thickness was studied in detail using LiF thin films of different thickness (10, 20, 40, 80, 160 and 265 nm) as shown in Fig. A2.2 [9]. It was found that the electronic sputtering decreases with increase in the film thickness as was observed in case of thin Au film. Two distinct regimes of thickness dependence of electronic sputtering were observed. In the regime I, up to 50 nm thickness film, the yield was of the order of 10^6 atoms/ion, whereas beyond this thickness the yield was an order less (10^5 atoms/ion). It was observed that the yield is more sensitive to thickness in regime I than in regime II. High sputtering yield in regime I is due to the combined effect of reduced thickness and grain size, whereas the change in yield in regime II is due to the change in grain size. Also the rate of change of the yield is higher in regime I as compared to regime II. For the dependence of grain size with film thickness, two distinct behaviours were also observed which indicates effective role of grain size on the sputtering, as discussed earlier.

A series of experiments were performed on CaF_2 and BaF_2 and results were compared with LiF thin films deposited on different substrates (glass, fused silica and Si) [10]. The film thicknesses were 10 and 100 nm for LiF and 100 nm for BaF_2 and CaF_2 films. The Li/F sputter yields, calculated from ERDA areal concentration versus fluence curves, were 2.3×10^6 and 6.2×10^4 atoms/ion from 10 and 100 nm LiF films deposited on Si substrates, whereas they were 7.4×10^6 and 1.9×10^5 for 10 and 100 nm LiF films deposited on glass substrates. The F and Ca sputter yields were 5.3×10^4 and 2.5×10^4 atoms/ion for CaF_2 deposited on Si substrates and 1.7×10^4 and 8.1×10^3 for film on glass substrate. The F and Ba sputter yields are 2.5×10^4 and 1.2×10^4 atoms/ion in BaF_2 films deposited on Si substrates, while they were 8.3×10^3 and 4.1×10^3 atoms/ion for films deposited on glass substrates, respectively. No significant difference in the sputtering yield was observed for films on glass and fused silica substrates. The sputtering yield for film deposited on glass substrate is nearly three times higher than that on Si substrate for all the materials. The observed yields for different materials were compared as a function of band gap of the bulk materials. The band gap of bulk LiF, CaF_2 and BaF_2 are 14.2, 12.1 and 9.2 eV, respectively. It is clear that the yield increases as the band gap of the material increases.

Insulating materials grown in form of thin films show higher sputtering than that in the corresponding bulk. In thin films, the sputtering has strong dependence on film growth and irradiation parameters. These parameters (grain size and thickness of the film, substrate, irradiation species, energy and temperature) greatly influence the rate of removal of material. The novelty of these experiments is to use equilibrium charge state of the ions for sputtering. For SHI induced surface and/or near-surface modifications of materials, the charge state of the incident ions should be equilibrium charge state i.e. of the order of effective charge defined in the electronic energy loss theory. To reach equilibrium charge state, all the projectiles from the accelerator having +9 charge states were passed through a thin carbon foil. The charge state +25, being the most probable in charge state distribution, was selected by dipole magnet for the experiment. The energy losses of these ions were 15.4, 15.8 and 16.5 keV/nm in LiF, BaF_2 and CaF_2 crystals, respectively, and contribution of nuclear energy loss is only up to 0.4%. During irradiation the layer thickness and stoichiometry were continuously monitored by ERDA in reflection geometry using a large area position sensitive detector telescope (LAPSDT) [11].

The on-line ERDA analysis performed to measure electronic sputtering of thin films by measuring the loss of material with the fluence and to quantify interface modification by quantifying the changes at the interface with fluence showed that the observations in electronic sputtering were qualitatively explainable by the thermal spike model, and that interface modification was quantitatively explained to be due to the diffusion of species during the transient melt phase. Diffusivity so obtained in the measurements is in the range of 10^{-8} to 10^{-6} $m^2 s^{-1}$. Such a high diffusivity is possible only for the molten state [12].

Dependence of Electronic Sputtering on Ion Velocity, Charge State and Substrate

Ion velocity, charge state and substrate dependence of electronic sputtering of fullerene (C_{60}) was studied with thin films deposited on Si and glass substrates using Au and Ag ions of different energies. Slower ion having same electronic energy deposition (S_e) as compared to its high velocity counterpart erodes more. C_{60} films deposited on more insulating substrate shows higher sputtering yield as compared to those deposited on Si substrate. No charge state effect was observed in the electronic sputtering yield within the experimental error of the set up [13].

Structural Effects on Electronic Sputtering

Structural effect on electronic sputtering of hydrogenated amorphous carbon films on bombardment with 150 MeV Ag^{13+} ions were studied with on-line elastic recoil detection analysis (ERDA) technique. A large erosion (~10^5 atoms/ion) of C and H from hydrogenated amorphous carbon films (a-C:H) was observed [14].

Angular Distribution of the Sputtering Yield

The angular distribution of the sputtering yield from highly oriented pyrolytic graphite sample irradiated with a 130 MeV Ag beam was studied using a high resolution ERDA set up. The maximum sputtering yield is observed at 53°, falling rapidly to almost zero at 90°, with an average sputter yield of 5.5×10^5 atoms/ion [15].

Electronic sputtering produces high sputtering yields from insulators, as the electronic excitations that cause sputtering are not immediately quenched, as they would be in a conductor. The yield of the electronically sputtered atoms is higher normal to the sample surface and as a function of the electronic energy loss (S_e) of the projectiles, the total sputter yield follows a S_e^4 law [16].

The angular dependence of electronic sputtering from HOPG with 120 MeV Au ion beam for three cases: from crystalline highly oriented pyrolytic graphite (HOPG) for (A) normal and (B) 70° incidence and from (C) amorphous carbon sample for normal incidence was studied in detail [17]. The sputtering yield shows an anisotropic distribution for all the three cases studied. However, the anomalous peak observed at 53° for normal incidence for HOPG sample is found to shift to 73° when the sample is tilted by 20°. Similar study with amorphous carbon sample shows no peak. The peaks observed in sputtering yield distribution have been attributed to the crystalline structure of the sample which allows the preferential release of pressure pulse along the crystal axis. The high exponent of over-cosine distribution ($n = 3.2$–3.8) signifies formation of intense pressure pulse induced jet-like sputtering.

Importance of Nuclear Energy Loss in Electronic Sputtering

Effect of 100 MeV Au irradiation on embedded Au nanoclusters in silica glass was analyzed using Rutherford backscattering spectrometry, transmission electron microscopy and optical absorption spectroscopy [18]. At lower irradiation fluence the high energy heavy ion irradiation has been found to result in a loss in Au due to an outward movement of the NCs together with a growth in size. At the highest irradiation fluence, almost 80% Au was lost, with only a few large NCs seen which had moved to the surface. These were found to be of a deformed non-spherical shape. The amount of Au lost was found to increase linearly with irradiation fluence indicating the movement of Au to be not dominated by diffusion. The enhanced sputtering of Au under SHI irradiation has been suggested to be due to collision cascades produced by nuclear energy loss S_n operating along with the inelastic scattering due to S_e [19].

Surface Modification by Electronic Sputtering

Swift heavy ions induced effects on optical (colour centres), structural and surface (electronic sputtering and morphology) modifications, in nano-grains LiF thin films were studied by glancing angle X-ray diffraction, optical absorption, photoluminescence and elastic recoil detection analysis techniques. Results show that grain size and irradiation temperature play a crucial role in materials modifications as a function of fluence for the selected ion beam parameters. Also for the first time, lamellae formation was observed in LiF thin films after a high fluence irradiation of 5×10^{13} ions cm^{-2} at liquid nitrogen temperature with 120 MeV Ag ions irradiation under grazing incidence (~10°) [20].

Surface Structuring through the Electronic Sputtering

Structure dependence of electronic sputtering of a-C:H films by 80 MeV Ni^{8+} and 150 MeV Ag^{13+} ion irradiations were analyzed from the characteristic graphitic (G) and disordered (D) modes of Raman vibration [21].

Scanning force microscopy studies on the organic single crystals irradiated with GeV energy ions reveal two different defect morphologies, i.e., either hillocks or craters. The defect morphology depends exclusively on the electronic energy loss of ions. For the same crystal namely, benzoyl glycine, hillocks are produced at an energy loss less than 4 keV/nm, and above this value up to 15 keV/nm craters are produced [22].

Formation of nanoscale metallic structures by the impact of 200 MeV Au^{15+} ions on cupric nitride thin film surface was studied with on-line elastic recoil detection analysis (ERDA) technique. A large depletion of N (~75% depletion) from the films due to electronic sputtering effect was observed whereas the copper content remains unchanged. The surface of the pristine film studied by atomic force microscope (AFM) shows nanodimensional grain formation. Conducting AFM (CAFM) measurements show that at certain regions (10–30 nm) of the irradiated film surface a rapid rise of current (~9000 pA) takes place. Enhancement of electron emission together with conducting AFM measurements lead to the conclusion that conductivity of the surface enhances due to formation of nanodimensional metallic zones under Au ion impact [23].

Electronic sputtering of different allotropes of carbon (diamond, graphite, fullerene, a-C and a-C:H) were studied under 200 MeV Au^{15+} ion irradiation. Erosion behaviour is distinctly different in different allotropes as observed by ERDA. Hardest known material diamond does not show any sputtering within the detection limit of the experimental set up, whereas the soft polymers like a-C:H shows highest sputtering yield (5.8×10^5 atoms/ion). Yields in case of other allotropes are 1×10^3 atoms/ion (graphite), 3×10^4 atoms/ion (fullerene), 1.8×10^4 atoms/ion (a-C), respectively [24].

Energy dependent sputtering of nanoclusters from a nanodisperse target of Au nanoparticles, prepared on Si substrate was studied at different ion energies to analyze the synergetic effects of nuclear stopping and electronic stopping [25].

Highly oriented pyrolytic graphite (HOPG) samples (Grade ZYB with grain size ~1 mm) were studied with 150 MeV Au beam using scanning tunnelling microscopy (STM) and scanning tunnelling spectroscopy (STS). The formation of hillocks is observed for the samples irradiated with fluences of 1×10^{11} ions cm^{-2}, 1×10^{12} ions cm^{-2} and 1×10^{13} ions cm^{-2} with typical diameters of 6.2, 2.2 and 1.5 nm, respectively. No hillocks are observed for the sample irradiated with fluence of 2×10^{13} ions cm^{-2}, though the formations of small craters were observed. The formation of hillocks is attributed to nuclear energy loss induced collision cascades near the surface. The reduction in hillocks size and formation of craters at higher fluence is attributed to the electronic sputtering from the surface. The STS studies of I–V characteristics show an increasing ohmic behaviour with fluence which is attributed to increasing metallic state for HOPG surface due to irradiation induced increase of carbon bond lengths [26].

Ejection of ZnS nanoparticles from ZnS film on Au irradiation was studied using transmission electron microscopy. No nanoparticle (NP) could be observed on irradiation with 35 keV Au ions. However, 2–7 nm size NPs were observed on MeV irradiations at room temperature. For particle sizes $\geq$ 3 nm, the distributions could be fitted to power law with decay exponents varying between 2 and 3.5. At 2 MeV, after correction for cluster breakup effects, the decay exponent was found to be close to 2, indicating shock waves induced ejection to be the dominant mechanism. The corrected decay exponent for the 100 MeV Au irradiation case was found to be about 2.6 [27].

The study of the influence of grain size on electronic sputtering of LiF thin showed a reduction in sputter yield with increasing grain size. The electrons liberated in different directions from the ion track have different diffusion length according to its energy. The motion of the electrons is affected by the smaller grain size due to grain boundary scattering resulting in reduction in the mean diffusion length of the electrons, which finally enhanced energy deposition inside the grains and thus the sputter yield. From ERDA measurements, reduction in the area concentration of F and Li due to ion bombardment with fluence was observed indicating the sputtering of Li and F from the film. It was found that the Li and F are nearly equally present in the film before and after sputtering, which showed the stoichiometric sputtering of LiF. The grain size decreased from 58 nm to 22 nm with the decrease in the substrate temperature from 500 K to 77 K, respectively, while the film thickness (150 nm) was kept constant for all the depositions. The reduction in sputter yield, from $\sim 5.5 \times 10^4$ to 7.1×10^3 atoms/ion, was observed with the increase in grain size of the film [28]. The sputtering from LiF single crystal were measured by Toulemonde et al. [29] using catcher technique and the yield of Li and F was found to be about 17530 and 15790 atoms/ion, respectively, for ions having electronic energy loss of 16.4 keV/nm.

Comparing the observed yield obtained in the films having different grain sizes, it was observed that the yield (1.6×10^4 atoms/ion for F) in case of the film having the grains of 46.8 nm is comparable to the existing results, while in other cases there is huge difference in the sputter yield with change in the grain size of the films.

The experimental results are assessed within the framework of thermal spike model along with size effect in thin films and influence of substrate. Reduction in the film thickness and the grain size can restrict the motion of the excited electrons because of the scattering from surface and interface and grain boundaries, respectively. So, the size effect can result in reduction of the mean diffusion length, λ, of the electrons resulting in increase in the deposition of energy in confined region, which finally can enhance the temperature spike in the thinner films/films having smaller grains. On the other hand, in smaller grains and thinner films, the duration of thermal spike will be more because of less out-diffusion. As a quantitative approach to explain the higher sputtering yield in case of insulator substrate, inelastic thermal spike model can be applied. As the range of ion (>15 μm for all the cases) is more than the film thickness, a thermal spike will be developed in the substrate also. This temperature spike can increase the temperature generated in the film resulting in higher sputtering. The temperature in Si substrate will smear out more efficiently due to its higher thermal conductivity (nearly 40 times more than that of glass/fused silica) and yield will be higher for glass/fused silica substrates. On the other hand, because of amorphous nature of glass/fused silica, the electron-phonon coupling strength will be stronger than that in Si, resulting in higher temperature rise in glass and silica substrates supported by thermal spike model. The contribution from temperature developed in substrate will enhance the yield and the sputtering will be lower in case of the film deposited on Si substrate than that for glass/fused silica.

Two exponents δ for the size distribution of n-atom clusters, $Y(n) \sim n^{-\delta}$, were found in Au clusters sputtered from embedded Au nanoparticles under swift heavy ion irradiation [30]. For small clusters, below 12.5 nm in size, δ was found to be 3/2, which can be due to a steady state aggregation process with size independent aggregation. For larger clusters, a δ value of 7/2 is suggested, which might come from a dynamical transition to another steady state where aggregation and evaporation rates are size dependent. In the present case, the observed decay exponents do not support any possibility of a thermodynamic liquid-gas-type phase transition taking place, resulting in cluster formation. The results imply that observables such as the sputtering yield may be used as signatures of the fast electron-lattice energy transfer in the electronic energy-loss regime [31].

REFERENCES

1. Sputtering by particle bombardments. Edited by R. Behrisch, Springer-Verlag, (1981) Berlin.
2. T. Schenkel, A.V. Hamza, A.V. Barnes, D.H. Schneider, J.C. Banks and B.L. Doyle, *Phys. Rev. Lett.*, **81** (1998) 2590.
3. Masahide Tona, Satoshi Takahashi, Kazuo Nagata, Nobuo Yoshiyasu, Chikashi, Yamada, Nobuyuki Nakamura and Shunsuke Ohtani, *App. Phys. Lett.*, **87** (2005) 224102.
4. G. Hayderer, S. Cernusca, M. Schmid, P. Varga, H.P. Winter, F. Aumayr, D. Niemann, V. Hoffmann, N. Stolterfoht, C. Lemell, L. Wirtz and J. Burgdörfer, *Phys. Rev. Lett.*, **86** (2001) 3530.
5. M.V.R. Murty, *Surf. Sci.*, **500** (2002) 523.
6. Robert E. Johnson and Bo U.R. Sundqvisr, *Physics Today*, March 1992, 27.
7. A. Gupta and D.K. Avasthi, *Phys. Rev.*, **B 64** (2001) 155407.
8. S. Ghosh, D.K. Avasthi, A. Tripathi, S.K. Srivastava, S.V.S. Nageswara Rao, T. Som, V.K. Mittal, F. Grüner and W. Assmann, *Nucl. Instr. and Meth.*, **B 190** (2002) 169.
9. Manvendra Kumar, S.A. Khan, Parasmani Rajput, F. Singh, A. Tripathi, D.K. Avasthi and A.C. Pandey, *J. Appl. Phys.*, **102** (2007) 083510.
10. Manvendra Kumar, Parasmani Rajput, S.A. Khan, D.K. Avasthi and A.C. Pandey, *Appl. Surf. Sci.*, **256** (2010) 2199.
11. S.A. Khan, Manvendra Kumar and D.K. Avasthi, *Nucl. Instr. and Meth.*, **B 266** (2008) 1912.
12. D.K. Avasthi, S. Ghosh, S.K. Srivastava and W. Assmann, *Nucl. Instr. and Meth.*, **B 219-220** (2004) 206.
13. S. Ghosh, D.K. Avasthi, T. Som, A. Tripathi, S.K. Srivastava, F. Grüner and W. Assmann, *Nucl. Instr. and Meth.*, **B 212** (2003) 431.
14. S. Ghosh, Alka Ingale, T. Som, D. Kabiraj, A. Tripathi, S. Mishra, S. Zhang, X. Hong and D.K. Avasthi, *Sol. Stat. Comm.*, **120** (2001) 445.
15. A. Tripathi, S.A. Khan, S.K. Srivastava, M. Kumar, S. Kumar, S.V.S.N. Rao, G.B.V.S. Lakshmi, Azhar M. Siddiqui, N. Bajwa, H.S. Nagaraja, V.K. Mittal, A. Szokefalvi, M. Kurth, A.C. Pandey, D.K. Avasthi and H.D. Carstanjen, *Nucl. Instr. and Meth.*, **B 212** (2003) 402.
16. M. Toulemonde, W. Assmann, C. Trautmann, F. Grüner, H.D. Mieskes, H. Kucal and Z.G. Wang, *Nucl. Instr. and Meth.*, **B 212** (2003) 346.
17. A. Tripathi, S.A. Khan, Manvendra Kumar, V. Baranwal, R. Krishna, Sarvesh Kumar, A.C. Pandey and D.K. Avasthi, *Nucl. Instr. and Meth.*, **B 266** (2008) 1265.

18. B. Joseph, J. Ghatak, H.P. Lenka, P.K. Kuiri, G. Sahu, N.C. Mishra and D.P. Mahapatra, *Nucl. Instr. and Meth.*, **B 256** (2007) 659.
19. H.D. Mieskes, W. Assmann, F. Grüner, H. Kucal, Z.G. Wang and M. Toulemonde, *Phys. Rev.*, **B 67** (2003) 155414.
20. M. Kumar, F. Singh, S.A. Khan, A. Tripathi, A. Gupta, D.K. Avasthi and A.C. Pandey, *Phys. Stat. Sol.*, **C 4** (2007) 1075.
21. S. Ghosh, D.K. Avasthi, T. Som, A. Tripathi, D. Kabiraj, A. Ingale, S. Mishra, V. Ganesan, S. Zhang and X. Hong, *Nucl. Instr. and Meth.*, **B 190** (2002) 164.
22. H.S. Nagaraja, R. Neumann and D.K. Avasthi, *Rad. Measurements*, **36** (2003) 729.
23. S. Ghosh, V.V. Sivakumar, A. Tripathi, S. Khan, V. Ganesan, A. Gupta, A. Nath and D.K. Avasthi, *Nucl. Instr. and Meth.*, **B 248** (2006) 71.
24. S. Ghosh, D.K. Avasthi, A. Tripathi, D. Kabiraj, S. Singh and D.S. Misra, *Nucl. Instr. and Meth.*, **B 219-220** (2004) 973.
25. B. Satpati, J. Ghatak, B. Joseph, T. Som, D. Kabiraj, B.N. Dev and P.V. Satyam, *Nucl. Instr. and Meth.*, **B 244** (2006) 278.
26. A. Tripathi, S.A. Khan, M. Kumar, V. Baranwal, R. Krishna and A.C. Pandey, *Nucl. Instr. and Meth.*, **B 244** (2006) 225.
27. P.K. Kuiri, J. Ghatak, B. Joseph, H.P. Lenka, G. Sahu, D.P. Mahapatra, A. Tripathi, D. Kanjilal and N.C. Mishra, *J. Appl. Phys.*, **101** (2007) 014313.
28. M. Kumar, S.A. Khan, F. Singh, A. Tripathi, D.K. Avasthi and A.C. Pandey, *Nucl. Instr. and Meth.*, **B 256** (2007) 328.
29. M. Toulemonde, W. Assmann, C. Trautmann and F. Grüner, *Phys. Rev. Lett.*, **88** (2002) 057602.
30. P.K. Kuiri, B. Joseph, H.P. Lenka, G. Sahu, J. Ghatak, D. Kanjilal and D.P. Mahapatra, *Phys. Rev. Lett.*, **100** (2008) 245501.
31. S. Mookerjee, M. Beuve, S.A. Khan, M. Toulemonde and A. Roy, *Phys. Rev.*, **B 78** (2008) 045435.

Appendix 3

Applications of the PIXE Technique

Proton Induced X-Ray Emission (PIXE), non-destructive elemental analysis technique, has proved to be of immense importance in diverse areas such as forensic science, bio-medical field, archaeological and environmental studies, geochemical prospecting etc.

Normally PIXE samples are mounted inside a vacuum chamber and exposed to the proton beam for measurements. It creates problems for archeological large samples and biological targets. It is possible to extract the beam outside the accelerator beam line and plan measurements at atmospheric presure with appropriate modifications, i.e. PIXE with External Beam.

Availability of micro-focussed proton beam has opened up a new horizon for exploiting PIXE technique in many disciplines. PIXE using micro-focussed beams, called Micro-PIXE, gives additional capability of microscopic analysis.

Trace Elements in Water

Elemental composition of water is of great importance as it is the prime source of the trace elements essential for the growth of living organisms. A systematic study of river water samples collected from various places in the Indo Gangetic valley (River Ganges) was done. The trace levels of several elements increase as one goes down the stream and pass through cities [1]. Concentrations of certain heavy elements are large near industrial areas and decrease downstream [2]. Samples from a hot spring, well known for its curative properties for skin diseases contain large S content, a main constituent of several skin medications. A comparison of the trace levels in various mineral water samples was carried out by Kennedy et al. [3].

A study of the drinking water from Salt Lake City, a residential locality in Kolkata, India, was done using a chelating agent (NaDDTC) for the pre-concentration of the trace elements. A large number of elements, namely Ca, Ti, Mn, Fe, Co, Ni, Cu, Zn, As, Sr, Ba, Tl and Pb were analyzed [4]. In a similar study at Chandigarh [5], the main pollutants in the river samples around Patiala District in Punjab were found to be K, Ca and S with the relative percentage of K as maximum.

Bolormaa et al. [6] investigated the effect of mining activity on the environment by studying the heavy metal contents in the Boroo River water samples collected in mining area of Mongolia. Similarly, Saleh [7] carried out investigations on trace metal contamination in drinking water in Jordan. Quinones et al. [8] used the PIXE technique with the objective of characterizing and monitoring of the trace elements in the water of São Francisco River, as well as to provide valuable information about the levels of metallic ions pollutants.

Analysis of Human Hair Samples

Hair is the most easily accessible biological tissue. It contains trace elements which reflect metabolic changes in the body over long periods of time. Also, continuous exposure to environment leaves an imprint of atmospheric pollutants. The elemental composition of hair vary greatly from individual to individual and with geographical location. Varier et al. [9] studied whether the elemental content exhibits any systemetic pattern for members of the same family and for different families in comparable environment and nutritional background. Hair samples from the different parts of the head of the same person and from different persons were subjected to PIXE analysis. The standard deviations in the estimated trace levels were about 17-28% in samples from same person and about 30% to 69% in samples from different individuals.

It was found that the zinc content remains constant along the length of the hair and is not governed by external pollutants but by internal metabolism. Concentration ratios of all elements in the samples relative to Zn were extracted from the data. Following are a few results from the study:

- Mn/Zn and Fe/Zn ratios are higher in the family from rural areas.
- Females were found to have higher Cu/Zn ratio.
- Married females were found to have higher Pb/Zn ratio.

No definite trend in the copper content of black hair (blackness due to a copper containing pigment called melanin) versus grey hair is found. The reason may be that other sources of copper content may mask the copper in melanin. It was observed that married women have more lead in their hair, probably due to the *kumkum* they apply on their foreheads.

Small scale miners in the mountainous regions of Benguet Province in the Philippines extract gold using a method which involves the use of mercury, via amalgamation. In the separation of gold from mercury the method involves the release of mercury vapour into the atmosphere. This is therefore expected to affect the people living in the nearby areas, which was investigated. This study involves the accumulation of baseline data on the extent of mercury contamination in humans through the analysis of their hair [10]. In 1989, Hursh et al. [11] studied human volunteers and found that uptake of mercury vapour through the skin is only about 1% of the uptake through inhalation [12]. In this light, any residual mercury which might have deposited in human hair is speculated to give an indication of how much mercury vapour the subject could have actually inhaled. A high concentration of mercury in the sample can therefore be indicative of the high rate of intake of the mercury vapour through inhalation.

Arsenic pollution in Bangladesh and Japan was studied using hair samples by Habib et al. [13]. The results show markedly higher levels of arsenic, manganese, iron and lead where the later three elements show a positive relation with arsenic in the case of Bangladeshi as compared to the samples from Japan. On the other hand, selenium concentrations show very low level in the Bangladeshi samples compared to Japanese, displaying an inverse relationship with arsenic. Sera et al. did quantitative analysis of untreated hair samples for monitoring human exposure to heavy metals. It was found that the concentration of mercury and arsenic in hairs taken from different parts of a body does not show significant difference, demonstrating that the concentration in hairs are good index for an estimation of human exposure to these toxic elements [14].

Forensic Science Studies

Elemental analysis has been found to be quite helpful in the identification of crime related samples, with possibility of identifying the criminal. For example, in cases of gunshot firing, the type of bullet and distance of firing are important parameters. The bullet carries along with it a part of the primer, gun powder and also a part of the material of the gun itself. Some of these get deposited around the bullet hole on the body of the victim. An analysis of the radial distribution of these elements (Sb, Ba, Cu, Pb and Fe) can be helpful in this respect. Laboratory simulated samples analysed for the gunshot residues by the PIXE technique shows clearly the dependence of the radial distributions of the various elements on the distance of firing [15]. Representative results for Pb are shown in Fig. A3.1.

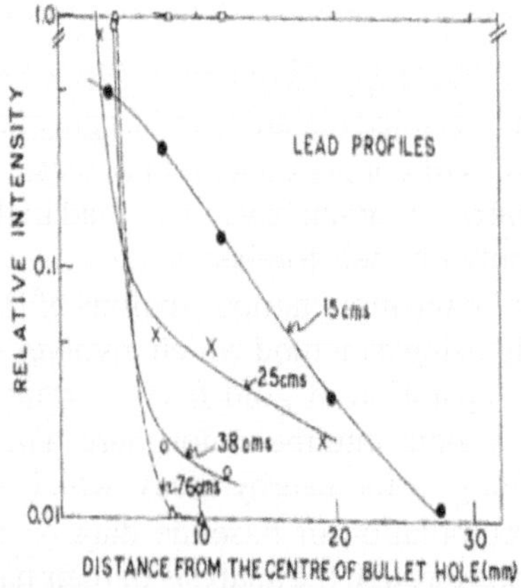

Figure A3.1: Relative intensities of Pb detected as a function of distance from the centre of bullet hole. Each curve corresponds to a different firing distance as specified. [Reprinted from S. Sen, K.M. Varier, G.K. Mehta, M.S. Rao, P. Sen and N. Panigrahi, *Nucl. Instr. and Meth.*, **181** (1981) 517 with permission from Elsevier.]

The non-destructive characteristic of PIXE has proven to be extremely valuable in the area of forensic applications. Samples can be analyzed maintaining their integrity. For example, bone samples from a victim in order to determine the presence of lead and samples returned to the court as evidence.

Warren et al. [16] used PIXE analysis in two forensic contexts: (1) case of cremation in which the nature of the remains is questioned and (2) cases of death by gunshot wound. In the first case, elemental analysis by PIXE revealed that the purported cremated remains are not bone, and in the second that radiopaque metallic residue embedded in bone is composed of lead from a projectile.

Other crime related samples subjected to the PIXE analysis include typewritten papers, nails, hair, blood stains etc. [17].

Geochemical Prospecting

Geochemical studies are largely concerned with establishing the distribution patterns of elements or group of elements, and classifying the laws governing these distribution patterns in natural systems. Mineral deposits represent anomalous concentrations of specific elements, usually within a relatively confined volume of the Earth's crust. Most mineral deposits include a central zone, or core, in which the valuable elements or minerals are concentrated, often in percentage quantities, to a degree sufficient to permit economic exploitation. The valuable elements surrounding this core generally decrease in concentration until they reach levels, measured in parts per million (ppm) or parts per billion (ppb), which appreciably exceed the normal background level of the enclosing rocks. These zones or halos afford means by which mineral deposits can be detected and traced; they are the geochemical anomalies being sought by all geochemical prospectors. The use of trace elements as indicators of geochemical processes provides an excellent way.

A large number of samples of the U.P. State Geology & Mining Department in India from eight regions spread over several kilometres in Kumaun Hills were analyzed [18]. It was concluded that Koirali region has good prospect for tungsten mining and indicated presence of uranium in Kuria Hill in Mirzapur district. A few samples from Orissa Mining Corporation indicated presence of Pt and Mo in the specimens from Sukinda Valley.

Ahmed [19] analyzed geochemical samples by the micro-PIXE technique of gold-bearing rocks, phosphorite ores and volcanic sediments. Elemental composition and distribution maps across single mineral grains, fluid inclusions, grain boundaries and matrices were measured. Frietas et al. [20] studied the serpentinophytes from north-east of Portugal for trace metal accumulation and to study its relevance to the management of mine environment. Ryan et al. [21, 22] carried out quantitative PIXE microanalysis of geological material using the CSIRO proton microprobe.

Archaeological Studies

Ancient cultures had developed highly sophisticated techniques manifested in beautiful artworks. For example pigments used in mural paintings provide very significant information. Ancient Indian coins made of Cu and Ag of Hindu Sahi dynasty (990–1015 AD) were analyzed. The presence of trace elements like Ti, Cr, Ni, Fe and Pb provides information about the source of Cu as from Khetri mines in Rajasthan. Analysis of metallic compositions of coins can provide valuable information regarding coin minting, methodology, provenance, art, culture and economics of the minting time [23, 24].

External proton induced X-ray emission (PIXE) is a good scientific method for nondestructive analysis of coins preserved in museum. Ten Kushana copper coins (3rd and 4th century AD) from the Orissa State Museum, Bhubaneswar were analyzed by using an external beam PIXE [25]. The study reveals that copper is the main constituent of the coins, with minor/trace elements like titanium, iron, nickel, zinc, lead and bismuth.

A quantitative analysis of an ancient statue was performed by external beam proton induced X-ray emission for the purpose of identifying its originality. The elemental composition of the statue is compared with that of several samples with definite ages. The analysed elements were Fe, Cu, Ag, Au and Hg for gold coating and Fe, Ni, Cu, Zn, As, Ag, Sn, Au, Pb and Bi for bronze body [26-28].

Sanchez del Reo et al. [29] applied the PIXE technique to the analysis of blue pigments contained in several Mesoamerican mural samples. Problems concerning the determination of technology and provenance of archaeological metals have been studied by PIXE and XRF by M.F. Guerra [30]. Bugoi et al. [31] have utilized a micro-PIXE set up for the study of the metal provenness of gold archaeological samples.

Bio-Medical Applications

The importance of trace elements in biomedical processes has been recognized but the details in which it manifests their biological role needs detailed study. Blood, tooth, nail, hair etc. have been investigated. However, most of these have been of exploratory type. As an example, blood samples of rheumatoid and non-rheumatoid group patients were analyzed as a part of a student project of the Medical College in Kanpur. It was found that Zn level is relatively high for non-rheumatoid group [32]. A patient lying in coma in the All India Institute of Medical Sciences in New Delhi was suspected of Hg poisoning but no technique was available to test the diagnosis. PIXE analysis done in IIT Kanpur helped the diagnosis and monitoring the progress of the treatment [32].

Quantitative analysis of lead levels in blood is important from the point of monitoring the effect of environment pollution on the human health. PIXE facility in the BARC laboratory in Mumbai was used to determine the blood lead levels in children admitted to hospital in Mumbai with suspected lead poisoning, ascribable to environmental pollution from heavy vehicular traffic and industrial sources. The lead concentration in the blood samples of the patients varied from 10 to 600 mg/dl.

The samples were grouped on the basis of clinical findings such as hypochromic microcytic anaemia, gastrointestinal symptomology, encephalopathy of unknown actiology, mental retardation and pica. The samples having significantly elevated (> 140 mg/dl) blood lead levels belonged mostly to the patients of encephalopathy, pica and anaemia groups [33].

Trace elemental analysis was carried out in the biological samples of cancer afflicted intestine [34]. It was found that the concentration of the elements Cr, Fe and Ni are higher in the cancerous tissue of the intestine than those observed in the normal tissue, whereas the concentration levels of the element Zn is slightly lower in the cancer tissue of intestine than that observed in the normal tissue. The concentrations of S, Cl, K, Ca, Ti, Mn, Co and Cu in the cancer tissue of the intestine are in agreement with those observed in the normal tissues within standard deviations. The present results support the belief that Ni and Cr are carcinogenic agents. The observed slightly low levels of zinc in the cancer tissue of the intestine suggest that zinc could possibly inhibit the tumour growth and development of neoplastic transformation. Trace elemental correlation studies in human malignant and normal tissues in different parts has been studied using PIXE technique [35-38].

Trace element changes in tissues of experimental animals (rats) as a result of liver necrosis or cirrhosis and the determination of the regional distribution of trace elements in the human brain have been done by Maenhaut et al. [39].

Trace elemental analysis was carried out in the tissue samples of normal, benign hypertrophic and carcinoma prostate using 3 MeV proton beam of Institute of Physics, Bhubaneswar. It was observed that in benign tissues the concentrations of Cl, K, Zn and Se are lower and those of Cr, Fe, Ni and Cu are higher than in normal tissues. The concentrations of K, Ca, Zn, Se and Br are lower and those of Ti, Cr, Mn, Fe, Ni and Cu are significantly higher in cancerous tissues than in normal tissues. Free radicals generated by elevated levels of Cr, Fe, Ni and Cu possibly initiate and promote prostate cancer by oxidative DNA damage. The excess Cu levels in cancerous tissues support the fact that Cu promotes cancer through angiogenesis. The higher levels of Fe observed in cancerous tissues might be a consequence of tumour growth through angiogenesis. Significantly higher levels of Ni and Cr observed in carcinoma tissues support the well established role of Ni and Cr as carcinogens. It is likely that the observed low levels of Zn and Se in cancerous tissues lead to the development of prostate cancer owing to a decrease in antioxidative defense capacity and impaired immune function of cells and also suggest that the inability to retain high levels of Zn and Se may possibly be an important factor in the development and progression of malignant prostate cells [40, 41]. However, in order to substantiate the observed elevated or deficient levels of trace elements in initiating, promoting and inhibiting prostate cancer, several cellular and molecular studies are required.

The role of some trace elements in the formation of gallstones was investigated. It was observed that 14 minor/trace elements, namely S, Cl, K, Ca, Ti, V, Cr, Mn, Fe, Ni, Cu, Zn, Br and Pb, were present in the cholesterol stone samples. The average concentration of Fe in south Indian (Chennai region) gallstone samples (503.4 ppm) is about 2.5 times more than that of the east Indian gallstone samples (205.0 ppm), whereas the concentration of Fe is still higher in other parts of south India (848.2 ppm) [42]. The higher concentration of Cu (in some parts of south India except the Chennai region) and Fe in south Indian cholesterol stone samples may be due to the intake of tamarind (*Garcinia camborginia*) as their regular food. The thermogravimetry curves provided information on the thermal decompositions of cholesterol stones.

The iron content in human alveolar macrophages has been studied by Corhe et al. [43] to investigate the clinical usefulness of the PIXE technique in occupational respiratory medicine and in various pulmonary diseases. Surgically excised malignant and normal tumours of breast tissue were subjected to PIXE investigations by Vatankhah et al. [44]. Statistically significant differences in the trace levels were noted between malignant and normal tumours.

PIXE using micro-focussed beams, called MicroPIXE, gives the additional capability of microscopic analysis. It can, for example, quantify the metal content of protein molecules. It is being utilized, for example, in Columbia University for trying to correlate trace element deficiencies with Alzheimer disease. Trace elements, such as Fe, Mn, Zn and Co, form the active centres of enzyme proteins and play important roles in their biochemical functions. Fluctuations in these elements affect the function of living tissues. Each tissue and organ has different patterns of trace elements. The micro-PIXE system at Tohoku University Japan [45], developed by Ishii's group, enables to obtain two-dimensional images of tracer elements in tissue slices with high spatial resolution. When combined with microbeam scanning of a sample surface, the PIXE method provides the spatial distribution of the elements in a cell. Two-dimensional maps of elemental compositions can be generated by scanning the microPIXE beam across the target.

REFERENCES

1. K.M. Varier, G.K. Mehta and S. Sen, *Nucl. Instr. and Meth.*, **181** (1981) 217.
2. V. John Kennedy, A. Augusthy, K.M. Varier, P. Magudapathy, S. Panchapakesan, C. Ramesh, K.G.M. Nair and V. Vijayan, *Nucl. Instr. and Meth.*, **B 134** (1998) 224.
3. V. John Kennedy, A. Augusthy, K.M. Varier, V. Vijayan and K.G.M. Nair, *Int. J. PIXE*, **8** (1998) 11.
4. M. Sudarshan, R.K. Dutta, V. Vijayan and S.N. Chintalapudi, *Nucl. Instr. and Meth.*, **B 168** (2000) 553.
5. I.M. Govil, *Particle Accelerator Conference*, **1720** (1993) 1753.

6. B. Oyuntsetseg, B. Jamsranjav, K. Katsunori, W. Makiko and H. Toshiyuki, *Nucl. Instr. and Meth.*, **B 243** (2006) 161.
7. N.S. Saleh, *Journal of Radioanalytical and Nuclear Chemistry*, **74** (1982) 257.
8. F.R. Espinoza-Quiñones, S.M. Palacio, R.M. Galante, F.L. Rossi, D.C. Zenatti, R.A. Pereira, R.A. Welter, N. Rossi, C.L. Obregón, J.M.T. De Abreu, M.A. Rizzutto, N. Added and M.H. Tabacniks, *Braz. J. Phys.*, **35** (2005) 757.
9. K.M. Varier, A.K. Nayak and G.K. Mehta, *IEEE Transactions on Nucl. Sc.*, **30** (1983) 1316.
10. E. Clemente, K. Sera, S. Futatsugawa and S. Murao, *Nucl. Instr. and Meth.*, **B 219-220** (2004) 161.
11. Hursh et al., *Arch. Environ. Health*, **44(2)** (1989) 120.
12. World Health Organization, IPCS, EHC118, Inorganic Mercury, Geneva, 1991, p. 42.
13. M.A. Habib, S. Mion, K. Sera and S. Futatsugawa, *International Journal of PIXE*, **12** (2002) 19.
14. K. Sera, S. Futatsugawa and S. Murao, *Nucl. Instr. and Meth.*, **189** (2002) 174.
15. S. Sen, K.M. Varier, G.K. Mehta, M.S. Rao, P. Sen and N. Panigrahi, *Nucl. Instr. and Meth.*, **181** (1981) 517.
16. M.W. Warren, A.B. Falsetti, I.I. Kravchenko, F.E. Dunnam and H.A. Van Rinsvelt, *Forensic Science International*, **125** (2002) 37.
17. P. Sen, N. Panigrahi, M.S. Rao, K.M. Varier, S. Sen and G.K. Mehta, *J. Forensic Sc., (U.S.A)*, **27** (1982) 330.
18. A.K. Nayak, K.M. Varier and G.K.Mehta, *Indian Mining and Engineering Journal*, May 1983.
19. M. Ahmed, *J. of Radioanalytical and Nucl. Chem.*, **265** (2005) 39.
20. H. Freitas, M.N.V. Prasad and J. Pratas, *Chemosphere*, **54** (2004) 1625.
21. C.G. Ryan, D.R. Cousens, S.H. Sie, W.L. Griffin, G.F. Suter and E. Clayton *Nucl. Instr. and Meth.*, **B 7** (1990) 55.
22. C.G. Ryan, D.R. Cousens, S.H. Sie and W.L. Griffin, *Nucl. Instr. and Meth.*, **B 49** (1990) 271.
23. M. Hajivaliei, M.L. Garg, K.P. Singh and I.M. Govil, *Nucl. Instr. and Meth.*, **168** (1999) 437.
24. M. Hajivaliei, M.L. Garg, D.K. Handa, K.L. Govil, T. Kakavand, V. Vijayan, K.P. Singh and I.M. Govil, *Nucl. Instr. and Meth.*, **B 150** (1999) 645.
25. S. Santra, Debasis Mitra, Manoranjan Sarkar, Dipan Bhattacharya, A. Denker, J. Pitz-Coutureau and J. Rauschenberg, *Nucl. Instr. and Meth.*, **B 229** (2005) 465.
26. V. Vijayan, R.K. Choudhary, B. Mallick, S. Sahu, S.K. Choudhary, H.P. Lenke, T.R. Rautray and P.K. Nayak, *Current Science*, **85** (2003) 772.
27. M. Ashok, S.N. Kalkura, V.J. Kennedy, A. Markwitz, V. Jayanthy, K.G.M. Nair and V. Vijayan, Fourth Bio PIXE Conference, Mexico, 2002.
28. B. Rout, S.K. Ghose, D.P. Mahapatra, B.N. Dev, H. Bhakru and A.W. Haberl, *Nucl. Instr. and Meth.*, **B 181** (2001) 110.
29. M. Sanchez del Rýo, P. Martinetto, A. Somogyi, C. Reyes-Valerio, E. Dooryhee, N. Peltier, L. Alianelli, B. Moignard, L. Pichon, T. Calligaro and J.-C. Dran, *Spectrochimica Acta*, **B 59** (2004) 1619.
30. Maria Filomena Guerra, *X-Ray Spectrometry*, **27** (1993) 73.
31. R. Bugoi, V. Cojocaru, B. Constantinescu, F. Constantin, D. Grambole and F. Herrmann, *Journal of Radioanalytical and Nuclear Chemistry*, **257** (2003) 375.

32. A.K. Nayak, K.M. Varier, S. Sen, G.K. Mehta and R. Popli, *IITK-Tech. Report* PHY/VDG/3/1983.
33. M. Lal, D. Joseph, R.K.Choudhury, H.N. Bajpai, Indu Gauba, M.R. Lokeshwar and C.S. Wagle, *The Sci. of the Total Environment* (Science Publisher, Amsterdam), **103** (1991) 2092.
34. Mitali Basu Chattopadhyay, Sutapa Mukherjee, Indira Kulkarni, V. Vijayan, Manika Doloi, N.B. Kanjilal and Malay Chatterjee, *Cancer Cell Int.*, **5** (2005) 16.
35. G.J. Naga Raju, P. Sarita, M. Ravi Kumar, A.V. Ramana Murty, B.S. Reddy, S. Lakshminarayana, V. Vijayan, P.V.B. Rama Lakshmi, S. Gavarasana and S. Bhuloka Reddy, *Nucl. Instr. and Meth.*, **B 247** (2006) 361.
36. G.J. Naga Raju, M.J. Charles, S.B. Reddy, P. Sarita, B.S. Reddy, P.V.B. Rama Lakshmi and V. Vijayan, *Nucl. Instr. and Meth.*, **B 229** (2005) 457.
37. A.K. Rath, G.S. Acharya and V. Vijayan, *Med. Phys.*, **18** (1993) 15.
38. A. Chakrabortty, S. Selvaraj, M. Sudarshan, R.K. Dutta, S.S. Ghugre and S.N. Chintalapudi, *Nucl. Instr. and Meth.*, **B 170** (2000) 156.
39. W. Maenhaut, J. Vandenhaute, H. Duflou and J.D. Reuck, *Nucl. Instr. and Meth.*, **B 22** (1987) 138.
40. S. Bhuloka Reddy, M. John Charles, G.J. Naga Raju, B.S. Deddy, T. Seshi Reddy, P.V.B. Rama Lakshmi and V. Vijayan, *Biological Trace Element Research*, **102** (2004) 265.
41. G.J. Naga Raju, P. Sarita, A.V. Ramana Murty, M. Ravi Kumar, B.S. Reddy, S. Lakshminarayana, K. Prema Chand, A. Durga Prasad, S. Bhuloka Reddy, V. Vijayan, P.V.B. Rama Lakshmi and S. Gavarasana, *European Journal of Cancer Prevention,* **16** (2007) 108.
42. R.T. Rautray, V. Vijayan and S. Panigrahi, *European Journal of Gastroenterology and Hepatology*, **18** (2006) 999.
43. J.L. Corhay, G. Weber, T. Bury, S. Mariz, I. Roelandts and M.F. Radermecker, *European Respir. Journal*, **5** (1992) 804.
44. S. Vatankhah, K. Moosavi, J. Salimi, L. Geranpayeh and H. Peyrovani, *Iran. J. Radiat. Res.*, **1** (2003) 23.
45. S. Matsuyama, K. Ishii, S. Abe, H. Ohtsu, H. Yamazaki, A. Suzuki, Ts. Amartivan, Y. Barbotteau, T. Yamaguchi, G. Momose, K. Inomata and Y. Watanabe, 10th International Conference on Particle Induced X-ray Emission and its Analytical Applications (PIXE 2004), Slovenia (2004) p. 931.1.

Appendix 4

Applications of RBS and ERDA

Introduction

The impingement of ions on solid results in scattering of ion, producing recoil, emission of photons such as X-rays, gamma rays and visible light, emission of gases for polymer targets, etc. All these events are utilized by developing techniques for the characterization of materials. The scattering of incident ion and the complementary event i.e. recoils are used in these techniques. Elastic recoil detection analysis (ERDA) is used for light elements depth profiling and Rutherford backscattering spectrometry for depth profiling of high Z elements. Some examples of these techniques are described here.

Rutherford Backscattering Spectrometry (RBS)

Compositional Analysis

The determination of stoichiometry of thin films is one of the major uses of RBS. It has been used very effectively in the determination of the metal content in metal dielectric nanocomposite thin films [1-5]. One such example is given here [6]. A typical RBS spectrum for the Au-silica nanocomposite thin film (Au nanoparticles embedded in silica) is shown in Fig. A4.1 for the films with two different Au content (10 at% and 20 at%). The peak in higher energy region of scattered α particle represents Au with two different contents. At lower energy, two edges correspond to the Si of the substrate and Si in silica of the nanocomposite thin film. A small peak riding over the plateau region corresponds to oxygen in silica in the nanocomposite thin film. The information of metal content in the noble metal silica nanocomposite thin film is useful in simulating the surface plasmon resonance, where metal content is one of the input parameters for simulation.

Ion Beam Mixing

RBS is one of the most effective tools for the interface analysis for the ion beam mixing experiments [7-18]. One example is discussed here [19].

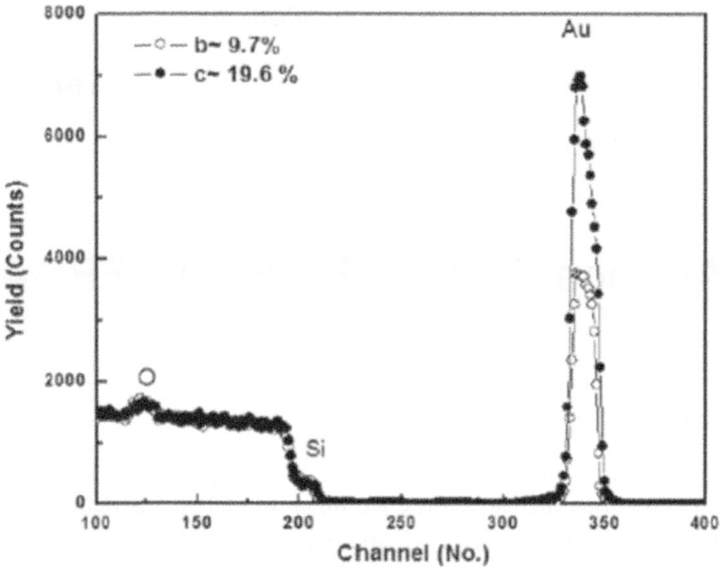

Figure A4.1: RBS spectra of Au-silica nanocomposite thin film. [Reprinted from Y.K. Mishra, S. Mohapatra, D.K. Avasthi, D. Kabiraj, N.P. Lalla, J.C. Pivin, H. Sharma, R. Kar and N. Singh, *Nanotechnology*, **18** (2007) 345606 with permission from Institute of Physics Publishing Ltd.]

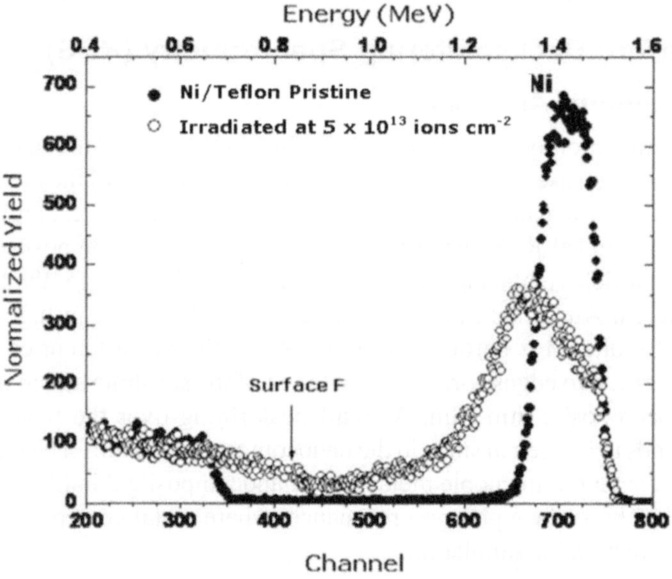

Figure A4.2: RBS spectra of pristine and SHI (120 MeV Au ions) irradiated sample of Ni deposited on Teflon. Asymmetrical broadening and shift of the low energy edge of Ni peak indicates strong mixing. [Reprinted from Jai Prakash, A. Tripathi, S.A. Khan, J.C. Pivin, F. Singh, Jalal Tripathi, Sarvesh Kumar and D.K. Avasthi, *Vacuum*, **84** (2010) 1275 with permission from Elsevier.]

RBS spectra of the pristine and the irradiated Ni/Teflon system (Ni thin film deposited on Teflon polymer sheet) are shown in Fig. A4.2 [19]. RBS spectrum of the pristine film has a low energy edge which correspond to F present in Teflon, whereas the peak at higher energy corresponds to Au. The RBS spectrum of irradiated sample has F edge extending as tail towards higher energy while the low energy edge of Au extends towards lower energy region which indicate that the irradiated sample had mixed region of Au and Teflon.

When interface in a multilayer is to be analyzed in which the layer thickness is smaller than the resolution (~10 nm) of RBS, one requires high resolution RBS. For example, the interface of SHI C_{60}/a-Si multilayer deposited on Si is examined by high resolution RBS using an electrostatic deflector. The high resolution RBS spectra recorded for pristine and SHI irradiated films are shown in Fig. A4.3 [20].

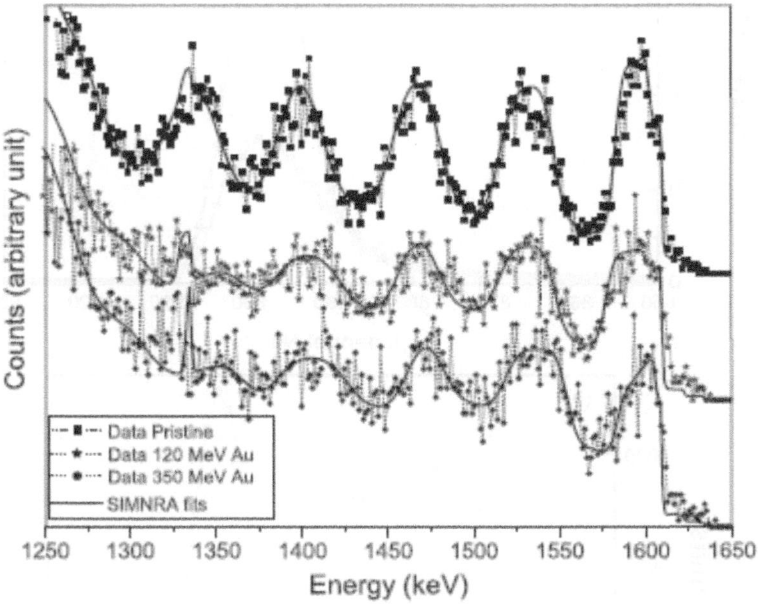

Figure A4.3: High resolution RBS spectra of pristine and SHI irradiated multilayer of C_{60}/Si. [Reprinted from S.K. Srivastava, D. Kabiraj, B. Schattat, H.D. Carstanjen and D.K.Avasthi, *Nucl. Instr. and Meth.*, **B 219-220** (2004) 815 with permission from Elsevier.]

Nuclear/Electronic Sputtering

The sputtering measurements of thin film are studied very effectively by measuring elemental composition of pristine and irradiated samples using RBS [21, 22]. The RBS spectra of pristine and 100 MeV Ag ion irradiated (at a fluence of 10^{13} ions cm^{-2}) Au film are shown in Fig. A4.4(a). The metal content clearly decreases with increase of SHI fluence, revealing SHI induced electronic sputtering. The decrease in film thickness with such low fluence is an indication of large electronic sputtering due to nano dimensional film thickness. RBS spectra for pristine (of the same thickness as in previous case) and 50 keV Si ion irradiated Au film at a fluence of 10^{16} ions cm^{-2} are shown in Fig. A4.4(b). The decrease in peak is due to loss of Au because of nuclear sputtering. The same film under SHI irradiation results in decrease in Au film thickness at a fluence of 10^{13} ions cm^{-2}, showing a large sputtering due to electronic energy loss. Thus electronic and nuclear sputtering are determined by RBS.

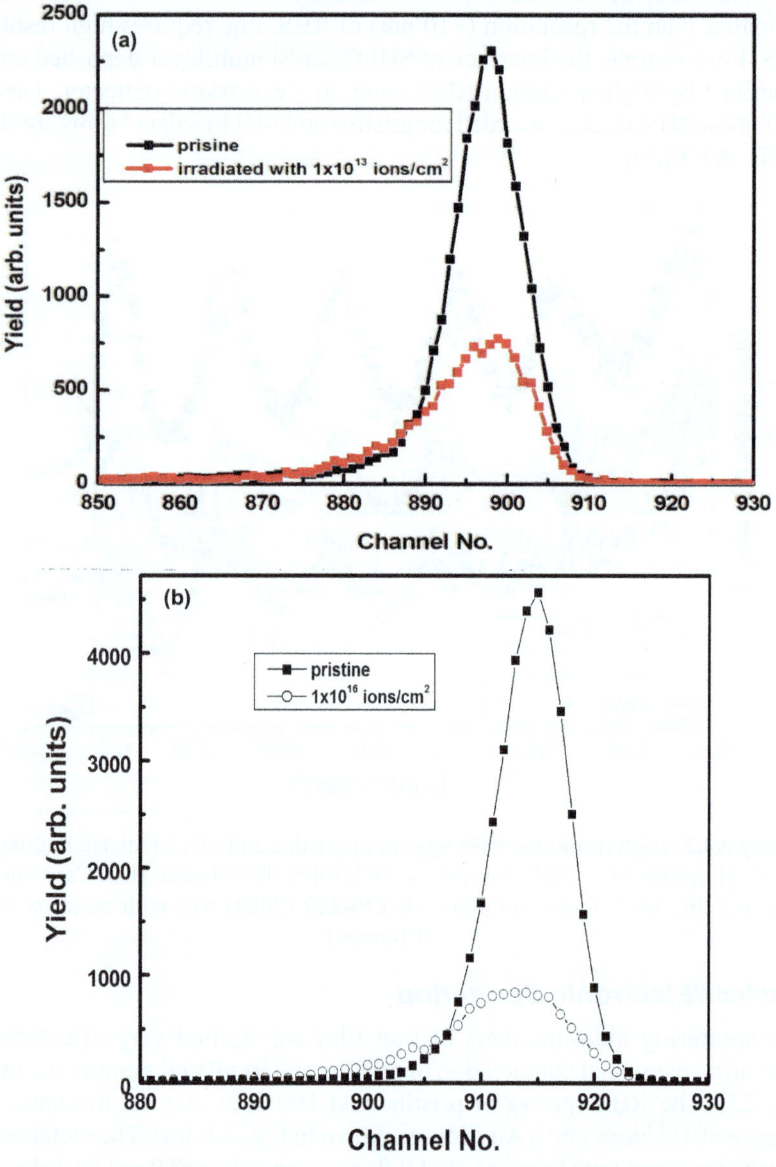

Figure A4.4: RBS spectra of the pristine and irradiated Au thin film on glass (a) for SHI with 100 MeV Ag ions and (b) for low energy with 50 keV Si ion.

RBS Channelling

RBS channelling is a unique way to measure strain in strained layer superlattice (SLS). An experiment to study the modification of strain in SHI irradiated hetrostructures (InGaAs/GaAs), reveals that the compressive strain decreases with ion fluence without loss of crystallinity [23]. The modification in strain with SHI irradiation results in controlled modification in optical properties. RBS channelling revealed that SHI irradiation can induce strain in an initially lattice matched system and can result in decrease in the compressive strain in an initially strained system without effecting the crystalline/interface quality [24]. Experiments of RBS channelling at room temperature and low temperature are capable of probing the atomic vibrations in lattice. RBS channelling is an effective tool to study ion induced epitaxial crystallization [25].

H Depth Profiling by Conventional ERDA

Correlation of H Content with the Microstructure

Hydrogen contributes to the carbon sp^3 or sp^2 bonding, which in turn affects the hybridization of carbon-carbon atoms during growth leading to graphitic or diamond-like character. Further, the microstructure of a-C:H films is intimately related to the total H content and the nature of C–C and C–H bonds in the films. ERDA provides an excellent technique for studying the correlation of hydrogen content with the microstructure of a-C:H films [26].

Dependence of Hardness on H Content

The depth profiling hydrogen in thin diamond-like carbon (DLC) films produced using dc glow discharge decomposition of acetylene show that high hardness DLC films have a low hydrogen content and higher thermal stability [27].

Ion Induced Delamination

Delamination of CVD diamond films deposited on Si substrates was studied with MeV He ions. It is found that film gets delaminated during 1.5 MeV He^+-ion irradiation. In-situ monitoring of hydrogen, during irradiation, using ERDA provides a way to estimate the threshold ion fluence for exfoliation to occur [28].

H Depth Profiling of Diamond Films

H profiling in diamond film grown by hot filament CVD and microwave CVD for different deposition pressure gave valuable information on the mechanism of growth of diamond films [29].

H Depletion from KH_2PO_4 Under He^+-ion Bombardment

Hydrogen depletion was probed in the optoelectronically important insulating material, KH_2PO_4 (KDP). It was found that depletion of hydrogen occurs under He^+-ion bombardment [30]. Investigation of the effect of Au of varying thicknesses showed that a 6 nm Au layer on KDP acts as a barrier and reduces the depletion of hydrogen from the sample by a considerable amount. These results provide insight into the ion beam induced structural and compositional changes in these materials to tailor their properties.

Ion Track Radius by H Loss Measurements

The diameter of the track produced by the ion is a quantity of interest for the understanding of basic ion insulator interaction. There have been attempts to measure the track diameters by scanning force microscopy and other state-of-the-art surface morphology probing equipments. A novel approach was demonstrated by the measurement of H loss during ion irradiation by the on-line elastic recoil detection analysis [31].

Swift heavy ion irradiation leads to decrease in hydrogen content in hydrogen containing materials. The hydrogen loss can be further correlated with other chemical transformations in these materials by the online-ERD measurements. For example, it was shown that 120 MeV Au irradiation of 400-500 nm thick films of methyltriethosilane on silicon substrates led to hydrogen loss from a narrow cylindrical zone which caused formation of C rich cylindrical zones [32].

Effect of Incident Ion Charge State on H Loss

The effect of charge state on hydrogen loss from polypropylene (PP) and polyethylene terephthalate (PET) foils was studied by online-ERD using 130 MeV Ag ions with 11^+, 14^+ and 25^+ charge states (q). It is known that the swift heavy ions with different charge states deposit different amount of energy in the first 10 nm or so while travelling in the material. This small thickness was not resolvable by the detection system and hence top 100 nm were considered for comparison. Hydrogen release cross-section from top 100 nm was found to vary as q^n, n being 2.98 and 1.94 for PP and PET respectively. The radii of ion tracks formed was also found to depend on the charge state [33].

H Depth Profiling in Pr-Pd Layers

The role of ion tracks formed in Pr-Pd layers on an all-round enhancement in the hydrogenation properties of these films was investigated with 120 MeV Ag^{+10} ions. Same ion beam was used to record hydrogen concentration in hydrogenation (absorption of hydrogen) and dehydrogenation (removal of hydrogen by creating vacuum) processes of these films by ERDA. The hydrogenation property has been found to be strongly influenced by the ion fluence. About 17.8% increase in the hydrogen stoichiometry value during hydrogenation, near maximal removal (about 31%) during dehydrogenation was observed for the highest ion fluence. Non-equilibrium structural changes during ion irradiation leading to the formation of nanotracks throughout the film thickness may provide two-way transport routes for H diffusion. This study provides a novel methodology of improving the hydrogenation properties of materials [34, 35].

The hydrogenation properties of nanoparticle Gd (grain size ~8 nm) and nanocrystalline Gd samples (grain size ~30 nm) were studied by ERDA [36]. The nanoparticle sample exhibited a larger difference in the stoichiometry ([H]/[Gd]) values (2.9 and 1.7) in comparison to polycrystalline sample (2.4 and 2.0) in the hydrogenated and dehydrogenated states respectively.

H Depth Profiling in NEG Strip

Hydrogen depth profiling of Non Evaporable Getter (NEG) material (an alloy of V, Fe and Zr) was performed in transmission geometry [37]. H recoil spectra of pristine and used NEG strip is shown in Fig. A4.5. NEG has the property of absorbing the gases after its activation and is, therefore, of interest in vacuum technology for its use as pumping device. Since hydrogen is one of the most dominant residual gases in vacuum, an experiment was performed to quantify the absorbed H in NEG strip used in vacuum pumping and in the pristine NEG strip.

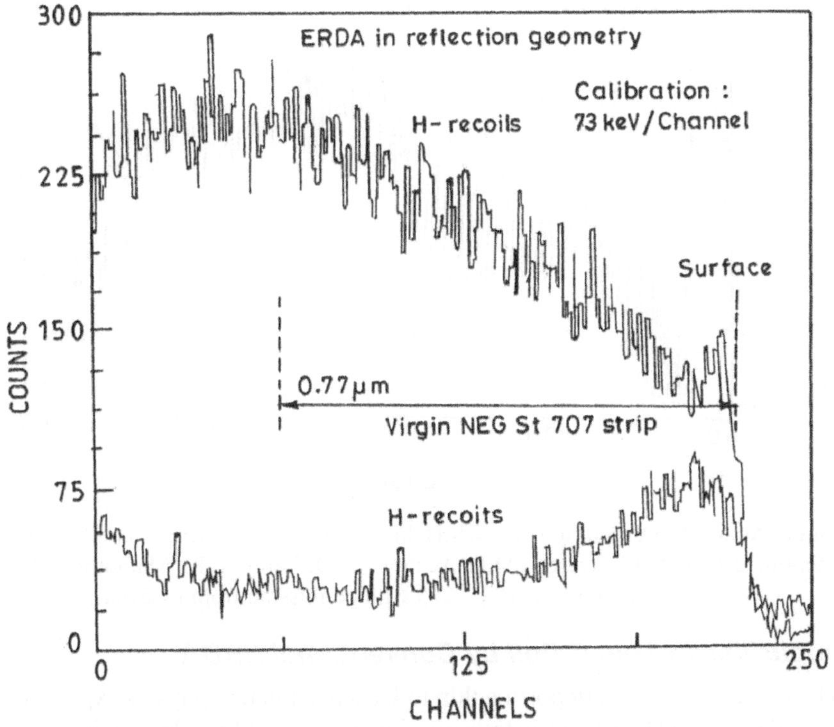

Figure A4.5: H recoil spectra of pristine and used (for vacuum pumping application) NEG strip. [Reprinted from D.K. Avasthi, *Material Science Forum*, **248-249** (1997) 405 with permission from Trans Tech Publications.]

Stoichiometric Analysis of a-SiN$_x$:H Thin Film by ERDA

Stoichiometric analysis of a thin a-SiN$_x$:H film on Fe substrate was performed [38]. A film on Fe substrate instead of Si substrate because of the interest to detect the content of Si along with other elements in the film and the recoils from Fe were stopped in stopper foil [38]. The sample was tilted at an angle of 20° and detector was kept at 34° with a stopper foil of 6 μm in front of it to stop unwanted scattered Ni ions and Fe recoils. The recoil spectrum is shown in Fig. A4.6 clearly distinguished recoils of H, N and Si and the film composition of H, N and Si were estimated as 3.8 at.%, 20.4 at.% and 75.7 at.% respectively.

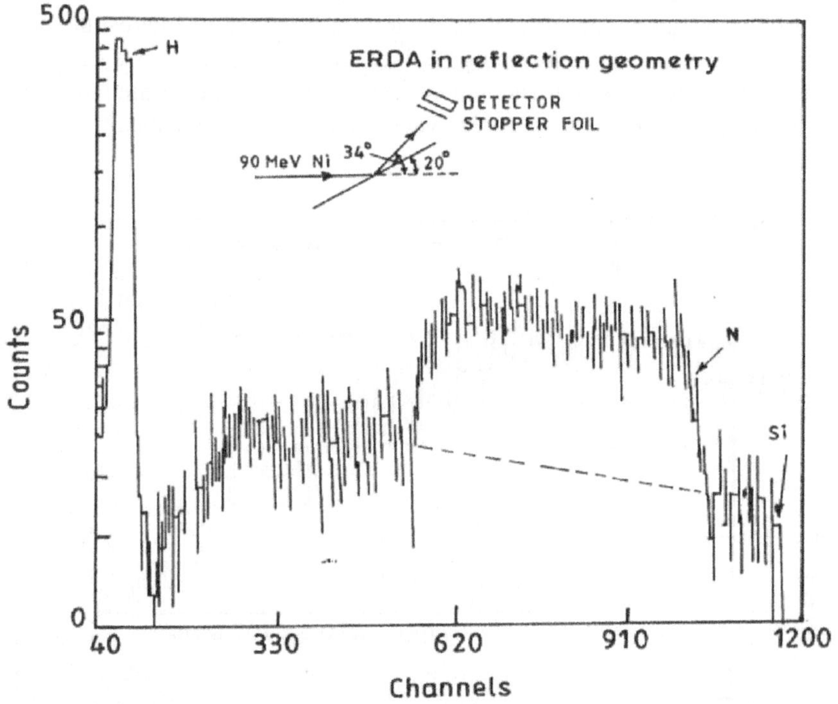

Figure A4.6: ERDA spectrum of a-SiN:H thin film using a stopper foil in tilted sample. [Reprinted from D.K. Avasthi, M.G. Acharya, R.D. Tarey, L.K. Malhotra and G.K. Mehta, *Vacuum*, **46** (1995) 265 with permission from Elsevier.]

Impurities in Thin C Foil by Conventional ERDA

The impurities in self supporting thin foil, used in nuclear physics experiments, were determined by ERDA in transmission geometry [39]. The recoil spectrum at 30° from a thin C foil for 10 MeV ^{127}I ions, indicating the impurities (N, O, Na and Cl) is shown in Fig. A4.7. The scattered ions do not reach the detector placed at 30° as the maximum scattering angle for ^{127}I on ^{12}C is 5.4°.

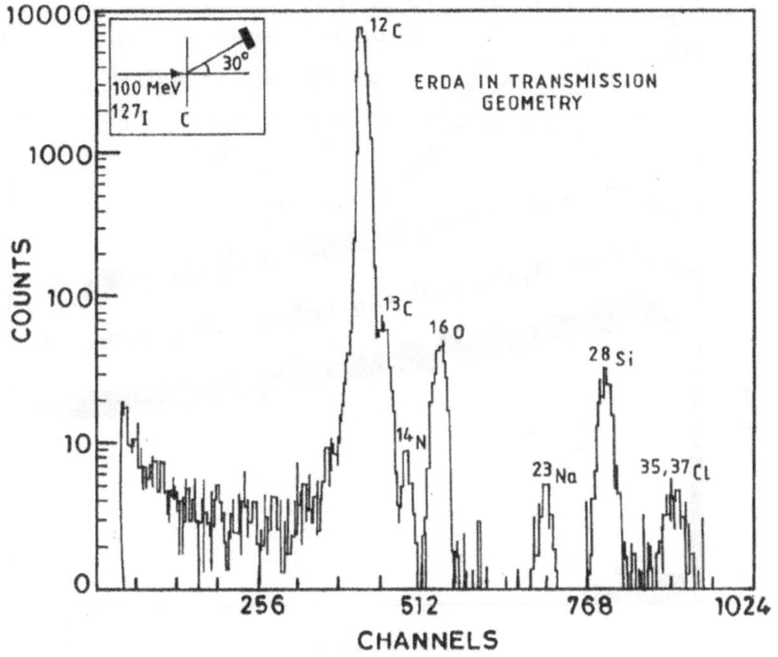

Figure A4.7: ERDA of a thin C foil using 100 MeV I ions, showing light mass impurities, which is possible only for self supporting thin film of a few tens of nm. [Reprinted from Jaipal, D. Kabiraj and D.K. Avasthi, *Nucl. Instr. and Meth.*, **A 334** (1993) 196 with permission from Elsevier.]

Simultaneous Detection of Several Light Elements ERDA Using Telescope Detector

ERDA with conventional detectors provide the required information when the sample has elements with well separated masses. If the sample has elements with neighbouring masses, the recoil energies overlap and it becomes difficult to distinguish them. In such a situation, the particle identification techniques are utilized for the discrimination of different elements [40-42]. There are different possible configuration for telescope detectors. Transmission type thin solid state detector as ΔE detector and thick solid state detector as E_{rest} detector is the simplest choice. But the solid state detectors are prone to radiation damage and thus have limited life in experiments. Second choice is to use a gaseous detector (which can be designed and fabricated according to the requirements) for ΔE along with a solid state detectors as the E_{rest} detector. Third choice is to use the gaseous detector for both the ΔE and E_{rest} detectors. Important feature and advantage of gaseous telescope detector is its being insensitive to radiation damage, rugged and can be fabricated indigenously. The telescope detectors are in use in major laboratories where SHIs are used for materials science such as LMU Munich [43], IUAC New Delhi [44], GSI Darmstadt [45], ANU

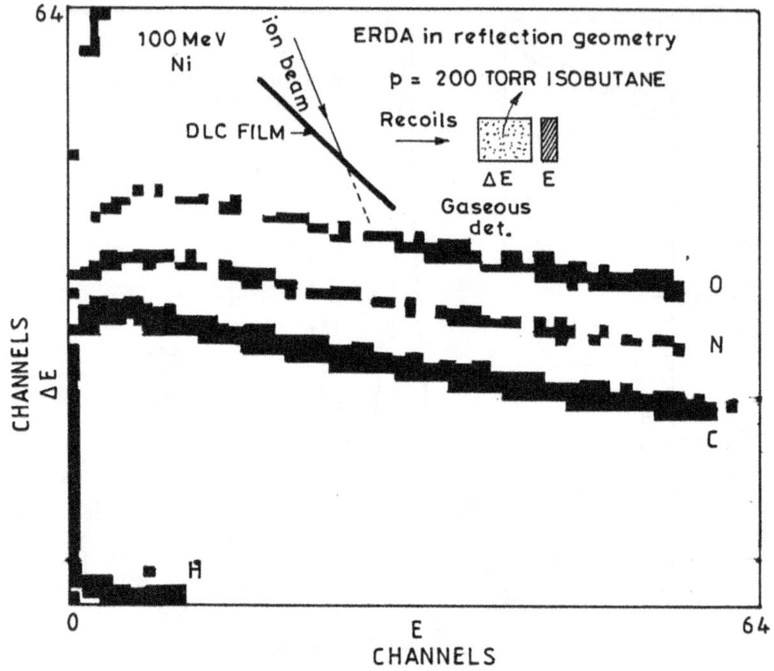

Figure A4.8: ERDA using a telescope detector showing the capability to distinguish light element in thin film of DLC on a Si substrate. [Reprinted from D.K. Avasthi, D. Kabiraj, A. Bhagwat, G.K. Mehta, V.D. Vankar and S.B. Ogale, *Nucl. Instr. and Meth.*, **B 93** (1994) 480 with permission from Elsevier.]

Canberra [46] and Rossendrof Dresden [47]. Figure A4.8 is a two dimensional (ΔE-E) ERDA spectrum [48] of a diamond-like carbon (DLC) film using such a telescope detector and 90 MeV Ni ions, which clearly shows the presence of N and O. The recoil energies from the surface of the sample for C, N and O recoils are well distinguished in the spectrum.

Need of Large Area Position Sensitive Telescope Detector in ERDA

Since there are possibilities of modification of sample in high energy heavy ion ERDA due to large S_e, it is desirable to record the recoil spectrum with sufficient statistics using only a small fluence. To accomplish this objective, it is necessary to increase the solid angle of the telescope detector. However, the increase in the solid angle of the detector results in larger kinematic broadening, which in turn hampers the depth resolution. This problem is overcome by using a large area position sensitive detector and making kinematic correction by software utilizing the position of detected recoils [49]. The advantage of this approach is to increase the sensitivity without compromising the depth resolution. A schematic sketch of telescope detector is shown in Fig. A4.9. The

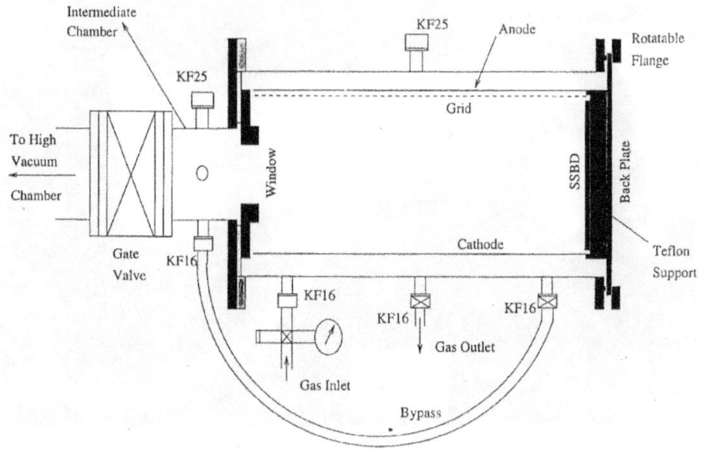

(a)

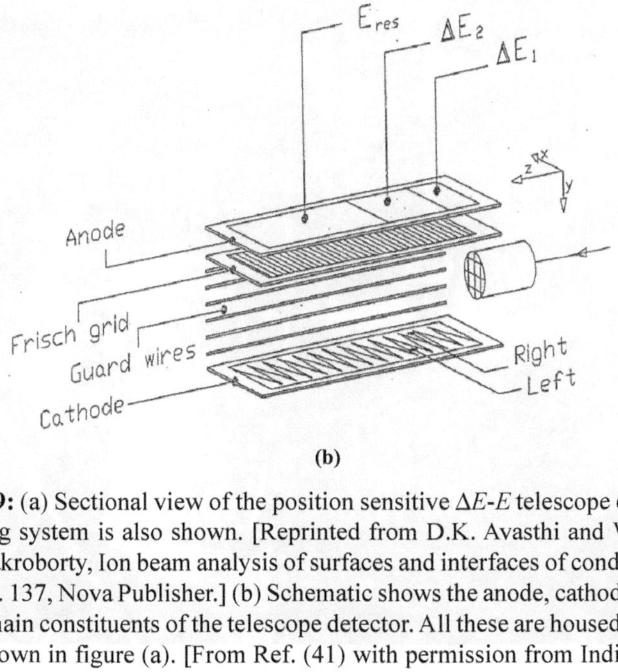

(b)

Figure A4.9: (a) Sectional view of the position sensitive ΔE-E telescope detector. The gas handling system is also shown. [Reprinted from D.K. Avasthi and W. Assmann, (Ed.) P. Chakroborty, Ion beam analysis of surfaces and interfaces of condensed matter systems. pp. 137, Nova Publisher.] (b) Schematic shows the anode, cathode Frisch grid which are main constituents of the telescope detector. All these are housed in a vacuum chamber shown in figure (a). [From Ref. (41) with permission from Indian Academy of Science.]

anode is split in two or three parts. The length of the anode is taken in such a way that the recoils of interest are stopped within this length when operating at a feasible gas pressure. Cathode is given a shape of backgammon so that the left and right side signals are created by the passage of recoil in the gas medium of detector to provide the information of position. The photograph of the on-line ERDA facility is shown in Fig. A4.10.

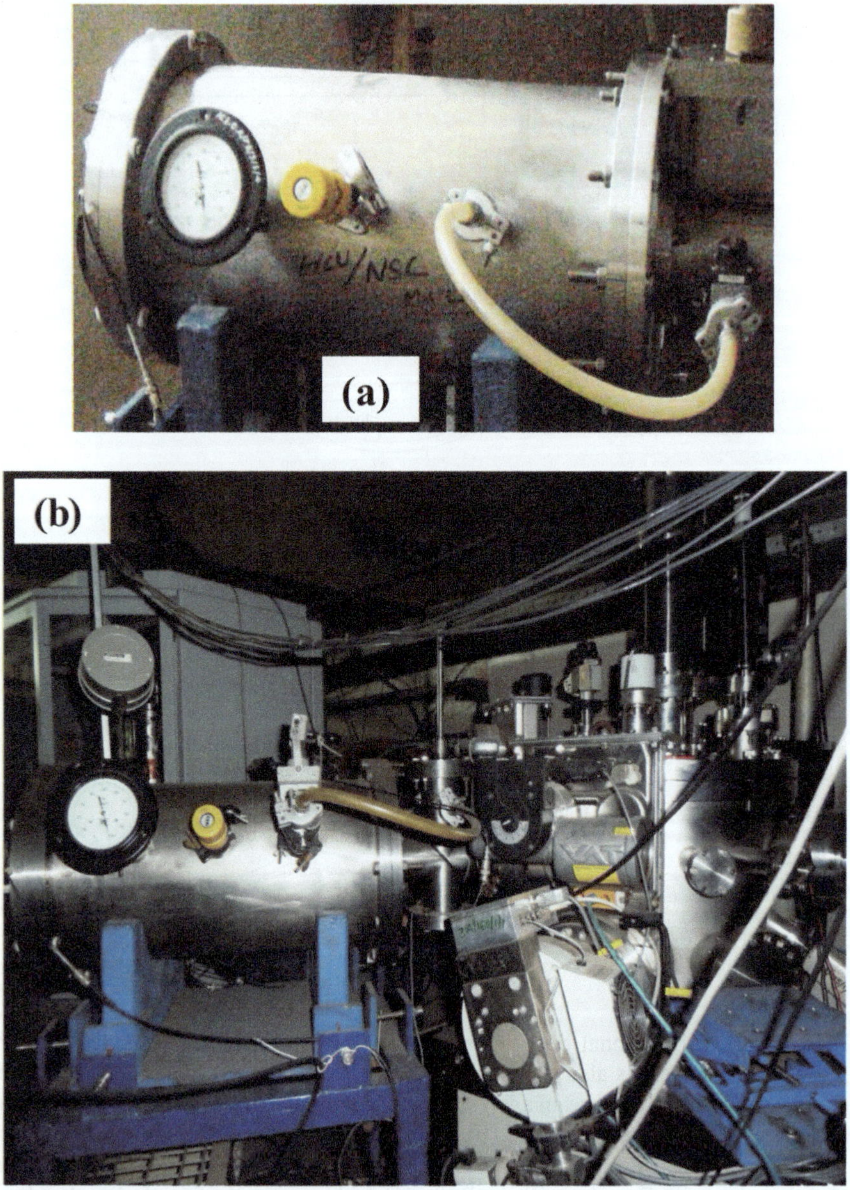

Figure A4.10: Photograph of the experimental facility of telescope detector in beam line. Bare telescope detecope detector (a), installed in experimental beam line as shown in (b).

On-line Monitoring of Ion Induced Modifications

On-line monitoring of ion induced modifications using a large area position sensitive detector [50], is one of the most interesting aspects of the ERDA. SHI is capable of producing modification at the surface, bulk and the interface of thin film and substrate. Electronic sputtering is an area which can be investigated by on-line monitoring in specific cases. The stoichiometric changes in the film, especially in the case of hydrogen and light element constituents, are also investigated by on-line ERDA. Modifying the sample by SHI irradiation and probing it at the same time is a unique feature of on-line ERDA.

N Depletion Studies

N loss in copper nitride film leads to the formation of nanoscale metallic structures in cupric nitride thin films by the impact of 200 MeV Au ions [51]. ERDA results show that the nitrogen content is reduced by 4.5 times due to irradiation to a fluence of 1.8×10^{13} ions cm^{-2}. Conducting AFM showed the presence of conducting regions in the irradiated films, which could be due to production of copper-rich region as a result of the large nitrogen loss.

Oxygen Content Measurements

Majority of oxide materials studied by on-line ERD [52, 53] show that although there may be significant electronic sputtering but there is no preferential sputtering of oxygen. The studies show that 250 nm thick film of ZnO has sputter yield of 400 atoms/ion on irradiation with 100 MeV Au but the zinc to oxygen atomic ratio remains nearly constant throughout [54]. Fe_2O_3 [55], $NiMn_{0.05}Ti_{0.2}MgFeO_4$ [56], $Li_{0.25}Mg_{0.5}Mn_{0.1}Fe_{2.15}O_4$ [57], CuO [58] and nickel oxide [59] also do not show any preferential depletion of oxygen. Similar result is obtained in the case of $GeO_{1.7}$ thin films under 100 MeV Au ion irradiation [60]. Even though this system shows Ge phase separation due to SHI irradiation with the same beam [61]. However, sub-stoichiometric indium oxide film irradiated with 120 MeV Ag ions show preferential decrease of oxygen as well as phase separation leading to indium clusters of 35-45 nm size [62].

Electronic Sputtering Measurements

The large area gaseous telescope detector is used to measure the elemental content in the film at different fluences, which in turn gives the electronic sputtering [49, 50, 64-67] or desorption yield. A two dimensional ΔE-E spectrum is shown in Fig. A4.11 for the recoils from thin CaF_2 deposited on Si substrate, when 100 MeV Au ions are incident on the sample [63]. Integrated counts of Ca and F at different fluences are used to determine the electronic sputtering. At these high energies the contribution to sputtering due to nuclear energy loss is negligible and the sputtering in totally mediated by electronic energy loss. The desorption of carbon and hydrogen from amorphous C film was found to be dependent on the structural properties of the film. The electronic sputtering of LiF thin films [63, 68-70] were studied extensively by on-line ERDA. Typical spectra of LiF and CaF thin films on Si substrate are shown [63] in Fig. A4.11. The decrease in the number of recoils in individual elements with fluence allowed the measurement of electronic sputtering of thin halide films. A series

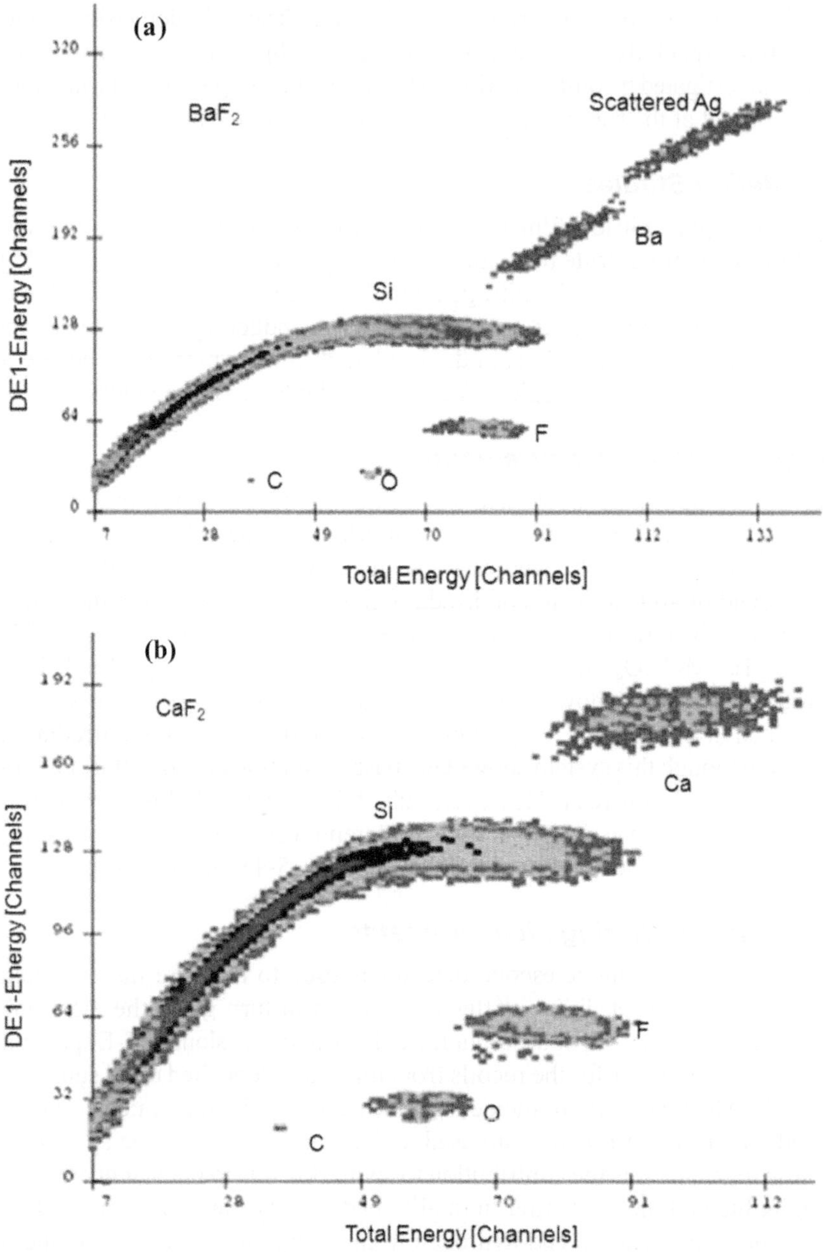

Figure A4.11: (a) A two dimensional ΔE-E spectrum for the recoils from BaF_2 thin film on Si substrate. (b) A two dimensional ΔE-E spectrum for the recoils from CaF_2 thin film on Si substrate. [Reprinted from Manvendra Kumar, Parasmani Rajput, S.A. Khan, D.K. Avasthi and A.C. Pandey, *Applied Surface Science*, **256** (2010) 2199 with permission from Elsevier.]

of experiments were performed on LiF, CaF_2 and BaF_2 thin films deposited on different substrates (glass, fused silica and Si) [63, 68-70]. Thickness of the films was 100 nm for all the films and the sputter yield of both the elements in the film was found to be stoichiometric. The total sputter yield, determined from ERDA areal concentration versus fluence curves, were 1.3×10^5, 2.5×10^4 and 1.2×10^4 atoms/ion, respectively from LiF, CaF_2 and BaF_2 films deposited on Si substrates, whereas they were 3.8×10^5, 7.8×10^4 and 3.7×10^4 atoms/ion respectively for LiF, CaF_2 and BaF_2 films deposited on glass substrates. No significant difference in the sputtering yield is observed for films on glass and fused silica substrates. The sputtering yield for film deposited on glass substrate is nearly three times higher than that on Si substrate for these halide thin films. The observed yields for different materials were compared as a function of band gap of the bulk materials. The band gap of bulk LiF, CaF_2 and BaF_2 are 14.2, 12.1 and 9.2 eV, respectively. It is clear that the yield increase as the band gap of the materials increases.

On-line Monitoring of Mixing at Interface

SHI produces mixing at the interface and the recoils provide information about the changes at the interface [49, 71]. For example, Cu recoil spectra indicated the mixing at the interface in an online ERDA measurement of CuO film on glass using 210 MeV I ion beam. The recoils detected in a large area position sensitive telescope detector and different masses appeared as different bands which were gated by software and the Cu recoil spectra at different fluences (in the beginning of irradiation and in the end of irradiation) was constructed, as shown in Fig. A4.12. The low energy region of the spectra represents the

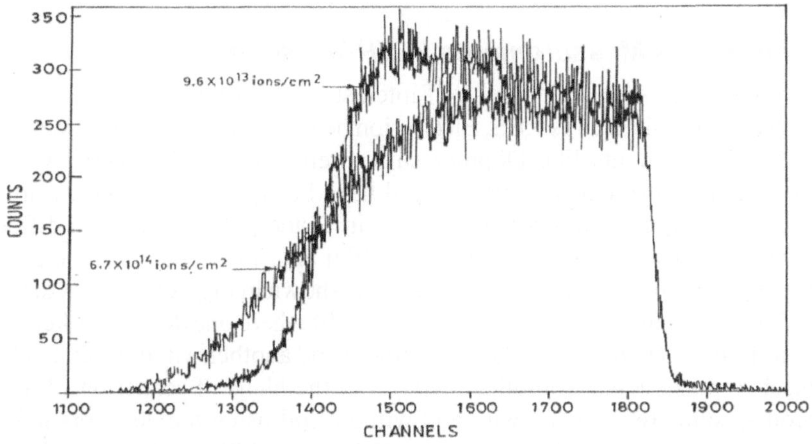

Figure A4.12: Recoil spectra of Fe (extracted from two dimensional ΔE-E spectra) obtained when 210 MeV I ion beam is incident on thin film of CuO deposited on glass. [Reprinted from D.K. Avasthi, W. Assmann, H. Nolte, H.D. Mieskes, H. Huber, E.T. Subramaniyam, A. Tripathi and S. Ghosh, *Nucl. Instr. and Meth.*, **B 156** (1999) 143 with permission from Elsevier.]

interface region. The change in the interface region is indication of the mixing caused by SHI's at the interface. In another study [71], the mixing induced by 230 MeV Au ions incident on a thin Fe film deposited on a Si substrate was monitored on-line. With the application of kinematic correction and subsequent improvement in depth resolution, the detection system allowed the on-line study of ion beam mixing in Fe/Ti bilayer system using 135 MeV Au projectiles at room temperature. The decrease in the slope of the recoil spectra corresponding to Fe/Ti interface indicates mixing. For this system, the mixing rate was found to be 147±9 nm^4 [72].

ERDA Channelling and Blocking Measurements

Channelling ERDA experiments with energetic heavy ion beams are performed by detecting the recoils in forward direction. The experiment is carried out with the aligned sample to get the channelling scan by recording recoils (for a fixed incidence charge) at different angles. The channelling ERDA technique has been used to measure the strain [73] at the interface of $CoSi_2$ and Si crystal. When the recoils get blocked in crystallographic directions (by manipulating the sample using a goniometer), one records these recoils in the axial direction by a two dimensional position sensitive detector and a 'shadow' of the crystal axis appears. The shadow pattern is referred to as blocking pattern. Sample with high crystallinity gives rise to a blocking pattern with a very sharp contrast of the axes and the surrounding region. The blocking pattern for an amorphous sample will not show any axes. The contrast of the axis with the background gives the content of the crystallinity of the sample. ERD blocking has been used [74, 75] to investigate the radiation damage induced by swift heavy ions in semiconductor crystals.

Energy Loss Measurements in ERDA Mode

Energy loss (dE/dx) is a quantity of interest in characterization of sample by ion beam analysis such as RBS, ERDA, ion beam based cancer therapy, nuclear physics experiments like Doppler shift attenuation method (DSAM), etc. A large number of experiments [76-86] have been performed to measure the dE/dx of light and heavy recoils upto Si in C and polymer films in different energy regimes. The recoils from a thin film act as ion beam for energy loss measurement in a foil of known thickness as shown in Fig. A4.13(a). A detector holding arrangement is shown in Fig. A4.13(b) where one detector is kept with the foil (in which dE/dx is to be determined) and another without foil. Both the detectors make equal azimuthal angle from the plane of scattering [77]. The recoil spectra are recorded with and without foil using the detectors holding arrangement shown in Fig. A4.13(b). Other possibility is to use a detector covered half by the foil in which energy loss measurement is required.

The usefulness of RBS and RBS channelling is shown for determining the content of elements, sputtering, ion beam mixing and determination of strained layer superlattice. The use of detector telescope in ERDA allowed depth profiling of a wide range of elements up to mass 150 simultaneously. The large

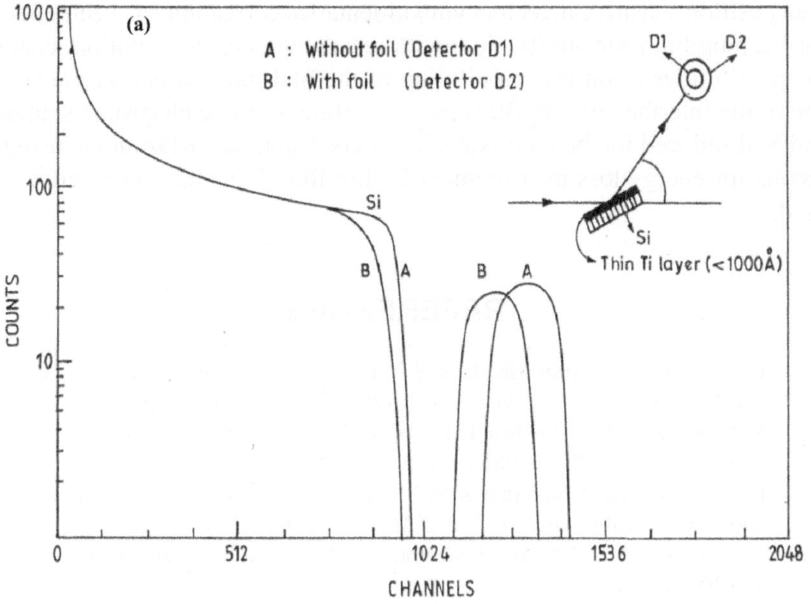

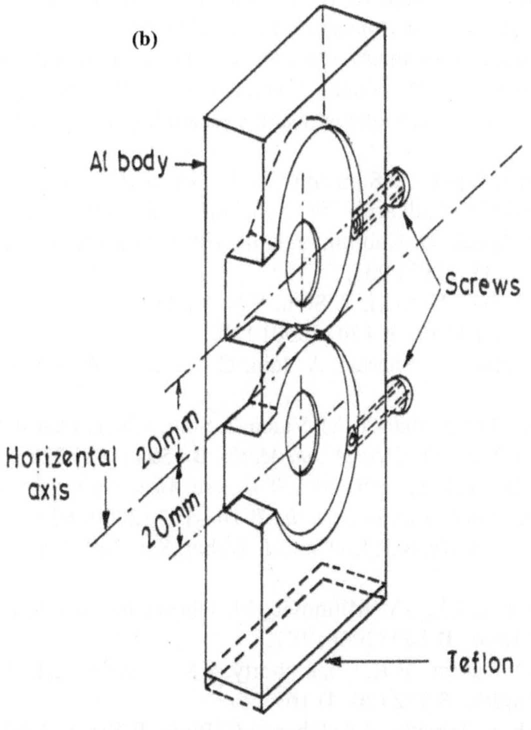

Figure A4.13: (a) Schematic sketch of the energy loss measurements using the recoils. (b) Au arrangement to mount two detector. [Reproduction from A. Bhagwat and D.K. Avasthi, *Jpn. J. Appl. Phys.*, **35** (1996) 313 with permission from Japan Society of Applied Physics.]

area position sensitive detectors with kinematics correction has been effective for keeping high sensitivity alongwith reasonable depth resolution, enabling the possibilities of on-line monitoring of the SHI induced chances at surface and at the interface of thin films on a substrate to probe electronic sputtering and SHI induced ion beam mixing. The experiment in ERDA mode using the recoils for energy loss measurement in thin foils/films have been effectively used.

REFERENCES

1. D. Kabiraj, S.R. Abhilash, Lionel Vanmarcke, Nicolas Cinausero, J.C. Pivin and D.K. Avasthi, *Nucl. Instr. and Meth.*, **B 244** (2006) 100.
2. S. Mohapatra, Y.K. Mishra, D.K. Avasthi, D. Kabiraj, J. Ghatak and S. Varma, *Appl. Phys. Lett.*, **92** (2008) 103105.
3. Hardeep Kumar, Y.K. Mishra, S. Mohapatra, D. Kabiraj, J.C. Pivin, S. Ghosh and D.K. Avasthi, *Nucl. Instr. and Meth.*, **B 266** (2008) 1511.
4. D.K. Avasthi, Y.K. Mishra, D. Kabiraj, N.P. Lalla and J.C. Pivin, *Nanotechnology*, **18** (2007) 125604.
5. Y.K. Mishra, S. Mohapatra, D. Kabiraj, B. Mohanta, N.P. Lalla, J.C. Pivin and D.K. Avasthi, *Scripta Materialia*, **56** (2007) 629.
6. Y.K. Mishra, S. Mohapatra, D.K. Avasthi, D. Kabiraj, N.P. Lalla, J.C. Pivin, H. Sharma, R. Kar and N. Singh, *Nanotechnology,* **18** (2007) 345606.
7. S. Dhar, T. Som, Y.N. Mohapatra and V.N. Kulkarni, *Appl. Phys. Lett.*, **67** (1995) 1700.
8. Sankar Dhar, Tapobrata Som and V.N. Kulkarni, *J. Appl. Phys.*, **83** (1998) 2363.
9. S. Dhar and V.N. Kulkarni, *Thin Solid Films*, **333** (1998) 20.
10. T. Som, P. Ayyub, D. Kabiraj, N. Kulkarni, V.N. Kulkarni and D.K. Avasthi, *J. Appl. Phys.*, **93** (2003) 903.
11. S.K. Sinha, D.C. Kothari, T. Som, V.N. Kulkarni, K.G.M. Nair and M. Natali, *Nucl. Inst. and Meth.*, **B 170** (2000) 120.
12. Garima Agarwal, P. Sharma, A. Jain, C. Lal and I.P. Jain, *Vacuum*, **83** (2009) 397.
13. S.K. Sinha, D.C. Kothari, K.M. Vigen, T. Som, V.N. Kulkarni, S. Panchapakesan and K.G.M. Nair, *Nucl. Instr. and Meth.*, **B 159** (1999) 227.
14. J. Sun, W. Bolse, K.P. Lieb and A. Traverse, *Mater. Sci. Eng.*, **A 196** (1995) 229.
15. D.M. Ruck, *Nucl. Instr. and Meth.*, **B 166-167** (2000) 602.
16. J. Conrad, W. Bolse, K.P. Lieb and T. Weber, *Suf. And Coat. Tech.*, **74-75** (1995) 941.
17. K. Zhang, K.P. Lieb, V. Milinovic, M. Uhrmacher and S. Klaumunzer, *Nucl. Instr. and Meth.*, **B 249** (2006) 167.
18. K. Diva, D. Kabiraj, B.R. Chakraborty, S.M. Shivaprasad, D.K. Avasthi, *Nucl. Instr. and Meth.*, **B 222** (2004) 169.
19. Jai Prakash, A. Tripathi, S.A. Khan, J.C. Pivin, F. Singh, Jalaj Tripathi, Sarvesh Kumar and D.K. Avasthi, *Vacuum*, **84** (2010) 1275.
20. S.K. Srivastava, D. Kabira, B. Schattat, H.D. Carstanjen and D.K. Avasthi, *Nucl. Instr. and Meth.*, **B 219-220** (2004) 815.

21. B. Joseph, J. Ghatak, H.P. Lenka, P.K. Kuiri, G. Sahu, N.C. Mishra and D.P. Mahapatra, *Nucl. Instr. and Meth.*, **B 256** (2007) 659.
22. B. Satpati, J. Ghatak, B. Joseph, T. Som, D. Kabiraj, B.N. Dev and P.V. Satyam, *Nucl. Instr. and Meth.*, **B 244** (2006) 278.
23. S. Dhamodaran, N. Sathish, A.P. Pathak, S.V.S.N. Rao, A.M. Siddiqui, S.A. Khan, D.K. Avasthi, T. Srinivasan, R. Murlidharan, C. Muntele, D. Ila and D. Emfietoglou, *Nucl. Instr. and Meth.*, **B 242** (2006) 538.
24. S.V.S. Nageswara Rao, Anand P. Pathak, Azher M. Siddiqui, D.K. Avasthi, Claudiu Muntele, Daryush Ila, B.N. Dev, R. Muralidharan, F. Eichhorn, R. Groetzschel and A. Turos, *Nucl. Instr. and Meth.*, **B 212** (2003) 442.
25. P.K. Sahoo, S. Gupta, A. Prdhan and V.N. Kulkarni, *Nucl. Instr. Meth. Phys. Res.*, **B 216** (2004) 313.
26. T. Som, M. Malhotra, V.N. Kulkarni and Satyendra Kumar, *Physica*, **B 355** (2005) 72.
27. M. Malhotra, T. Som, V.N. Kulkarni and S. Kumar, *Vacuum*, **47** (1996) 1265.
28. T. Som, S. Bhargava, M. Malhotra, H.D. Bist, V.N. Kulkarni and Satyendra Kumar, *Appl. Phys. Lett.*, **72** (1998) 3014.
29. Tarun Sharda, D.S. Misra, D.K. Avasthi and G.K Mehta, *Solid State Comm.*, **98** (1996) 879.
30. Tapobrata Som, Sankar Dhar, Shiraz N. Minawalla and Vishwas N. Kulkarni, *Nucl. Instr. and Meth.*, **B 122** (1997) 244.
31. V.K. Mittal, S. Lotha and D.K. Avasthi, *Radiation Eff. Defects Solids*, **147** (1999) 199.
32. A. Kumar, F. Singh, S.A. Khan, D.C. Agarwal, A. Tripathi, D.K. Avasthi and J.C. Pivin, *Nucl. Instr. and Meth.*, **B 244** (2006) 23.
33. D.P. Gupta, R.S. Chauhan, Shyam Kumar, P.K. Diwan, S.A. Khan, Ambuj Tripathi, Fouran Singh, Santanu Ghosh, D.K. Avasthi and V.K. Mittal, *Rad. Effects and Defects in Solids*, **161** (2006) 331.
34. S. Kala, B.R. Mehta, S.A. Khan and D.K. Avasthi, *Appl. Phys. Lett.*, **90** (2007) 153121.
35. Manika Khanuja, B.R. Mehta, Pragya Agar, P.K. Kulriya and D.K. Avasthi, *J. Appl. Phys.*, **106** (2009) 093515.
36. I. Aruna, B.R. Mehta, L.K. Malhotra, S.A. Khan and D.K. Avasthi, *J. Nanosci. Nanotech.*, **5** (2005) 1728.
37. D.K. Avasthi, *Material Science Forum*, **248-249** (1997) 405.
38. D.K. Avasthi, M.G. Acharya, R.D. Tarey, L.K. Malhotra and G.K. Mehta, *Vacuum*, **46** (1995) 265.
39. Jaipal, D. Kabiraj and D.K. Avasthi, *Nucl. Instr. and Meth.*, **A 334** (1993) 196.
40. D. Bromeley (Ed.), Treatise on Heavy Ion Science, Plenum Press, New York, 1984.
41. D.K. Avasthi and W. Assmann, *Current Sci.*, **80** (2001) 1532.
42. D.K. Avasthi and W. Assmann, ERDA: A tool for surface and near-surface studies. *In:* P. Chakroborty (Ed.), Ion beam analysis of surfaces and interfaces of condensed matter systems, pp. 137, Nova Science Publishers (2003).
43. W. Assmann, P. Hartung, H. Huber, P. Staat, H. Steffens and Ch. Steinhausen, *Nucl. Instr. and Meth.*, **B 85** (1994) 726.
44. D.K. Avasthi, *Nucl. Instr. and Meth.*, **B 136-138** (1994) 729.

45. M. Bender, H. Kollmus, M.C. Bellachioma and W. Assmann, *Nucl. Instr. and Meth.*, **B 268** (2010) 1986.
46. Santosh K. Shrestha and Heiko Timmers, *Nucl. Instr. and Meth.*, **B 249** (2006) 257.
47. R. Grötzschel, Michael Mäder, Ulrich Kreissig, Stephan Grigull, Walter Assmann and Stefan Parhofer, *Nucl. Instr. and Meth.*, **B 118** (1996) 139.
48. D.K. Avasthi, D. Kabiraj, A. Bhagwat, G.K. Mehta, V.D. Vankar and S.B. Ogale, *Nucl. Instr. and Meth.*, **B 93** (1994) 480.
49. S.A. Khan, Manvendra Kumar and D.K. Avasthi, *Nucl. Instr. and Meth.*, **B 266** (2008) 1912.
50. D.K. Avasthi, W. Assmann, H. Huber, H.D. Mieskes and H. Nolte, *Nucl. Instr. and Meth.*, **B 142** (1998) 117.
51. S. Ghosh, V.V. Sivakumar, A. Tripathi, S. Khan, V. Ganesan, A. Gupta, A. Nath and D.K. Avasthi, *Nucl. Instr. And Meth.*, **B 248** (2006) 71.
52. W.M. Arnoldbik, P.A. Zeijlmans van Emmichoven, F.H.P.M. Habraken, *Phys. Rev. Lett.*, **94** (2005) 245504.
53. N. Matsunami, M. Sataka, A. Iwase, S. Okayasu, *Nucl. Instr. and Meth.*, **B 209** (2003) 288.
54. D.C. Agarwal, R.S. Chauhan, D.K. Avasthi, S.A. Khan, D. Kabiraj and I. Sulania, *J. of Appl. Phys.*, **104** (2008) 24304.
55. Yatendra S. Chaudhary, Saif A. Khan, Rohit Shrivastav, Vibha R. Satsangi, Satya Prakash, Umesh K. Tiwari, D.K. Avasthi, Navendu Goswami and Sahab Dass, *Thin Solid Film*, **492** (2005) 332.
56. Anjana Dogra, Ravi Kumar, S.A. Khan, V.V. Siva Kumar, N. Kumar and M. Singh, *Nucl. Instr. and Meth.*, **B 225** (2004) 283.
57. Sanjukta Ghosh, P. Ayyub, N. Kumar, S.A. Khan, D.K. Avasthi, R. Chander and D. Banerjee, *Nucl. Instr. and Meth.*, **B 212** (2003) 510.
58. Yatendra S. Chaudhury, Saif A. Khan, Rohit Shrivastav, Vibha R. Satsangi, Satya Prakash, D.K. Avasthi and Sahab Dass, *Nucl. Instr. and Meth.*, **B 225** (2003) 291.
59. U.S. Joshi, S.J. Trivedi, K.H. Bhavsar, U.N. Trivedi, S.A. Khan and D.K. Avasthi, *J. of Appl. Phys.*, **105** (2009) 73704.
60. Y. Batra, S.A. Khan, D. Kabiraj, S. Kumar and D. Kanjilal, *Nucl. Instr. and Meth.*, **B 266** (2008) 1697.
61. Y. Batra, D. Kabiraj, S. Kumar and D. Kanjilal, *J. Phys. D: Appl. Phys.*, **40** (2007) 4568.
62. Neeti Tripathi, Shyama Rath, P.K. Kulriya, S.A. Khan, D. Kabiraj and D.K. Avasthi, *Nucl. Instr. and Meth.*, **B 268** (2010) 3335.
63. Manvendra Kumar, Parasmani Rajput, S.A. Khan, D.K. Avasthi and A.C. Pandey, *Applied Surface Science*, **256** (2010) 2199.
64. D.K. Avasthi, S. Ghosh, S.K. Srivastava and W. Assmann, *Nucl. Instr. and Meth.*, **B 219-220** (2004) 206.
65. S. Ghosh, A. Tripathi, T. Som, D. Kabiraj, A. Ingale, S. Mishra, V. Ganesan, S. Zhang and Z. Hong, *Nucl. Instr. and Meth.*, **B 190** (2002)164.
66. S. Ghosh, Alka Ingale, T. Som, D. Kabiraj, A. Tripathi, S. Mishra, S. Zhang, X. Hong and D.K. Avasthi, *Sol. State Commu.*, **120** (2001) 445.

67. S. Ghosh, D.K. Avasthi, A. Tripathi, S.K. Srivastava, S.V.S. Nageswara Rao, T. Som, V.K. Mittal, F. Gruner and W. Assmann, *Nucl. Instr. and Meth.*, **B 190** (2002) 169.
68. M. Kumar, S.A. Khan, F. Singh, A. Tripathi, D.K. Avasthi and A.C. Pandey, *Nucl. Instr. and Meth.*, **B 256** (2007) 328.
69. M. Kumar, F. Singh, S.A. Khan, V. Baranwal, S. Kumar, D.C. Agarwal, A.M. Siddiqui, A. Tripathi, A. Gupta, D.K. Avasthi and A.C. Pandey, *J. Phys. D: Applied Phys.*, **38** (2005) 637.
70. Manvendra Kumar, S.A. Khan, Parasmani Rajput, F. Singh, A. Tripathi, D.K. Avasthi and A.C. Pandey, *J. Appl. Phys.*, **102** (2007) 083510.
71. D.K. Avasthi, W. Assmann, H. Nolte, H.D. Mleskes, H. Huber, E.T. Subramaniyam, A. Tripathi and S. Ghosh, *Nucl. Instr. and Meth.*, **B 156** (1999) 143.
72. Sarvesh Kumar, R.S. Chauhan, S.A. Khan, W. Bolse and D.K. Avasthi, *Nucl. Instr. and Meth.*, **B 244** (2006) 194.
73. H. Nolte, W. Assmann, H. Huber, S.A. Karamian and H.D. Mieskes, *Nucl. Instr. and Meth.*, **B 136-138** (1998) 587.
74. D. Plachke, H.D. Carstanjen, R.M. Emrick, H. Nolte, H. Huber, W. Assmann and R. Grötzschel, *Nucl. Instr. and Meth.*, **B 136-138** (1998) 816.
75. W. Assmann, *Annual Report Tandem Accelerator Laboratory*, 1997.
76. N. Nath, O.P. Dhaniwal, A. Bhagwat, D.K.Avasthi, V. Hari Kumar and A.P. Pathak, *Surface and Coating Technology*, **66** (1994) 231.
77. A. Bhagwat and D.K.Avasthi, *Jpn. J. Appl. Phys.*, **35** (1996) 313.
78. S. Kumar, S.K. Sharma, N. Nath, V. Hari Kumar, A.P. Pathak, D. Kabiraj, D.K. Avasthi, *Radiat. Effects and Defects in Solids*, **139** (1996) 197.
79. V. Harikumar. A.P. Pathak, S.K. Sharma, Shyam Kumar, N. Nath, D. Kabiraj and D.K. Avasthi, *Nucl. Instr. and Meth.*, **B 108** (1996) 223.
80. V. Hari Kumar, A.P. Pathak, N. Nath, Shyam Kumar, S.K. Sharma, S.K. Hui and D.K. Avasthi, *Nucl. Instr. and Meth.*, **B 129** (1997) 143.
81. Shyam Kumar, S.K. Sharma, N. Nath, V. Harikumar, A.P. Pathak, S.K. Hui, D. Kabiraj and D.K. Avasthi, *Vacuum*, **48** (1997) 1027.
82. A. Sharma, Shyam Kumar, S.K. Sharma, N. Nath, V. Harikumar, A.P. Pathak, L.N.S. Prakash Goteti, S.K. Hui and D.K. Avasthi, *Journal of Physics G: Nucl. Part. Phys.*, **25** (1999) 135.
83. A. Sharma, S. Kumar, S.K. Sharma, P.K. Diwan, N. Nath, V.K. Mittal, S. Ghosh and D.K. Avasthi, *Nucl. Instr. and Meth. in Phys. Res.*, **B 170** (2000) 323.
84. A. Sharma, P.K. Diwan, S. Kumar, S.K. Sharma, V.K. Mittal, S.V.S. Nageswara Rao, B. Sannakki, S. Ghosh and D.K. Avasthi, *Nucl. Instr. and Meth. in Phys. Res.*, **B 194** (2002) 7.
85. P.K. Diwan, V. Sharma, S. Kumar, V.K. Mittal, S.A. Khan and D.K. Avasthi, *Nucl. Instr. and Meth.*, **B 258** (2007) 293.
86. P.K. Diwan, V. Sharma, S. Kumar and D.K. Avasthi, *Indian Journal of Pure & Applied Physics*, **43** (2005) 733.

Index

GPSR Compliance

The European Union's (EU) General Product Safety Regulation (GPSR) is a set of rules that requires consumer products to be safe and our obligations to ensure this.

If you have any concerns about our products, you can contact us on ProductSafety@springernature.com

In case Publisher is established outside the EU, the EU authorized representative is:

Springer Nature Customer Service Center GmbH
Europaplatz 3
69115 Heidelberg, Germany

Batch number: 10816734

Printed by Printforce, the Netherlands